Graduate Texts in Mathematics 256

For other titles published in this series, go to
www.springer.com/series/136

Gregor Kemper

A Course in Commutative Algebra

With 14 Illustrations

Gregor Kemper
Technische Universität
Zentrum Mathematik - M11
Boltzmannstr. 3
85748 Garching
Germany
kemper@ma.tum.de

ISSN 0072-5285
ISBN 978-3-642-26632-4 ISBN 978-3-642-03545-6 (eBook)
DOI 10.1007/978-3-642-03545-6
Springer Heidelberg Dordrecht London New York

Cover design: WMXDesign GmbH, Heidelberg, Germany

Printed on acid-free paper

Springer is part of Springer Science+Business Media (www.springer.com)

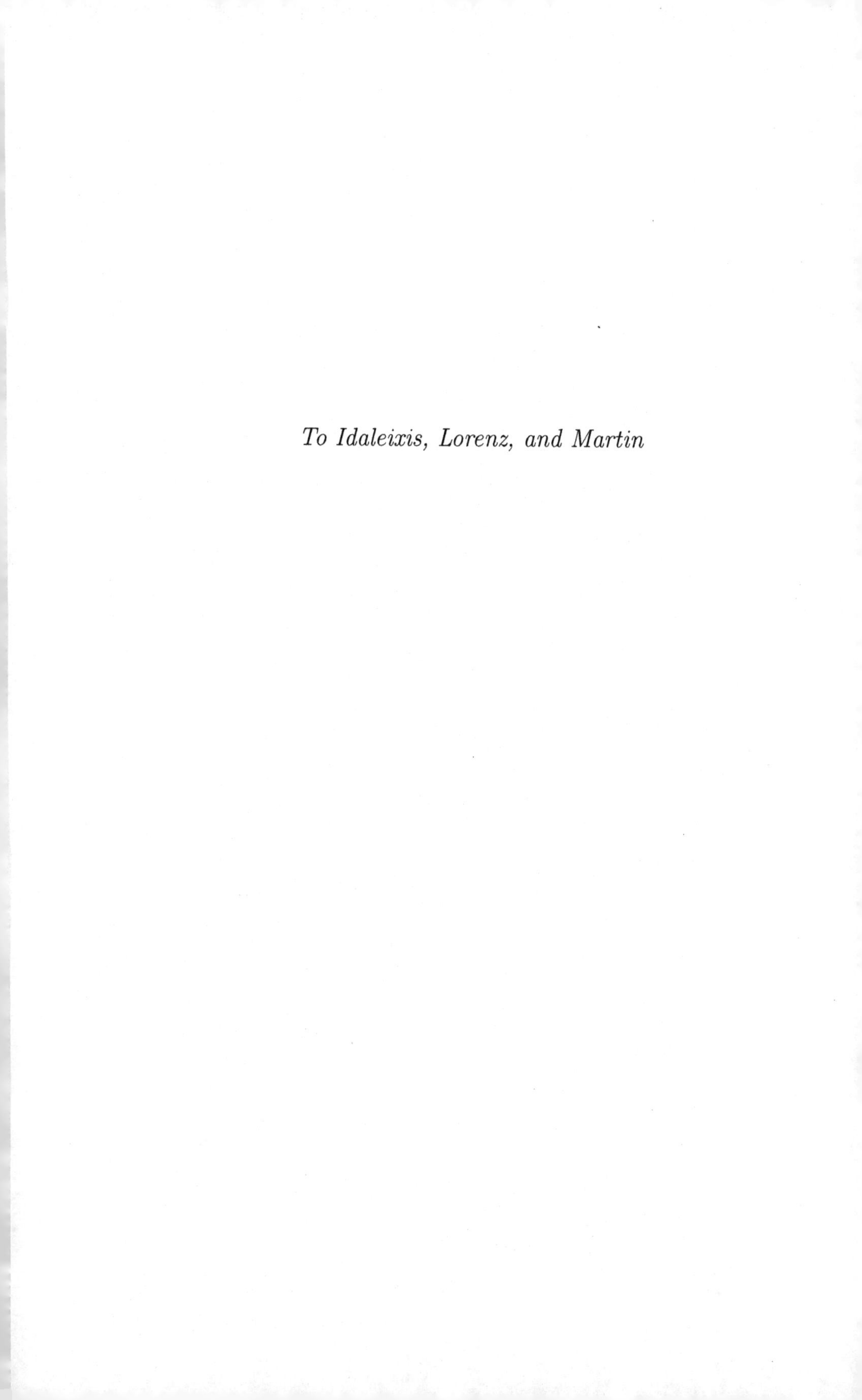

To Idaleixis, Lorenz, and Martin

Preface

This book has grown out of various courses in commutative algebra that I have taught in Heidelberg and Munich. Its primary objective is to serve as a guide for an introductory graduate course of one or two semesters, or for self-study. I have striven to craft a text that presents the concepts at the center of the field in a coherent, tightly knit way, with streamlined proofs and a focus on the core results. Needless to say, for an imperfect writer like me, such high-flying goals will always remain elusive. To introduce readers to the more recent algorithmic branch of the subject, one part of the book is devoted to computational methods. Virtually all concepts and results of commutative algebra have natural geometric interpretations. In fact, it is the geometric viewpoint that brings out the "true meaning" of the theory. This is why the first part of the book is entitled "The Algebra–Geometry Lexicon," and why I have tried to keep a focus on the geometric context throughout. I hope that this will make the theory more alive for readers, more meaningful, more visual, and easier to remember.

I welcome any comments, suggestions for improvements, and error reports from readers. Please send them to `kemper@ma.tum.de`.

Acknowledgments. First and foremost, I thank the students who attended the three courses on commutative algebra that I have taught at Heidelberg and Munich. This book has benefited greatly from their participation. Particularly fruitful was the last course, given in 2008, in which I awarded one euro for every mistake in the manuscript that the students reported. This method was so successful that it cost me a small fortune. I would like to mention Peter Heinig in particular, who brought to my attention innumerable mistakes and quite a few didactic subtleties.

I am also grateful to Gert-Martin Greuel, Bernd Ulrich, Robin Hartshorne, Viet-Trung Ngo, Dale Cutkosky, Martin Kohls, and Steve Gilbert for interesting conversations.

My interest in commutative algebra grew out of my main research interest, invariant theory. In particular, the books by Sturmfels [50] and Benson [4], although they do not concentrate on commutative algebra, first awakened my fascination for it. So my thanks go to Bernd Sturmfels and David Benson, too.

Last but not least, I am grateful to David Kramer for his outstanding job of copyediting the manuscript, to the anonymous referees, and to the people at Springer for the swift and efficient handling of the publication process.

Munich
November 2010

Gregor Kemper

Contents

Introduction

How To Use This Book

The main intention of this book is to provide material for an introductory graduate course of one or two semesters. The duration of the course clearly depends on such parameters as speed and teaching hours per week and on how much material is covered. In the book, I have indicated three options for skipping material. For example, one possibility is to omit Chapter 10 and most of Section 7.2. Another is to skip Chapters 9–11 almost entirely. But apart from these options, interdependencies in the text are close enough to make it hard to skip material without tearing holes into proofs that come later. So the instructor can best limit the amount of material by choosing where to stop. A relatively short course would stop after Chapter 8, while other natural stopping points are after Chapters 11 or 13.

The book contains a total of 143 exercises. Some of them deal with examples that illustrate definitions (such as an example of an Artinian module that is not Noetherian) or shed some light on the necessity of hypotheses of theorems (such as an example in which the principal ideal theorem fails for a non-Noetherian ring). Others give extensions to the theory (such as a series of exercises that deal with formal power series rings), and yet others invite readers to do computations on examples. These examples often come from geometry and also serve to illustrate the theory (such as examples of desingularization of curves). Some exercises depend on others, as is usually indicated in the hints for the exercise. However, no theorem, proposition, lemma, or corollary in the text depends on results from the exercises. I put a star by some exercises to indicate that I consider them more difficult. Solutions to all exercises are collected in a solutions manual, which is available for instructors.

Although the ideal way of using the book is to read it from beginning to end (every author desires such readers!), an extensive subject index should facilitate a less linear navigation.

G. Kemper, *A Course in Commutative Algebra*, Graduate Texts in Mathematics 256, DOI 10.1007/978-3-642-03545-6_1,

Prerequisites

Readers should have taken undergraduate courses in linear algebra and abstract algebra. Everything that is assumed is contained in Lang's book [33], but certainly not everything in that book is assumed. Specifically, readers should have a grasp of the following subjects:

- definition of a (commutative) ring,
- ideals, prime ideals, and maximal ideals,
- zero divisors,
- quotient rings (also known as factor rings),
- subrings and homomorphisms of rings,
- principal ideal domains,
- factorial rings (also known as unique factorization domains),
- polynomial rings in several indeterminates,
- finite field extensions, and
- algebraically closed fields.

In accordance with the geometric viewpoint of this book, it sometimes uses language from topology. Specifically, readers should know the definitions of the following terms:

- topological space,
- closure of a set,
- subspace topology, and
- continuous map.

All these can be found in any textbook on topology, for example Bourbaki [6].

Contents

The first four chapters of the book have a common theme: building the "Algebra–Geometry Lexicon," a machine that translates geometric notions into algebraic ones and vice versa. The opening chapter deals with Hilbert's Nullstellensatz, which translates between ideals of a polynomial ring and affine varieties. The second chapter is about the basic theory of Noetherian rings and modules. One result is Hilbert's basis theorem, which says that every ideal in a polynomial ring over a field is finitely generated. The results from Chapter 2 are used in Chapter 3 to prove that affine varieties are made up of finitely many irreducible components. That chapter also introduces the Zariski topology, another important element of our lexicon, and the notion of the spectrum of a ring, which allows us to interpret prime ideals as generalized points in a more abstract variant of geometry. Chapter 4 provides a summary of the lexicon.

In any mathematical theory connected with geometry, dimension is a central, but often subtle, notion. The four chapters making up the second part of the book relate to this notion. In commutative algebra, dimension is defined by the Krull dimension, which is introduced in Chapter 5. The main result of the chapter is that the dimension of an affine algebra coincides with its transcendence degree. Chapter 6 is an interlude introducing an important construction that is used throughout the book: localization. Along the way, the notions of local rings and height are introduced. Chapter 6 sets up the conceptual framework for proving Krull's principal ideal theorem in Chapter 7. That chapter also contains an investigation of fibers of morphisms, which leads to the nice result that forming a polynomial ring over a Noetherian ring increases the dimension by 1. Chapter 8 discusses the notions of integral ring extensions and normal rings. One of the main results is the Noether normalization theorem, which is then used to prove that all maximal chains of prime ideals in an affine domain have the same length.

The third part of the book is devoted to computational methods. Theoretical and algorithmic aspects go hand in hand in this part. The main computational tool is Buchberger's algorithm for calculating Gröbner bases, which is developed in Chapter 9. As a first application, Gröbner bases are applied to compute elimination ideals, which have important geometric interpretations. Chapter 10, the second chapter of this part, continues the investigation of fibers of morphisms started in Chapter 7. This chapter contains a constructive version of Grothendieck's generic freeness lemma. This is one of the main ingredients of an algorithm for computing the image of a morphism of affine varieties, probably a novelty. The chapter also contains Chevalley's result that the image of a morphism is a constructible set. The results of Chapter 10 are not used elsewhere in the book, so there is an option to skip that chapter and the parts of Chapter 7 that deal with fibers of morphisms. Finally, Chapter 11 deals with the Hilbert function and Hilbert series of an ideal in a polynomial ring. The main result, whose proof makes use of Noether normalization, states that the Hilbert function is eventually represented by a polynomial whose degree is the dimension of the affine algebra given by the ideal. This result leads to an algorithm for computing the dimension of an affine algebra, and it also plays an important role in Chapter 12 (which belongs to the fourth part of the book). Nevertheless, it is possible to skip the third part of the book almost entirely by modifying some parts of the text, as indicated in an exercise.

The fourth and last part of the book deals with local rings. Geometrically, local rings relate to local properties of varieties. Chapter 12 introduces the associated graded ring and presents a new characterization of the dimension of a local ring. Chapter 13 studies regular local rings, which correspond to nonsingular points of a variety. An important result is the Jacobian criterion for calculating the singular locus of an affine variety. A consequence is that an affine variety is nonsingular almost everywhere. The final chapter deals with topics related to rings of dimension one. The starting point is the observation that a Noetherian local ring of dimension one is regular if and only if it is normal. From this it follows that affine curves can be desingularized. After an excursion to multiplicative ideal theory for more general rings, the attention is focused to Dedekind domains, which are characterized as “rings with a perfect multiplicative ideal theory.” The chapter closes with an application that explains how the group law on an elliptic curve can be defined by means of multiplicative ideal theory.

Further Reading

The contents of a book may also be described by what is missing. Since this book is relatively short and concentrates on the central issues, it pays a price in comprehensiveness. Homological concepts and methods should probably appear at the top of the list of what is missing. In particular, the book does not treat syzygies, resolutions, and Tor and Ext functors. As a consequence, depth and the Cohen–Macaulay property cannot be dealt with sensibly (and would require much more space in any case), so only one exercise touches on Cohen–Macaulay rings. Flat modules are another topic that relates to homological methods and is not treated. The subject of completion is also just touched on. I have decided not to include associated primes and primary decomposition in the book, although these topics are often regarded as rather basic and central, because they are not needed elsewhere in the book.

All the topics mentioned above are covered in the books by Matsumura [37] and Eisenbud [17], which I warmly recommend for further reading. Of these books, [37] presents the material in a more condensed way, while [17] shares the approach of this book in its focus on the geometric context and in its inclusion of Gröbner basis methods. Eisenbud's book, more than twice as large as this one, is remarkable because it works as a textbook but also contains a lot of material that appeals to experts.

Apart from deepening their knowledge in commutative algebra, readers of this book may continue their studies in different directions. One is algebraic geometry. Hartshorne's textbook [26] still seems to be the authoritative source on the subject, but Harris [25] and Smith et al. [47] (to name just two) provide more recent alternatives. Another possible direction to go in is computational commutative algebra. A list of textbooks on this appears at the beginning of Chapter 9 of this book. I especially recommend the book by Cox et al. [12], which does a remarkable job of blending aspects of geometry, algebra, and computation.

Part I
The Algebra–Geometry Lexicon

Chapter 1
Hilbert's Nullstellensatz

Hilbert's Nullstellensatz may be seen as the starting point of algebraic geometry. It provides a bijective correspondence between affine varieties, which are geometric objects, and radical ideals in a polynomial ring, which are algebraic objects. In this chapter, we give proofs of two versions of the Nullstellensatz. We exhibit some further correspondences between geometric and algebraic objects. Most notably, the coordinate ring is an affine algebra assigned to an affine variety, and points of the variety correspond to maximal ideals in the coordinate ring.

Before we get started, let us fix some conventions and notation that will be used throughout the book. By a **ring** we will always mean a commutative ring with an identity element 1. In particular, there is a ring $R = \{0\}$, the **zero ring**, in which $1 = 0$. A ring R is called an **integral domain** if R has no zero divisors (other than 0 itself) and $R \neq \{0\}$. A **subring** of a ring R must contain the identity element of R, and a homomorphism $R \to S$ of rings must send the identity element of R to the identity element of S.

If R is a ring, an **R-algebra** is defined to be a ring A together with a homomorphism $\alpha: R \to A$. In other words, by an algebra we will mean a commutative, associative algebra with an identity element. A **subalgebra** of an algebra A is a subring that contains the image $\alpha(R)$. If A and B are R-algebras with homomorphisms α and β, then a map $\varphi: A \to B$ is called a **homomorphism of (R-)algebras** if φ is a ring homomorphism, and $\varphi \circ \alpha = \beta$. If A is a nonzero algebra over a field K, then the map α is injective, so we may view K as a subring of A. With this identification, a homomorphism of nonzero K-algebras is just a ring homomorphism fixing K elementwise.

One of the most important examples of an R-algebra is the ring of polynomials in n indeterminates with coefficients in R, which is written as $R[x_1, \dots, x_n]$. If A is any R-algebra and $a_1, \dots, a_n \in A$ are elements, then there is a unique algebra homomorphism $\varphi: R[x_1, \dots, x_n] \to A$ with $\varphi(x_i) = a_i$, given by applying α to the coefficients of a polynomial and substituting x_i by a_i. Clearly the image of φ is the smallest subalgebra of A

G. Kemper, *A Course in Commutative Algebra*, Graduate Texts in Mathematics 256, DOI 10.1007/978-3-642-03545-6_2,

containing all a_i, i.e., the subalgebra of A generated by the a_i. We write this image as $R[a_1, \dots, a_n]$, which is consistent with the notation $R[x_1, \dots, x_n]$ for a polynomial ring. We say that A is **finitely generated** if there exist $a_1, \dots, a_n$ with $A = R[a_1, \dots, a_n]$. Thus an algebra is finitely generated if and only if it is isomorphic to the quotient ring $R[x_1, \dots, x_n]/I$ of a polynomial ring by an ideal $I \subseteq R[x_1, \dots, x_n]$. By an **affine ($K$-)algebra** we mean a finitely generated algebra over a field K. An **affine (K-)domain** is an affine K-algebra that is an integral domain.

Recall that the definition of a module over a ring is identical to the definition of a vector space over a field. In particular, an ideal in a ring R is the same as a submodule of R viewed as a module over itself. Recall that a module does not always have a basis (= a linearly independent generating set). If it does have a basis, it is called **free**. If M is an R-module and $S \subseteq M$ is a subset, we write $(S)_R = (S)$ for the submodule of M generated by S, i.e., the set of all R-linear combinations of S. (The index R may be omitted if it is clear which ring we have in mind.) If $S = \{m_1, \dots, m_k\}$ is finite, we write $(S)_R = (m_1, \dots, m_k)_R = (m_1, \dots, m_k)$. In particular, if $a_1, \dots, a_k \in R$ are ring elements, then $(a_1, \dots, a_k)_R = (a_1, \dots, a_k)$ denotes the ideal generated by them.

1.1 Maximal Ideals

Let $a \in A$ be an element of a nonzero algebra A over a field K. As in field theory, a is said to be **algebraic** (over K) if there exists a nonzero polynomial $f \in K[x]$ with $f(a) = 0$. We say that A is **algebraic** (over K) if every element from A is algebraic. Almost everything that will be said about affine algebras in this book has its starting point in the following lemma.

Lemma 1.1 (Fields and algebraic algebras). *Let A be an algebra over a field K.*

(a) If A is an integral domain and algebraic over K, then A is a field.
(b) If A is a field and is contained in an affine K-domain, then A is algebraic.

Proof. (a) We need to show that every $a \in A \setminus \{0\}$ is invertible in A. For this, it suffices to show that $K[a]$ is a field. We may therefore assume that $A = K[a]$. With x an indeterminate, let $I \subseteq K[x]$ be the kernel of the map $K[x] \to A$, $f \mapsto f(a)$. Then $A \cong K[x]/I$. Since A is an integral domain, I is a prime ideal, and since a is algebraic over K, I is nonzero. Since $K[x]$ is a principal ideal domain, it follows that $I = (f)$ with $f \in K[x]$ irreducible, so I is a maximal ideal. It follows that $A \cong K[x]/I$ is a field.

(b) By way of contradiction, assume that A has an element a_1 that is not algebraic. By hypothesis, A is contained in an affine K-domain $B = K[a_1, \dots, a_n]$ (we may include a_1 in the set of generators). We can reorder $a_2, \ldots, a_n$ in such a way that $\{a_1, \ldots, a_r\}$ forms a maximal K-algebraically independent subset of $\{a_1, \ldots, a_n\}$. Then the field of fractions $\mathrm{Quot}(B)$ of B is a finite field extension of the subfield $L := K(a_1, \ldots, a_r)$. For $b \in \mathrm{Quot}(B)$, multiplication by b gives an L-linear endomorphism of $\mathrm{Quot}(B)$. Choosing an L-basis of $\mathrm{Quot}(B)$, we obtain a map $\varphi\colon \mathrm{Quot}(B) \to L^{m \times m}$ assigning to each $b \in \mathrm{Quot}(B)$ the representation matrix of this endomorphism. Let $g \in K[a_1, \ldots, a_r] \setminus \{0\}$ be a common denominator of all the matrix entries of all $\varphi(a_i)$, $i = 1, \ldots, n$. So $\varphi(a_i) \in K[a_1, \ldots, a_r, g^{-1}]^{m \times m}$ for all i. Since φ preserves addition and multiplication, we obtain

$$\varphi(B) \subseteq K[a_1, \ldots, a_r, g^{-1}]^{m \times m}.$$

$K[a_1, \ldots, a_r]$ is isomorphic to a polynomial ring and therefore factorial (see, for example, Lang [33, Chapter V, Corollary 6.3]). Take a factorization of g, and let $p_1, \ldots, p_k$ be those irreducible factors of g that happen to lie in $K[a_1]$. Let $p \in K[a_1]$ be an arbitrary irreducible element. Then $p^{-1} \in A \subseteq B$ since $K[a_1] \subseteq A$ and A is a field. Applying φ to p^{-1} yields a diagonal matrix with all entries equal to p^{-1}, so there exists a nonnegative integer s and an $f \in K[a_1, \ldots, a_r]$ with $p^{-1} = g^{-s} \cdot f$, so $g^s = p \cdot f$. By the irreducibility of p, it follows that p is a K-multiple of one of the p_i. Since this holds for all irreducible elements $p \in K[a_1]$, every element from $K[a_1] \setminus K$ is divisible by at least one of the p_i. But none of the p_i divides $\prod_{i=1}^k p_i + 1$. This is a contradiction, so all elements of A are algebraic. □

The following proposition is an important application of Lemma 1.1. A particularly interesting special case of the proposition is that $A \subseteq B$ is a subalgebra and φ is the inclusion.

Proposition 1.2 (Preimages of maximal ideals). *Let $\varphi\colon A \to B$ be a homomorphism of algebras over a field K, and let $\mathfrak{m} \subset B$ be a maximal ideal. If B is finitely generated, then the preimage $\varphi^{-1}(\mathfrak{m}) \subseteq A$ is also a maximal ideal.*

Proof. The map $A \to B/\mathfrak{m}$, $a \mapsto \varphi(a) + \mathfrak{m}$, has kernel $\varphi^{-1}(\mathfrak{m}) =: \mathfrak{n}$. So $A/\mathfrak{n}$ is isomorphic to a subalgebra of $B/\mathfrak{m}$. By Lemma 1.1(b), $B/\mathfrak{m}$ is algebraic over K. Hence the same is true for the subalgebra $A/\mathfrak{n}$, and $A/\mathfrak{n}$ is also an integral domain. By Lemma 1.1(a), $A/\mathfrak{n}$ is a field and therefore $\mathfrak{n}$ is maximal. □

Example 1.3. We give a simple example that shows that intersecting a maximal ideal with a subring does not always produce a maximal ideal. Let $A = K[x]$ be a polynomial ring over a field and let $B = K(x)$ be the rational function field. Then $\mathfrak{m} := \{0\} \subset B$ is a maximal ideal, but $A \cap \mathfrak{m} = \{0\}$ is not maximal in A. ◁

Before drawing a "serious" conclusion from Proposition 1.2 in Proposition 1.5, we need a lemma.

Lemma 1.4. *Let K be a field and $P = (\xi_1, \dots, \xi_n) \in K^n$ a point in K^n. Then the ideal*

$$\mathfrak{m}_P := (x_1 - \xi_1, \dots, x_n - \xi_n) \subseteq K[x_1, \dots, x_n]$$

in the polynomial ring $K[x_1, \dots, x_n]$ is maximal.

Proof. It is clear from the definition of $\mathfrak{m}_P$ that every polynomial $f \in K[x_1, \dots, x_n]$ is congruent to $f(\xi_1, \dots, \xi_n)$ modulo $\mathfrak{m}_P$. It follows that $\mathfrak{m}_P$ is the kernel of the homomorphism φ: $K[x_1, \dots, x_n] \to K$, $f \mapsto f(\xi_1, \dots, \xi_n)$, so $K[x_1, \dots, x_n]/\mathfrak{m}_P \cong K$. This implies the result. □

Together with Lemma 1.4, the following proposition describes all maximal ideals in a polynomial ring over an algebraically closed field. Recall that a field K is called *algebraically closed* if every nonconstant polynomial in $K[x]$ has a root in K.

Proposition 1.5 (Maximal ideals in a polynomial ring). *Let K be an algebraically closed field, and let $\mathfrak{m} \subset K[x_1, \dots, x_n]$ be a maximal ideal in a polynomial ring over K. Then there exists a point $P = (\xi_1, \dots, \xi_n) \in K^n$ such that*

$$\mathfrak{m} = (x_1 - \xi_1, \dots, x_n - \xi_n).$$

Proof. By Proposition 1.2, the intersection $K[x_i] \cap \mathfrak{m}$ is a maximal ideal in $K[x_i]$ for each $i = 1, \dots, n$. Since $K[x_i]$ is a principal ideal domain, $K[x_i] \cap \mathfrak{m}$ has the form $(p_i)_{K[x_i]}$ with p_i an irreducible polynomial. Since K is algebraically closed, we obtain $(p_i)_{K[x_i]} = (x_i - \xi_i)_{K[x_i]}$ with $\xi_i \in K$. So there exist $\xi_1, \dots, \xi_n \in K$ with $x_i - \xi_i \in \mathfrak{m}$. With the notation of Lemma 1.4, it follows that $\mathfrak{m}_P \subseteq \mathfrak{m}$, so $\mathfrak{m} = \mathfrak{m}_P$ by Lemma 1.4. □

We make a definition before giving a refined version of Proposition 1.5.

Definition 1.6. *Let $K[x_1, \dots, x_n]$ be a polynomial ring over a field.*

(a) For a set $S \subseteq K[x_1, \dots, x_n]$ of polynomials, the **affine variety** *given by S is defined as*

$$\mathcal{V}(S) = \mathcal{V}_{K^n}(S) := \{(\xi_1, \dots, \xi_n) \in K^n \mid f(\xi_1, \dots, \xi_n) = 0 \text{ for all } f \in S\}.$$

The index K^n is omitted if no misunderstanding can occur.

(b) A subset $X \subseteq K^n$ is called an **affine (K-)variety** *if X is the affine variety given by a set $S \subseteq K[x_1, \dots, x_n]$ of polynomials.*

Remark. In the literature, affine varieties are sometimes assumed to be irreducible. Moreover, the definition of an affine variety is sometimes made only in the case that K is algebraically closed. ◁

Theorem 1.7 (Correspondence points–maximal ideals). *Let K be an algebraically closed field and $S \subseteq K[x_1, \dots, x_n]$ a set of polynomials. Let $\mathcal{M}_S$ be the set of all maximal ideals $\mathfrak{m} \subset K[x_1, \dots, x_n]$ with $S \subseteq \mathfrak{m}$. Then the map*

$$\Phi\colon \mathcal{V}(S) \to \mathcal{M}_S,\ (\xi_1,\ldots,\xi_n) \mapsto (x_1-\xi_1,\ldots,x_n-\xi_n)$$

is a bijection.

Proof. Let $P := (\xi_1,\ldots,\xi_n) \in \mathcal{V}(S)$. Then $\Phi(P)$ is a maximal ideal by Lemma 1.4. All $f \in S$ satisfy $f(P) = 0$, so $f \in \Phi(P)$. It follows that $\Phi(P) \in \mathcal{M}_S$. On the other hand, let $\mathfrak{m} \in \mathcal{M}_S$. By Proposition 1.5, $\mathfrak{m} = (x_1-\xi_1,\ldots,x_n-\xi_n)$ with $(\xi_1,\ldots,\xi_n) \in K^n$, and $S \subseteq \mathfrak{m}$ implies $(\xi_1,\ldots,\xi_n) \in \mathcal{V}(S)$. This shows that Φ is surjective.

To show injectivity, let $P = (\xi_1,\ldots,\xi_n)$ and $Q = (\eta_1,\ldots,\eta_n)$ be points in $\mathcal{V}(S)$ with $\Phi(P) = \Phi(Q) =: \mathfrak{m}$. For each i, we have $x_i - \xi_i \in \mathfrak{m}$ and also $x_i - \eta_i \in \mathfrak{m}$, so $\xi_i - \eta_i \in \mathfrak{m}$. This implies $\xi_i = \eta_i$, since otherwise $\mathfrak{m} = K[x_1,\ldots,x_n]$. □

Corollary 1.8 (Hilbert's Nullstellensatz, first version). *Let K be an algebraically closed field and let $I \subsetneqq K[x_1,\ldots,x_n]$ be a proper ideal in a polynomial ring. Then*

$$\mathcal{V}(I) \neq \emptyset.$$

Proof. Consider the set of all proper ideals $J \subsetneqq K[x_1,\ldots,x_n]$ containing I. Using Zorn's lemma, we conclude that this set contains a maximal element $\mathfrak{m}$. (Instead of Zorn's lemma, we could also use the fact that $K[x_1,\ldots,x_n]$ is Noetherian (see Corollary 2.13). But even then, the axiom of choice, which is equivalent to Zorn's lemma, would have to be used to do the proof without cheating. See Halmos [24] to learn more about Zorn's lemma and the axiom of choice.) So $\mathfrak{m}$ is a maximal ideal with $I \subseteq \mathfrak{m}$. Now $\mathcal{V}(I) \neq \emptyset$ follows by Theorem 1.7. □

Remark. (a) To see that the hypothesis that K is algebraically closed cannot be omitted from Corollary 1.8, consider the example $K = \mathbb{R}$ and $I = (x^2+1) \subsetneqq \mathbb{R}[x]$.

(b) Hilbert's Nullstellensatz is really a theorem about systems of polynomial equations. Indeed, let $f_1,\ldots,f_m \in K[x_1,\ldots,x_n]$ be polynomials. If there exist polynomials $g_1,\ldots,g_m \in K[x_1,\ldots,x_n]$ such that

$$\sum_{i=1}^{m} g_i f_i = 1, \tag{1.1}$$

then obviously the system of equations

$$f_i(\xi_1,\ldots,\xi_n) = 0 \quad \text{for} \quad i = 1,\ldots,m \tag{1.2}$$

has no solutions. But the existence of $g_1,\ldots,g_m$ satisfying (1.1) is equivalent to the condition $(f_1,\ldots,f_m) = K[x_1,\ldots,x_n]$. So Hilbert's Nullstellensatz says that if the obvious obstacle (1.1) to solvability does not exist, and if K is algebraically closed, then indeed the system (1.2) is solvable. In other words, for algebraically closed fields, the obvious obstacle to the solvability of systems of polynomial equations is the only one! In Chapter 9 we will see how it can be checked algorithmically whether the obstacle (1.1) exists (see (9.4) on page 123). ◁

1.2 Jacobson Rings

The main goal of this section is to prove the second version of Hilbert's Nullstellensatz (Theorem 1.17). We start by defining the spectrum and the maximal spectrum of a ring.

Definition 1.9. *Let R be a ring.*

(a) The **spectrum** *of R is the set of all prime ideals in R:*

$$\operatorname{Spec}(R) := \{P \subset R \mid P \text{ is a prime ideal}\}.$$

(b) The **maximal spectrum** *of R is the set of all maximal ideals in R:*

$$\operatorname{Spec}_{\max}(R) := \{P \subset R \mid P \text{ is a maximal ideal}\}.$$

(c) We also define the **Rabinowitsch spectrum** *of R as the set*

$$\operatorname{Spec}_{\mathrm{rab}}(R) := \{R \cap \mathfrak{m} \mid \mathfrak{m} \in \operatorname{Spec}_{\max}(R[x])\},$$

where $R[x]$ is the polynomial ring over R. This is an ad hoc definition, which is not found in the standard literature and will be used only within this section.

Remark. The idea of using an additional indeterminate for proving the second version of Hilbert's Nullstellensatz goes back to J. L. Rabinowitsch [45], and is often referred to as Rabinowitsch's trick. This made my student Martin Kohls suggest that the set from Definition 1.9(c) be called the Rabinowitsch spectrum. ◁

We have the inclusions

$$\operatorname{Spec}_{\max}(R) \subseteq \operatorname{Spec}_{\mathrm{rab}}(R) \subseteq \operatorname{Spec}(R).$$

Indeed, the second inclusion follows since for any prime ideal $P \subset S$ in a ring extension S of R, the intersection $R \cap P$ is a prime ideal in R. The first inclusion is proved in Exercise 1.3. Only the second inclusion will be used in this book. Exercise 1.4 gives an example in which both inclusions are strict. The importance of the Rabinowitsch spectrum is highlighted by Proposition 1.11.

Recall that for an ideal $I \subseteq R$ in a ring R, the **radical ideal** of I is defined as

$$\sqrt{I} := \left\{f \in R \mid \text{there exists a positive integer } k \text{ with } f^k \in I\right\}.$$

I is called a **radical ideal** if $\sqrt{I} = I$. For example, a nonzero ideal $(a) \subseteq \mathbb{Z}$ is radical if and only if a is square-free. Recall that every prime ideal is a radical ideal.

Lemma 1.10. *Let R be a ring, $I \subseteq R$ an ideal, and $\mathcal{M} \subseteq \operatorname{Spec}(R)$ a subset. Then*

$$\sqrt{I} \subseteq \bigcap_{\substack{P \in \mathcal{M}, \\ I \subseteq P}} P.$$

If there exist no $P \in \mathcal{M}$ with $I \subseteq P$, the intersection is to be interpreted as R.

Proof. Let $a \in \sqrt{I}$, so $a^k \in I$ for some k. Let $P \in \mathcal{M}$ with $I \subseteq P$. Then $a^k \in P$. Since P is a prime ideal, it follows that $a \in P$. □

Proposition 1.11 (The raison d'être of the Rabinowitsch spectrum). *Let $I \subseteq R$ be an ideal in a ring. Then*

$$\sqrt{I} = \bigcap_{\substack{P \in \operatorname{Spec}_{\mathrm{rab}}(R), \\ I \subseteq P}} P.$$

If there exist no $P \in \operatorname{Spec}_{\mathrm{rab}}(R)$ with $I \subseteq P$, the intersection is to be interpreted as R.

Proof. The inclusion "$\subseteq$" follows from Lemma 1.10 and the fact that $\operatorname{Spec}_{\mathrm{rab}}(R) \subseteq \operatorname{Spec}(R)$.

To prove the reverse inclusion, let a be in the intersection of all $P \in \operatorname{Spec}_{\mathrm{rab}}(R)$ with $I \subseteq P$. Consider the ideal

$$J := (I \cup \{ax - 1\})_{R[x]} \subseteq R[x]$$

generated by I and by $ax - 1$. Assume that $J \subsetneqq R[x]$. By Zorn's lemma, there exists $\mathfrak{m} \in \operatorname{Spec}_{\max}(R[x])$ with $J \subseteq \mathfrak{m}$. We have $I \subseteq R \cap J \subseteq R \cap \mathfrak{m} \in \operatorname{Spec}_{\mathrm{rab}}(R)$, so by hypothesis, $a \in \mathfrak{m}$. But also $ax - 1 \in \mathfrak{m}$, so $\mathfrak{m} = R[x]$. This is a contradiction, showing that $J = R[x]$. In particular, we have

$$1 = \sum_{j=1}^{n} g_j b_j + g(ax - 1) \tag{1.3}$$

with $g, g_1, \dots, g_n \in R[x]$ and $b_1, \dots, b_n \in I$. Let $R[x, x^{-1}]$ be the ring of Laurent polynomials and consider the map $\varphi\colon R[x] \to R[x, x^{-1}]$, $f \mapsto f(x^{-1})$. Applying φ to both sides of (1.3) and multiplying by some x^k yields

$$x^k = \sum_{j=1}^{n} h_j b_j + h(a - x) \quad \text{with} \quad h_j = x^k \varphi(g_j) \quad \text{and} \quad h = x^{k-1}\varphi(g).$$

For $k \geq \max\{\deg(g_1), \dots, \deg(g_n), \deg(g)+1\}$, all h_j and h lie in $R[x]$, so we may substitute $x = a$ in the above equation and obtain

$$a^k = \sum_{j=1}^{n} h_j(a) b_j \in I,$$

so $a \in \sqrt{I}$. This completes the proof. □

We get the following important consequence.

Corollary 1.12 (Intersecting prime ideals). *Let R be a ring and $I \subseteq R$ an ideal. Then*

$$\sqrt{I} = \bigcap_{\substack{P \in \mathrm{Spec}(R),\\ I \subseteq P}} P.$$

If there exist no $P \in \mathrm{Spec}(R)$ with $I \subseteq P$, the intersection is to be interpreted as R.

Proof. This follows from Lemma 1.10 and Proposition 1.11. □

Theorem 1.13 (Intersecting maximal ideals). *Let A be an affine algebra and $I \subseteq A$ an ideal. Then*

$$\sqrt{I} = \bigcap_{\substack{\mathfrak{m} \in \mathrm{Spec}_{\max}(A),\\ I \subseteq \mathfrak{m}}} \mathfrak{m}.$$

If there exist no $\mathfrak{m} \in \mathrm{Spec}_{\max}(A)$ with $I \subseteq \mathfrak{m}$, the intersection is to be interpreted as A.

Proof. The inclusion "$\subseteq$" again follows from Lemma 1.10.

Let $P \in \mathrm{Spec}_{\mathrm{rab}}(A)$. Then $P = A \cap \mathfrak{m}$ with $\mathfrak{m} \in \mathrm{Spec}_{\max}(A[x])$. But $A[x]$ is finitely generated as an algebra over a field, so by Proposition 1.2 it follows that $P \in \mathrm{Spec}_{\max}(A)$. We conclude that

$$\mathrm{Spec}_{\mathrm{rab}}(A) \subseteq \mathrm{Spec}_{\max}(A).$$

(In fact, equality holds, but we do not need this.) Now the inclusion "$\supseteq$" follows from Proposition 1.11. □

We pause here to make a definition, which is inspired by Theorem 1.13.

Definition 1.14. *A ring R is called a* **Jacobson ring** *if for every proper ideal $I \subsetneq R$ the equality*

$$\sqrt{I} = \bigcap_{\substack{\mathfrak{m} \in \mathrm{Spec}_{\max}(R),\\ I \subseteq \mathfrak{m}}} \mathfrak{m}$$

holds.

So Theorem 1.13 says that every affine algebra is a Jacobson ring. A further example is the ring $\mathbb{Z}$ of integers (see Exercise 1.6). So one wonders whether the polynomial ring $\mathbb{Z}[x]$ is Jacobson, too. This is indeed the case. It is an instance of the general fact that every finitely generated algebra A over a Jacobson ring R is again a Jacobson ring. A proof is given in Eisenbud [17, Theorem 4.19]. There we also find the following: If α is the homomorphism making A into an R-algebra, then for every $\mathfrak{m} \in \operatorname{Spec}_{\max}(A)$ the preimage $\alpha^{-1}(\mathfrak{m})$ is also maximal. This is in analogy to Proposition 1.2.

A typical example of a non-Jacobson ring is the formal power series ring $K[[x]]$ over a field K (see Exercise 1.2). A similar example is the ring of all rational numbers with odd denominator.

We can now prove the second version of Hilbert's Nullstellensatz. To formulate it, a bit of notation is useful.

Definition 1.15. *Let K be a field and $X \subseteq K^n$ a set of points. The* **(vanishing) ideal** *of X is defined as*

$$\begin{aligned}\mathcal{I}(X) &= \mathcal{I}_{K[x_1,\ldots,x_n]}(X)\\ &:= \{f \in K[x_1,\ldots,x_n] \mid f(\xi_1,\ldots,\xi_n) = 0 \text{ for all } (\xi_1,\ldots,\xi_n) \in X\}.\end{aligned}$$

The index $K[x_1,\ldots,x_n]$ is omitted if no misunderstanding can occur.

Remark 1.16. It is clear from the definition that the ideal of a set of points is always a radical ideal. ◁

Theorem 1.17 (Hilbert's Nullstellensatz, second version). *Let K be an algebraically closed field and let $I \subseteq K[x_1,\ldots,x_n]$ be an ideal in a polynomial ring. Then*

$$\mathcal{I}(\mathcal{V}(I)) = \sqrt{I}.$$

Proof. We start by showing the inclusion "$\supseteq$", which does not require K to be algebraically closed. Let $f \in \sqrt{I}$, so $f^k \in I$ for some k. Take $(\xi_1,\ldots,\xi_n) \in \mathcal{V}(I)$. Then $f(\xi_1,\ldots,\xi_n)^k = 0$, so $f(\xi_1,\ldots,\xi_n) = 0$. This shows that $f \in \mathcal{I}(\mathcal{V}(I))$.

For the reverse inclusion, assume $f \in \mathcal{I}(\mathcal{V}(I))$. In view of Theorem 1.13, we need to show that f lies in every $\mathfrak{m} \in \mathcal{M}_I$, where

$$\mathcal{M}_I = \{\mathfrak{m} \in \operatorname{Spec}_{\max}(K[x_1,\ldots,x_n]) \mid I \subseteq \mathfrak{m}\}.$$

So let $\mathfrak{m} \in \mathcal{M}_I$. By Theorem 1.7, $\mathfrak{m} = (x_1 - \xi_1,\ldots,x_n - \xi_n)_{K[x_1,\ldots,x_n]}$ with $(\xi_1,\ldots,\xi_n) \in \mathcal{V}(I)$. This implies $f(\xi_1,\ldots,\xi_n) = 0$, so $f \in \mathfrak{m}$. This completes the proof. □

The following corollary is the heart of what we call the algebra–geometry lexicon. We need an (easy) lemma.

Lemma 1.18. *Let K be a field and $X \subseteq K^n$ an affine variety. Then*

$$\mathcal{V}(\mathcal{I}(X)) = X.$$

Proof. By assumption, $X = \mathcal{V}(S)$ with $S \subseteq K[x_1, \ldots, x_n]$. So $S \subseteq \mathcal{I}(X)$, and applying $\mathcal{V}$ yields

$$\mathcal{V}(\mathcal{I}(X)) \subseteq \mathcal{V}(S) = X \subseteq \mathcal{V}(\mathcal{I}(X)).$$

The lemma follows. □

Corollary 1.19 (Ideal–variety correspondence). *Let K be an algebraically closed field and n a positive integer. Then there is a bijection between the sets*

$$\mathcal{A} := \{I \subseteq K[x_1, \ldots, x_n] \mid I \text{ is a radical ideal}\}$$

and

$$\mathcal{B} := \{X \subseteq K^n \mid X \text{ is an affine variety}\},$$

given by

$$\mathcal{A} \to \mathcal{B}, \quad I \mapsto \mathcal{V}(I)$$

and the inverse map

$$\mathcal{B} \to \mathcal{A}, \quad X \mapsto \mathcal{I}(X).$$

Both maps reverse inclusions, i.e., if $I, J \in \mathcal{A}$, then

$$I \subseteq J \quad \Longleftrightarrow \quad \mathcal{V}(J) \subseteq \mathcal{V}(I),$$

and the corresponding statement holds for the inverse map.

Proof. If $I \in \mathcal{A}$ is a radical ideal, it follows from the Nullstellensatz (Theorem 1.17) that $\mathcal{I}(\mathcal{V}(I)) = I$. On the other hand, take $X \in \mathcal{B}$. Then $\mathcal{I}(X) \in \mathcal{A}$ by Remark 1.16, and $\mathcal{V}(\mathcal{I}(X)) = X$ by Lemma 1.18. This shows that the given maps are inverses to each other. The last statement follows since $I \subseteq J$ implies $\mathcal{V}(J) \subseteq \mathcal{V}(I)$ for $I, J \in \mathcal{A}$, and $X \subseteq Y$ implies $\mathcal{I}(Y) \subseteq \mathcal{I}(X)$ for $X, Y \in \mathcal{B}$. Now apply $\mathcal{I}$ and $\mathcal{V}$ to get the converse implications. □

1.3 Coordinate Rings

The next part of the algebra–geometry lexicon is provided by assigning to an affine variety X an affine algebra, the coordinate ring $K[X]$, which encodes the properties of X.

Definition 1.20. *Let K be a field and $X \subseteq K^n$ an affine variety. Let $I := \mathcal{I}(X) \subseteq K[x_1, \ldots, x_n]$ be the ideal of X. Then the* **coordinate ring** *of X is the quotient ring*

$$K[X] := K[x_1, \ldots, x_n]/I.$$

The coordinate ring is sometimes also called the **ring of regular functions** *on X.*

Remark 1.21. (a) Every element of the coordinate ring $K[X]$ of an affine variety is a class $f + I$ with $f \in K[x_1, \dots, x_n]$. Such a class yields a well-defined function $X \to K$, given by $(\xi_1, \dots, \xi_n) \mapsto f(\xi_1, \dots, \xi_n)$, and different classes yield different functions. So $K[X]$ can be identified with an algebra of functions $X \to K$. The functions from $K[X]$ are called **regular functions**. They are precisely those functions $X \to K$ that are given by polynomials.

(b) If $X = \mathcal{V}(J)$ with $J \subseteq K[x_1, \dots, x_n]$ an ideal, then it is not necessarily true that $K[X] = K[x_1, \dots, x_n]/J$. However, if K is algebraically closed, then $K[X] = K[x_1, \dots, x_n]/\sqrt{J}$ by the Nullstellensatz (Theorem 1.17). ◁

The following lemma compares ideals in a quotient ring R/I to ideals in R. It is rather boring and elementary, but very important.

Lemma 1.22 (Ideals in quotient rings). *Let R be a ring and let $I \subseteq R$ be an ideal. Consider the sets*

$$\mathcal{A} := \{J \subseteq R \mid J \text{ is an ideal and } I \subseteq J\}$$

and

$$\mathcal{B} := \{\mathcal{J} \subseteq R/I \mid \mathcal{J} \text{ is an ideal}\}.$$

The map

$$\Phi\colon \mathcal{A} \to \mathcal{B},\ J \mapsto \{a + I \mid a \in J\} = J/I$$

is an inclusion-preserving bijection with inverse map

$$\Psi\colon \mathcal{B} \to \mathcal{A},\ \mathcal{J} \mapsto \{a \in R \mid a + I \in \mathcal{J}\}.$$

If $J \in \mathcal{A}$, then

$$R/J \cong (R/I)\big/\Phi(J), \tag{1.4}$$

and there are equivalences

$$J \text{ is a prime ideal} \iff \Phi(J) \text{ is a prime ideal}$$

and

$$J \text{ is a maximal ideal} \iff \Phi(J) \text{ is a maximal ideal.}$$

Moreover, if $J = (a_1, \dots, a_n)_R$ with $a_i \in R$, then $\Phi(J) = (a_1 + I, \dots, a_n + I)_{R/I}$.

Proof. It is easy to check that Φ and Ψ are inclusion-preserving maps and that $\Psi \circ \Phi = \mathrm{id}_{\mathcal{A}}$ and $\Phi \circ \Psi = \mathrm{id}_{\mathcal{B}}$. The isomorphism (1.4) follows since $\Phi(J)$ is the kernel of the epimorphism $R/I \to R/J,\ a + I \mapsto a + J$. Both equivalences follow from (1.4). The last statement is also clear. □

If $X \subseteq K^n$ is an affine variety, then a **subvariety** is a subset $Y \subseteq X$ that is itself an affine variety in K^n. We can now prove a correspondence between subvarieties of a variety and radical ideals in the coordinate ring.

Theorem 1.23 (Correspondence subvarieties–radical ideals). *Let X be an affine variety over an algebraically closed field K. Then there is an inclusion-reversing bijection between the set of subvarieties $Y \subseteq X$ and the set of radical ideals $J \subseteq K[X]$. The bijection is given by mapping a subvariety $Y \subseteq X$ to $\mathcal{I}(Y)/\mathcal{I}(X) \subseteq K[X]$, and mapping an ideal $J \subseteq K[X]$ to*

$$\mathcal{V}_X(J) := \{x \in X \mid f(x) = 0 \text{ for all } f \in J\}.$$

If $J \subseteq K[X]$ is the ideal corresponding to a subvariety Y, then

$$K[Y] \cong K[X]/J,$$

with an isomorphism given by $K[X]/J \to K[Y]$, $f + J \mapsto f|_Y$.

Restricting our bijection to subvarieties consisting of one point yields a bijection

$$X \to \operatorname{Spec}_{\max}(K[X]), \ x \mapsto \mathcal{I}(\{x\})/\mathcal{I}(X).$$

Proof. All claims are shown by putting Corollary 1.19 and Lemma 1.22 together. □

Another correspondence between points and algebraic objects that relates to the coordinate ring is given in Exercise 1.11. The next theorem tells us which types of rings occur as coordinate rings of affine algebras. To state it, we need a definition.

Definition 1.24. *Let R be a ring.*

(a) *An element $a \in R$ is called* **nilpotent** *if there exists a positive integer k with $a^k = 0$.*
(b) *The set of all nilpotent elements is called the* **nilradical** *of R, written as* $\operatorname{nil}(R)$. *(So the nilradical is equal to the radical ideal $\sqrt{\{0\}}$ of the zero ideal, which by Corollary 1.12 is the intersection of all prime ideals.)*
(c) *R is called* **reduced** *if* $\operatorname{nil}(R) = \{0\}$. *(In particular, every integral domain is reduced.)*

Theorem 1.25 (Coordinate rings and reduced algebras). *Let K be a field.*

(a) *For every affine K-variety X, the coordinate ring $K[X]$ is a reduced affine K-algebra.*
(b) *Suppose that K is algebraically closed, and let A be a reduced affine K-algebra. Then there exists an affine K-variety X with $K[X] \cong A$.*

Proof. (a) With $I = \mathcal{I}(X)$, we have $K[X] = K[x_1, \ldots, x_n]/I$, so $K[X]$ is an affine algebra, and it is reduced since I is a radical ideal.
(b) Choose generators $a_1, \ldots, a_n$ of A. Then the epimorphism

$$\varphi\colon K[x_1, \ldots, x_n] \to A, \ f \mapsto f(a_1, \ldots, a_n)$$

yields $A \cong K[x_1,\ldots,x_n]/I$ with $I = \ker(\varphi)$. Since A is reduced, I is a radical ideal. Set $X := \mathcal{V}(I)$. By the Nullstellensatz (Theorem 1.17), $I = \mathcal{I}(X)$, so $A \cong K[X]$. □

Remark. The affine variety X in Theorem 1.25(b) is not uniquely determined. In fact, in the proof we have given, X depends on the choice of the generators of A. However, given the correct concept of an isomorphism of varieties (see Definition 3.4), it can be shown that all affine varieties with coordinate ring A are isomorphic. In fact, we get a bijective correspondence between isomorphism classes of affine K-varieties and isomorphism classes of reduced affine K-algebras. ◁

Exercises for Chapter 1

1.1 (Some counterexamples). Give examples which show that none of the hypotheses in Lemma 1.1(a) and (b) and in Proposition 1.2 can be omitted.

1.2 (Formal power series ring). Consider the formal power series ring

$$K[[x]] := \left\{ \sum_{i=0}^{\infty} a_i x^i \mid a_i \in K \right\}$$

over a field K.

(a) Show that $K[[x]]$ is an integral domain.
(b) Show that all power series $f = \sum_{i=0}^{\infty} a_i x^i$ with $a_0 \neq 0$ are invertible in $K[[x]]$. Assuming for a moment that K is only a ring, show that f is invertible if and only if a_0 is invertible in K.
(c) Show that $K[[x]]$ has exactly one maximal ideal $\mathfrak{m}$, i.e., $K[[x]]$ is a local ring (see Definition 6.7).
(d) Show that $K[[x]]$ is not a Jacobson ring.
(e) Show that the ring

$$L := \left\{ \sum_{i=k}^{\infty} a_i x^i \mid k \in \mathbb{Z},\ a_i \in K \right\}$$

of formal Laurent series is a field. The field L of formal Laurent series is often written as $K((x))$.
(f) Is $K[[x]]$ finitely generated as a K-algebra?

1.3 (Maximal spectrum and Rabinowitsch spectrum). Let R be a ring. Show that

$$\operatorname{Spec}_{\mathrm{max}}(R) \subseteq \operatorname{Spec}_{\mathrm{rab}}(R).$$

(Solution on page 217)

***1.4 (Three types of spectra).** Let $R = K[[y]]$ be the formal power series ring over a field K, and let $S = R[z]$ be a polynomial ring over R. Show that

$$\operatorname{Spec}_{\max}(S) \subsetneqq \operatorname{Spec}_{\text{rab}}(S) \subsetneqq \operatorname{Spec}(S).$$

Hint: Consider the ideals $(y)_S$ and $(z)_S$.

1.5 (Jacobson rings). Show that for verifying that a ring R is a Jacobson ring it is enough to check that every prime ideal $P \in \operatorname{Spec}(R)$ is an intersection of maximal ideals.

1.6 ($\mathbb{Z}$ is a Jacobson Ring). Show that the ring $\mathbb{Z}$ of integers is a Jacobson ring.

1.7 (Explicit computations with a variety). Consider the ideal

$$I = \left(x_1^4 + x_2^4 + 2x_1^2x_2^2 - x_1^2 - x_2^2\right) \subseteq \mathbb{R}[x_1, x_2].$$

(a) Determine $X := \mathcal{V}(I) \subseteq \mathbb{R}^2$ and draw a picture.
(b) Is I a prime ideal? Is I a radical ideal?
(c) Does Hilbert's Nullstellensatz (Theorem 1.17) hold for I?

1.8 (Colon ideals). If I and $J \subseteq R$ are ideals in a ring, the **colon ideal** is defined as

$$I : J := \{a \in R \mid a \cdot b \in I \text{ for all } b \in J\}.$$

In this exercise we give a geometric interpretation of the colon ideal.

(a) Set $\mathcal{M} := \{P \in \operatorname{Spec}(R) \mid I \subseteq P \text{ and } J \not\subseteq P\}$ and show that

$$\sqrt{I} : J = \bigcap_{P \in \mathcal{M}} P.$$

(b) Let K be a field and $X, Y \subseteq K^n$ such that Y is an affine variety. Show that

$$\mathcal{I}(X) : \mathcal{I}(Y) = \mathcal{I}(X \setminus Y).$$

1.9 (A generalization of Hilbert's Nullstellensatz). Let K be a field and $\overline{K}$ its algebraic closure. Let $I \subseteq K[x_1, \ldots, x_n]$ be an ideal in a polynomial ring. Show that

$$\mathcal{I}_{K[x_1,\ldots,x_n]}\left(\mathcal{V}_{\overline{K}^n}(I)\right) = \sqrt{I}.$$

1.10 (Order-reversing maps). This exercise puts Corollary 1.19 and its proof in a more general framework. Let $\mathcal{A}'$ and $\mathcal{B}'$ be two partially ordered sets. Let $\varphi\colon \mathcal{A}' \to \mathcal{B}'$ and $\psi\colon \mathcal{B}' \to \mathcal{A}'$ be maps satisfying the following properties:

(1) If $a_1, a_2 \in \mathcal{A}'$ with $a_1 \leq a_2$, then $\varphi(a_1) \geq \varphi(a_2)$;
(2) if $b_1, b_2 \in \mathcal{B}'$ with $b_1 \leq b_2$, then $\psi(b_1) \geq \psi(b_2)$;
(3) if $a \in \mathcal{A}'$, then $\psi(\varphi(a)) \geq a$;
(4) if $b \in \mathcal{B}'$, then $\varphi(\psi(b)) \geq b$.

Set $\mathcal{A} := \psi(\mathcal{B}')$ and $\mathcal{B} := \varphi(\mathcal{B}')$, and show that the restriction

$$\varphi|_{\mathcal{A}} \colon \mathcal{A} \to \mathcal{B}$$

is a bijection with inverse map $\psi|_{\mathcal{B}}$.
Remark: In the light of this exercise, all that is needed for the proof of Corollary 1.9 is that all radical ideals in $K[x_1, \ldots, x_n]$ occur as vanishing ideals of sets of points in K^n (which is a consequence of Theorem 1.17). Another typical situation in which this exercise applies is the correspondence between subgroups and intermediate fields in Galois theory.

1.11 (Points of a variety and homomorphisms). Let K be a (not necessarily algebraically closed) field and X a K-variety. Construct a bijection between X and the set

$$\mathrm{Hom}_K\left(K[X], K\right) := \{\varphi \colon K[X] \to K \mid \varphi \text{ is an algebra homomorphism}\}.$$

Remark: In the language of affine schemes, an algebra homomorphism $K[X] \to K$ induces a morphism $\mathrm{Spec}(K) \to \mathrm{Spec}(K[X])$. Such a morphism is called a K-rational point of the affine scheme associated to X.

Chapter 2
Noetherian and Artinian Rings

In this chapter we develop the theory of Noetherian and Artinian rings. In the first section, we will see that the Artin property, although in complete formal analogy to the Noether property, implies the Noether property and is, in fact, much more special (see Theorem 2.8). Both properties will also be considered for modules. In the second section, we concentrate on the Noether property. The most important results are Hilbert's basis theorem (Corollary 2.13) and its consequences. Using the Noether property often yields elegant but nonconstructive proofs. The most famous example is Hilbert's proof [27] that rings of invariants of GL_n and SL_n are finitely generated, which for its nonconstructive nature drew sharp criticism from Gordan, the "king of invariant theory" at the time, who exclaimed, "Das ist Theologie und nicht Mathematik!"[1]

2.1 The Noether and Artin Properties for Rings and Modules

Definition 2.1. *Let R be a ring and M an R-module.*

(a) *M is called* **Noetherian** *if the submodules of M satisfy the* ascending chain condition, *i.e., for submodules $M_1, M_2, M_3, \ldots \subseteq M$ with $M_i \subseteq M_{i+1}$ for all positive integers i, there exists an integer n such that $M_i = M_n$ for all $i \geq n$. In other words, every strictly ascending chain of submodules is finite.*
(b) *R is called* **Noetherian** *if R is Noetherian as a module over itself. In other words, R is Noetherian if the ideals of R satisfy the ascending chain condition.*
(c) *M is called* **Artinian** *if the submodules of M satisfy the* descending chain condition, *i.e., for submodules $M_1, M_2, M_3, \ldots \subseteq M$ with $M_{i+1} \subseteq M_i$ for*

[1] "This is theology and not mathematics."

G. Kemper, *A Course in Commutative Algebra*, Graduate Texts in Mathematics 256, DOI 10.1007/978-3-642-03545-6_3,

all positive integers i, there exists an integer n such that $M_i = M_n$ for all $i \geq n$.

(d) *R is called* **Artinian** *if R is Artinian as a module over itself, i.e., if the ideals of R satisfy the descending chain condition.*

Example 2.2. (1) The ring $\mathbb{Z}$ of integers is Noetherian, since ascending chains of ideals correspond to chains of integers $a_1, a_2, \ldots$ with a_{i+1} a divisor of a_i. So the well-ordering of the natural numbers yields the result.

(2) By the same argument, a polynomial ring $K[x]$ over a field is Noetherian. More trivially, every field is Noetherian.

(3) Let X be an infinite set and K a field (in fact, any nonzero ring will do). The set $R := K^X$ of all functions from X to K forms a ring with pointwise operations. For every subset $Y \subseteq X$, the set

$$I_Y := \{f \in R \mid f \text{ vanishes on } Y\}$$

is an ideal of R. Since there are infinite strictly descending chains of subsets of X, there are also infinite strictly ascending chains of ideals in R. So R is not Noetherian.

(4) The rings $\mathbb{Z}$ and $K[x]$ considered above are not Artinian.

(5) Every field and every finite ring or module is Artinian.

(6) The ring K^X, as defined in (3), is Artinian if and only if X is a finite set.

(7) Let $R := K[x]$ be a polynomial ring over a field. Then $S := R/(x^2)$ is Artinian, and S is also Artinian as an R-module. ◁

The ring from Example 2.2(3) is a rather pathological example of a non-Noetherian ring. In particular, it is not an integral domain. The following provides a less pathological counterexample.

Example 2.3. Let $S := K[x, y]$ be the polynomial ring in two indeterminates over a field K. Consider the subalgebra

$$R := K + S \cdot x = K[x, xy, xy^2, xy^3, \ldots].$$

It is shown in Exercise 2.1 that R is not Noetherian. ◁

The following proposition shows that the Noether property and the Artin property behave well with submodules and quotient modules.

Proposition 2.4 (Submodules and quotient modules). *Let M be a module over a ring R, and let $N \subseteq M$ be a submodule. Then the following statements are equivalent:*

(a) *M is Noetherian.*

(b) *Both N and the quotient module M/N are Noetherian.*

In particular, every quotient ring of a Noetherian ring is Noetherian.

All statements of this proposition hold with "Noetherian" replaced by "Artinian."

Proof. First assume that M is Noetherian. It follows directly from Definition 2.1 that N is Noetherian, too. To show that M/N is Noetherian, let $U_1, U_2, \ldots \subseteq M/N$ be an ascending chain of submodules. With $\varphi: M \to M/N$ the canonical epimorphism, set $M_i := \varphi^{-1}(U_i)$. This yields an ascending chain of submodules of M. By hypothesis, there exists an n with $M_i = M_n$ for $i \geq n$. Since $\varphi(M_i) = U_i$, it follows that $U_i = U_n$ for $i \geq n$. So we have shown that (a) implies (b).

Now assume that (b) is satisfied. To show (a), let $M_1, M_2, \ldots \subseteq M$ be an ascending chain of submodules. We obtain an ascending chain $\varphi(M_1), \varphi(M_2), \ldots \subseteq M/N$ of submodules of M/N. Moreover, the intersections $N \cap M_i \subseteq N$ yield an ascending chain of submodules of N. By hypothesis, there exists an n such that for $i \geq n$ we have $\varphi(M_i) = \varphi(M_n)$ and $N \cap M_i = N \cap M_n$. We claim that also $M_i = M_n$ for all $i \geq n$. Indeed, let $m \in M_i$. Then there exists an $m' \in M_n$ with $\varphi(m) = \varphi(m')$, so

$$m - m' \in N \cap M_i = N \cap M_n \subseteq M_n.$$

We conclude that $m = m' + (m - m') \in M_n$. So the equivalence of (a) and (b) is proved.

To prove the statement on quotient rings, observe that the ideals of a quotient ring R/I are precisely the submodules of R/I viewed as an R-module.

To get the proof for the case of Artinian modules, replace every occurrence of the word "ascending" in the above argument by "descending," and exchange "M_i" and "M_n" in the proof of $M_i = M_n$. □

We need the following definition to push the theory further.

Definition 2.5 (Ideal product). *Let R be a ring, $I \subseteq R$ and ideal, and M an R-module.*

(a) The product of I and M is defined to be the abelian group generated by all products $a \cdot m$ of elements from I and elements from M. So

$$IM = \left\{ \sum_{i=1}^{n} a_i m_i \,\middle|\, n \in \mathbb{N},\ a_i \in I,\ \text{and } m_i \in M \right\}.$$

Clearly $IM \subseteq M$ is a submodule.

(b) An interesting special case is that in which $M = J$ is another ideal of R. Then the product IJ is called the **ideal product**. *Clearly the formation of the ideal product is commutative and associative, and the rules*

$$IJ \subseteq I \cap J \quad \text{and} \quad \sqrt{IJ} = \sqrt{I \cap J}$$

hold (check this!).

(c) For $n \in \mathbb{N}_0$, I^n denotes the product of n copies of I, with $I^0 := R$.

The following lemma gives a connection between ideal powers and radical ideals.

Lemma 2.6 (Ideal powers and radical ideals). *Let R be a ring and $I, J \subseteq R$ ideals. If I is finitely generated, then*

$$I \subseteq \sqrt{J} \iff \text{there exists} \quad k \in \mathbb{N}_0 \quad \text{such that} \quad I^k \subseteq J.$$

Proof. We have $I = (a_1, \ldots, a_n)$. Suppose that $I \subseteq \sqrt{J}$. Then there exists $m > 0$ with $a_i^m \in J$ for $i = 1, \ldots, n$. Set $k := n \cdot (m-1) + 1$. We need to show that the product of k arbitrary elements from I lies in J. So let $x_1, \ldots, x_k \in I$ and write

$$x_i = \sum_{j=1}^{n} r_{i,j} a_j \quad \text{with} \quad r_{i,j} \in R.$$

When we multiply out the product $x_1 \cdots x_k$, we find that every summand has some a_j^m as a subproduct. Therefore $x_1 \cdots x_k \in J$. This shows that $I^k \subseteq J$.

The converse statement is clear (and does not require finite generation of I). □

Theorem 2.8, which we start proving now, gives a comparison between the Noether property and the Artin property for rings. Readers who are mainly interested in the Noether property can continue with reading Section 2.2. Theorem 2.8 will not be used before Chapter 7.

Lemma 2.7. *Let R be a ring and $\mathfrak{m}_1, \ldots, \mathfrak{m}_n \in \operatorname{Spec}_{\max}(R)$ maximal ideals (which are not assumed to be distinct) such that the ideal product $\mathfrak{m}_1 \cdots \mathfrak{m}_n$ is zero. Then R is Artinian if and only if it is Noetherian. Moreover,*

$$\operatorname{Spec}(R) = \{\mathfrak{m}_1, \ldots, \mathfrak{m}_n\}.$$

Proof. Setting

$$I_i := \mathfrak{m}_1 \cdots \mathfrak{m}_i,$$

we get a chain

$$\{0\} = I_n \subseteq I_{n-1} \subseteq \cdots \subseteq I_2 \subseteq I_1 \subseteq I_0 := R$$

of ideals. Applying Proposition 2.4 repeatedly, we see that R is Noetherian (Artinian) if and only if every quotient module I_{i-1}/I_i is Noetherian (Artinian). But $\mathfrak{m}_i \cdot (I_{i-1}/I_i) = \{0\}$, so I_{i-1}/I_i is a vector space over the field $K_i := R/\mathfrak{m}_i$, and a subset of I_{i-1}/I_i is an R-submodule if and only if it is a K_i-subspace. So both the Noether and the Artin property for I_{i-1}/I_i are equivalent to $\dim_{K_i}(I_{i-1}/I_i) < \infty$. This yields the claimed equivalence.

To prove the second claim, take $P \in \operatorname{Spec}(R)$. By hypothesis, $\mathfrak{m}_1 \cdots \mathfrak{m}_n \subseteq P$. From the primality of P and the definition of the ideal product, we conclude that there exists i with $\mathfrak{m}_i \subseteq P$, so $P = \mathfrak{m}_i$. □

Theorem 2.8 (Artinian and Noetherian rings). *Let R be a ring. Then the following statements are equivalent:*

(a) R is Artinian.
(b) R is Noetherian and every prime ideal of R is maximal.

Using the concept of dimension as defined in Definition 5.1, the condition (b) in Theorem 2.8 can be rephrased as, "R is Noetherian and has dimension 0 or -1" (where -1 occurs if and only if R is the zero ring). We prove only the implication "(a) $\Rightarrow$ (b)" here and postpone the proof of the converse to the end of Chapter 3 (see page 42).

Proof of "(a) $\Rightarrow$ (b)". Suppose that R is Artinian. The first claim is that R has only finitely many maximal ideals. Assume the contrary. Then there exist infinitely many pairwise distinct maximal ideals $\mathfrak{m}_1, \mathfrak{m}_2, \mathfrak{m}_3, \ldots \in \operatorname{Spec}_{\max}(R)$. Setting $I_i := \bigcap_{j=1}^{i} \mathfrak{m}_j$ yields a descending chain of ideals, so by hypothesis there exists n such that $I_{n+1} = I_n$. This implies $\bigcap_{j=1}^{n} \mathfrak{m}_j \subseteq \mathfrak{m}_{n+1}$, so there exists $j \leq n$ with $\mathfrak{m}_j = \mathfrak{m}_{n+1}$, a contradiction. We conclude that there exist finitely many maximal ideals $\mathfrak{m}_1, \ldots, \mathfrak{m}_k$. Setting

$$I := \mathfrak{m}_1 \cdots \mathfrak{m}_k,$$

we obtain a descending chain of ideals I^i, $i \in \mathbb{N}_0$, so there exists $n \in \mathbb{N}_0$ with

$$I^i = I^n =: J \quad \text{for} \quad i \geq n. \tag{2.1}$$

By way of contradiction, assume $J \neq \{0\}$. Then the set

$$\mathcal{M} := \{J' \subseteq R \mid J' \text{ is an ideal and } J'J \neq \{0\}\}$$

is nonempty. There exists a minimal element $\widehat{J}$ in $\mathcal{M}$, since otherwise $\mathcal{M}$ would contain an infinite, strictly descending chain of ideals. Pick an $x \in \widehat{J}$ with $xJ \neq \{0\}$. Then $\widehat{J} = (x)$ by the minimality. Moreover, (2.1) implies that $J^2 = J$, so

$$(x)J \cdot J = (x)J^2 = (x)J \neq \{0\},$$

so $(x)J = (x)$ again by the minimality of $\widehat{J}$. Therefore there exists $y \in J$ with $xy = x$. By the definition of J, y lies in every maximal ideal of R, and so $y - 1$ lies in *no* maximal ideal. This means that $y - 1$ is invertible, and $(y - 1)x = 0$ implies $x = 0$. This contradicts $xJ \neq \{0\}$. We conclude that $J = \{0\}$. So we can apply Lemma 2.7 and get that R is Noetherian and that every prime ideal is maximal. □

Theorem 2.8 raises the question whether it is also true that every Artinian module over a ring is Noetherian. This is answered in the negative by Exercise 2.2.

2.2 Noetherian Rings and Modules

The following theorem gives an alternative definition of Noetherian modules. There is no analogue for Artinian modules.

Theorem 2.9 (Alternative definition of Noetherian modules). *Let R be a ring and M an R-module. The following statements are equivalent:*

(a) M is Noetherian.
(b) For every subset $S \subseteq M$ there exist finitely many elements $m_1, \dots, m_k \in S$ such that

$$(S)_R = (m_1, \dots, m_k)_R.$$

(c) Every submodule of M is finitely generated.

In particular, R is Noetherian if and only if every ideal of R is finitely generated, and then every generating set of an ideal contains a finite generating subset.

Proof. Assume that M is Noetherian, but there exists $S \subseteq M$ that does not satisfy (b). We define finite subsets $S_i \subseteq S$ $(i = 1, 2, \dots)$ recursively, starting with $S_1 = \emptyset$. Suppose S_i has been defined. Since S does not satisfy (b), there exists $m_{i+1} \in S \setminus (S_i)_R$. Set

$$S_{i+1} := S_i \cup \{m_{i+1}\}.$$

(In fact, the axiom of choice is needed to make this definition precise.) By construction we have $(S_i)_R \subsetneqq (S_{i+1})_R$ for all i, contradicting (a). So (a) implies (b), and it is clear that (b) implies (c).

So suppose that (c) holds, and let $M_1, M_2, \dots \subseteq M$ be an ascending chain of submodules. Let $N := \cup_{i \in \mathbb{N}} M_i$ be the union. It is easy to check that N is a submodule, so by (c) we have $N = (m_1, \dots, m_k)_R$ with $m_j \in N$. Each m_j lies in some M_{i_j}. Let $n := \max\{i_1, \dots, i_k\}$. Then all m_j lie in M_n, so for $i \geq n$ we have

$$M_i \subseteq N = (m_1, \dots, m_k)_R \subseteq M_n \subseteq M_i,$$

which implies equality. Therefore (a) holds. □

Theorem 2.9 implies that every Noetherian module over a ring is finitely generated. This raises the question whether the converse is true, too. But this is clearly false in general: If R is a non-Noetherian ring, then R is not Noetherian as a module over itself, but it is finitely generated (with 1 the only generator). The following theorem shows that if the converse does not go wrong in this very simple way, then in fact it holds.

Theorem 2.10 (Noetherian modules and finite generation). *Let R be a Noetherian ring and M an R-module. Then the following statements are equivalent:*

(a) M is Noetherian.
(b) M is finitely generated.

In particular, every submodule of a finitely generated R-module is also finitely generated.

Proof. We need to show only that (b) implies (a), since the converse implication is a consequence of Theorem 2.9. So let $M = (m_1, \dots, m_k)_R$. We use induction on k. There is nothing to show for $k = 0$, so assume $k > 0$. Consider the submodule

$$N := (m_1, \dots, m_{k-1})_R \subseteq M.$$

By induction, N is Noetherian. The homomorphism

$$\varphi\colon R \to M/N, \ a \mapsto am_k + N,$$

is surjective, so $M/N \cong R/\ker(\varphi)$. By hypothesis and by Proposition 2.4, $R/\ker(\varphi)$ is Noetherian, so M/N is Noetherian, too. Applying Proposition 2.4 again shows that M is Noetherian. □

The following theorem is arguably the most important result on Noetherian rings.

Theorem 2.11 (Polynomial rings over Noetherian rings). *Let R be a Noetherian ring. Then the polynomial ring $R[x]$ is Noetherian, too.*

Proof. Let $I \subseteq R[x]$ be an ideal. By Theorem 2.9, we need to show that I is finitely generated. For a nonnegative integer i, set

$$J_i := \left\{a_i \in R \mid \text{there exist } a_0, \dots, a_{i-1} \in R \text{ such that } \sum_{j=0}^{i} a_j x^j \in I\right\}.$$

Clearly $J_i \subseteq R$ is an ideal. Let $a_i \in J_i$ with $f = \sum_{j=0}^{i} a_j x^j \in I$. Then $I \ni xf = \sum_{j=0}^{i} a_j x^{j+1}$, so $a_i \in J_{i+1}$. It follows that the J_i form an ascending chain of ideals of R. By hypothesis, there exists an n such that for $i \geq n$ we have $J_i = J_n$. Again by hypothesis, every J_i is finitely generated, so

$$J_i = (a_{i,1}, \dots, a_{i,m_i})_R \quad \text{for} \quad i \leq n \tag{2.2}$$

and

$$J_i = J_n = (a_{n,1}, \dots, a_{n,m_n})_R \quad \text{for} \quad i > n. \tag{2.3}$$

By the definition of J_i, there exist polynomials $f_{i,j} \in I$ of degree at most i whose ith coefficient is $a_{i,j}$. Set

$$I' := \left(f_{i,j} \mid i = 0, \dots, n, \ j = 1, \dots, m_i\right)_{R[x]} \subseteq I.$$

We claim that $I = I'$. To prove the claim, consider a polynomial $f = \sum_{i=0}^{d} b_i x^i \in I$ with $\deg(f) = d$. We use induction on d. We first consider the case $d \leq n$. Since $b_d \in J_d$, we can use (2.2) and write $b_d = \sum_{j=1}^{m_d} r_j a_{d,j}$ with $r_j \in R$. Then

$$\widetilde{f} := f - \sum_{j=1}^{m_d} r_j f_{d,j}$$

lies in I and has degree less than d, so by induction $\widetilde{f} \in I'$. This implies $f \in I'$. Now assume $d > n$. Then we can use (2.3) and write $b_d = \sum_{j=1}^{m_n} r_j a_{n,j}$ with $r_j \in R$. So

$$\widetilde{f} := f - \sum_{j=1}^{m_n} r_j x^{d-n} f_{n,j}$$

lies in I and has degree less than d, so by induction $\widetilde{f} \in I'$. Again we conclude that $f \in I'$. So indeed $I = I'$ is a finitely generated ideal. □

The corresponding statement for formal power series rings is contained in Exercise 2.4. By applying Theorem 2.11 repeatedly and using the second statement of Proposition 2.4, we obtain the following corollary.

Corollary 2.12 (Finitely generated algebras). *Every finitely generated algebra over a Noetherian ring is Noetherian. In particular, every affine algebra is Noetherian.*

A special case is the celebrated basis theorem of Hilbert.

Corollary 2.13 (Hilbert's basis theorem). *Let K be a field. Then the polynomial ring $K[x_1, \ldots, x_n]$ is Noetherian. In particular, every ideal in $K[x_1, \ldots, x_n]$ is finitely generated.*

The name *basis theorem* comes from the fact that generating sets of ideals are sometimes called *bases*. One consequence is that every affine variety $X \subseteq K^n$ is the solution set of a *finite* system of polynomial equations: $X = \mathcal{V}(f_1, \ldots, f_m)$.

Exercises for Chapter 2

2.1 (A non-Noetherian ring providing many counterexamples). Consider the polynomial ring $S = K[x, y]$ and the subalgebra $R := K + S \cdot x$ given in Example 2.3. Show that R is not Noetherian. Conclude that R is not finitely generated as an algebra. Explain why this provides an example for the following caveats:

- Subrings of Noetherian rings need not be Noetherian
- Subalgebras of finitely generated algebras need not be finitely generated

In Exercise 7.4 we will also see that Krull's principal ideal theorem (Theorem 7.4) fails for R. In Exercise 2.6 we explore whether Example 2.3 is, in some sense, the smallest of its kind.

2.2 (An Artinian module that is not Noetherian). Let $p \in \mathbb{N}$ be a prime number and consider the $\mathbb{Z}$-modules

$$\mathbb{Z}_p := \left\{a/p^n \in \mathbb{Q} \mid a, n \in \mathbb{Z}\right\} \subset \mathbb{Q} \quad \text{and} \quad M := \mathbb{Z}_p/\mathbb{Z}.$$

Show that M is Artinian but not Noetherian.

2.3 (Modules over an Artinian ring). Show that a finitely generated module M over an Artinian ring R is Artinian.

2.4 (The Noether property for formal power series rings). Let R be a Noetherian ring and

$$R[[x]] := \left\{\sum_{i=0}^{\infty} a_i x^i \mid a_i \in R\right\}$$

the formal power series ring over R. Show that $R[[x]]$ is Noetherian.

2.5 (Separating subsets). Let K be a field and $A \subseteq K[x_1, \ldots, x_n]$ a subalgebra of a polynomial algebra (which, as we have seen in Example 2.3, need not be finitely generated). Every polynomial from A defines a function $K^n \to K$. A subset $S \subseteq A$ is called **(A-)separating** if the following condition holds for all points $P_1, P_2 \in K^n$:

> If there exists $f \in A$ with $f(P_1) \neq f(P_2)$, then there exists $f \in S$ with $f(P_1) \neq f(P_2)$.

(Loosely speaking, this means that S has the same capabilities of separating points as A.)

(a) Show that if $S \subseteq A$ generates A as an algebra, then S is separating. (In other words, "separating" is a weaker condition than "generating." It is seen in (b) and (c) that it is in fact substantially weaker.)
*(b) Show that A has a finite separating subset.
(c) Exhibit a finite R-separating subset of the algebra $R \subset K[x, y]$ from Example 2.3.

(Solution on page 217)

***2.6 (Subalgebras of $K[x]$).** Let K be a field and $K[x]$ a polynomial ring in one indeterminate. Is every subalgebra of $K[x]$ finitely generated? Give a proof or a counterexample.

2.7 (Graded rings). A ring R is called **graded** if it has a direct sum decomposition

$$R = R_0 \oplus R_1 \oplus R_2 \oplus \cdots = \bigoplus_{d \in \mathbb{N}_0} R_d$$

(as an abelian group) such that for all $a \in R_i$ and $b \in R_j$ one has $ab \in R_{i+j}$. Then an element from R_d is called **homogeneous** of degree d. A standard example is $R = K[x_1, \ldots, x_n]$ with R_d the space of all homogeneous polynomials of degree d (including the zero polynomial). Let R be graded and set

$$I = \bigoplus_{d \in \mathbb{N}_{>0}} R_d,$$

which obviously is an ideal. Sometimes I is called the *irrelevant ideal.* Prove the equivalence of the following statements.

(a) R is Noetherian.
(b) R_0 is Noetherian and I is finitely generated.
(c) R_0 is Noetherian and R is finitely generated as an R_0-algebra.

Remark: By Corollary 2.12, a finitely generated algebra over a Noetherian rings is Noetherian. However, Noetherian algebras are not always finitely generated. So graded rings constitute a special case in which this converse holds.

2.8 (The Noether property and subrings). In Exercise 2.1 we have seen that in general the Noether property does not go down to subrings. In this exercise we look at a situation in which it does.

(a) Let S be a Noetherian ring and $R \subseteq S$ a subring such that there exists a homomorphism $\varphi\colon S \to R$ of R-modules with $\varphi|_R = \mathrm{id}_R$. Show that R is Noetherian, too.
(b) Show that for a ring R, the following three statements are equivalent: (i) R is Noetherian; (ii) $R[x]$ is Noetherian; (iii) $R[[x]]$ is Noetherian.

2.9 (Right or wrong?). Decide whether each of following statements is true or false. Give reasons for your answers.

(a) Every finitely generated module over an Artinian ring is Artinian.
(b) Every Artinian module is finitely generated.
(c) Every ring has a module that is both Noetherian and Artinian.
(d) The set of all ideals of a ring, together with the ideal sum and ideal product, forms a commutative semiring (i.e., we have an additive and a multiplicative commutative monoid, and a distributive law).

2.10 (The ring of analytic functions). Let R be the ring of all analytic functions $\mathbb{R} \to \mathbb{R}$, i.e., all functions that are given by power series that converge on all of $\mathbb{R}$. Show that R is not Noetherian.

Can your argument be used for showing that other classes of functions $\mathbb{R} \to \mathbb{R}$ form non-Noetherian rings, too?

Chapter 3
The Zariski Topology

In this chapter we will put a topology on K^n and on affine varieties. This topology is quite weak, but surprisingly useful. We will define an analogous topology on $\operatorname{Spec}(R)$. In both cases, there are correspondences between closed sets and radical ideals. As a consequence of some general topological considerations, affine varieties can be decomposed into irreducible components. Another consequence is that a Noetherian ring contains only finitely many minimal prime ideals.

Readers who are unfamiliar with the language of topology can find all that is needed for this book in any textbook on topology (for example Bourbaki [6]), usually on the first few pages.

3.1 Affine Varieties

In this section we define the Zariski topology on K^n and on its subsets. We first need a proposition.

Proposition 3.1 (Unions and intersections of affine varieties). *Let $K[x_1, \ldots, x_n]$ be a polynomial ring over a field K.*

(a) Let $I, J \subseteq K[x_1, \ldots, x_n]$ be ideals. Then

$$\mathcal{V}(I) \cup \mathcal{V}(J) = \mathcal{V}(I \cap J).$$

(b) Let $\mathcal{M}$ be a nonempty set of subsets of $K[x_1, \ldots, x_n]$. Then

$$\bigcap_{S \in \mathcal{M}} \mathcal{V}(S) = \mathcal{V}\Big(\bigcup_{S \in \mathcal{M}} S\Big).$$

Proof. We first prove (a). It is clear that $\mathcal{V}(I) \cup \mathcal{V}(J) \subseteq \mathcal{V}(I \cap J)$. To prove the reverse inclusion, let $P \in \mathcal{V}(I \cap J)$. Assume $P \notin \mathcal{V}(I)$, so there exists

G. Kemper, *A Course in Commutative Algebra*, Graduate Texts in Mathematics 256, DOI 10.1007/978-3-642-03545-6_4,

$f \in I$ with $f(P) \neq 0$. We need to show that $P \in \mathcal{V}(J)$, so let $g \in J$. Then $fg \in I \cap J$, so $f(P)g(P) = 0$. But this implies $g(P) = 0$.

Part (b) is clear. □

Proposition 3.1 tells us that finite unions and arbitrary intersections of affine varieties in K^n are again affine varieties. Since K^n and $\emptyset$ are also affine varieties, this suggests that we can define a topology using the affine varieties as closed sets. This is exactly what we will do.

Definition 3.2. *Let K be a field and n a positive integer. Then the* **Zariski topology** *is defined on K^n by saying that a subset $X \subseteq K^n$ is (Zariski) closed if and only if X is an affine variety. On a subset $Y \subseteq K^n$, we define the Zariski topology to be the subset topology induced by the Zariski topology on K^n, i.e., the closed subsets in Y are the intersections of closed subsets in K^n with Y.*

We make a few remarks.

Remark 3.3. (a) By definition, the closed subsets of K^n have the form $\mathcal{V}(S)$ with $S \subseteq K[x_1, \ldots, x_n]$ a subset. By Lemma 1.18, we may substitute S by $\mathcal{I}(X)$, i.e., we may assume S to be an ideal, and in fact even a radical ideal.

(b) For a subset $X \subseteq K^n$, the topological closure (also called the Zariski closure) is

$$\overline{X} = \mathcal{V}\left(\mathcal{I}(X)\right).$$

(c) If $Y \subseteq K^n$ is an affine variety, then by definition the Zariski topology on Y has the subvarieties of Y as closed sets.

(d) On $\mathbb{R}^n$ and $\mathbb{C}^n$, the Zariski topology is coarser than the usual Euclidean topology.

(e) Every finite subset of K^n is Zariski closed. In other words, K^n is a T_1 space. This also applies to every subset $Y \subseteq K^n$.

(f) On the "affine line" K^1, the closed subsets are precisely the finite subsets, and all of K^1. So the Zariski topology is the coarsest topology for which singletons (i.e., sets with one element) are closed. This illustrates how much coarser the Zariski topology is compared to the usual topology on $\mathbb{R}$ or $\mathbb{C}$.

(g) All polynomials $f \in K[x_1, \ldots, x_n]$, viewed as functions $K^n \to K$, are continuous with respect to the Zariski topology. In fact, the Zariski topology is the coarsest topology such that all polynomials are continuous (assuming that $\{0\} \subset K^1$ is closed).
On the other hand, there exist continuous functions $K^n \to K$ that are not polynomials, e.g., the function $\mathbb{C} \to \mathbb{C}$, $x \mapsto \overline{x}$ (complex conjugation).

(h) The Zariski-open subsets of K^n are unions of solution sets of polynomial inequalities.

(i) Recall that a Hausdorff space (also called a T_2 space) is a topological space in which for any two distinct points $P_1 \neq P_2$ there exist disjoint open sets U_1 and U_2 with $P_i \in U_i$. If K is an infinite field, then K^n with the Zariski topology is never Hausdorff. In fact, it is not hard to see that two nonempty open subsets $U_1, U_2 \subseteq K^n$ always intersect. This is extended in Exercise 3.7, where it is shown that no infinite subset of K^n is Hausdorff. ◁

Further examples of continuous maps are morphisms of varieties, which we deal with now.

Definition 3.4. *Let K be a field and let $X \subseteq K^m$ and $Y \subseteq K^n$ be affine varieties. A map $f: X \to Y$ is called a* **morphism (of varieties)** *if there exist polynomials $f_1, \ldots, f_n \in K[x_1, \ldots, x_m]$ such that f is given by*

$$f(P) = (f_1(P), \ldots, f_n(P)) \quad \textit{for} \quad P \in X.$$

We write $\mathrm{Mor}(X,Y)$ for the set of all morphisms $X \to Y$. Since compositions of morphisms are obviously again morphisms, this definition makes the collection of affine K-varieties into a category.

A morphism $f: X \to Y$ is called an **isomorphism** *if there exists a morphism $g: Y \to X$ with $f \circ g = \mathrm{id}_Y$ and $g \circ f = \mathrm{id}_X$. In particular, every isomorphism is a homeomorphism (i.e., a topological isomorphism).*

In particular, the regular functions on X are precisely the morphisms $X \to K^1$ (see Remark 1.21 (a)).

Let $f: X \to Y$ be a morphism given by polynomials $f_1, \ldots, f_n$. Then we have a homomorphism of K-algebras $\varphi: K[Y] \to K[X]$ given as follows: If $K[X] = K[x_1, \ldots, x_m]/\mathcal{I}(X)$ and $K[Y] = K[y_1, \ldots, y_n]/\mathcal{I}(Y)$, then

$$\varphi(y_i + \mathcal{I}(Y)) := f_i + \mathcal{I}(X).$$

It is routine to check that this is well defined. The homomorphism φ is said to be **induced** from f. Assigning coordinate rings to affine varieties and assigning induced homomorphisms to morphisms provides a contravariant functor from the category of affine K-varieties to the category of affine K-algebras.

We have a reverse process. Indeed, if $\varphi: K[Y] \to K[X]$ is an algebra homomorphism, then we have polynomials $f_1, \ldots, f_n \in K[x_1, \ldots, x_m]$ with $\varphi(y_i + \mathcal{I}(Y)) = f_i + \mathcal{I}(X)$, and it is easy to check that these f_i define a morphism $f: X \to Y$, which does not depend on the choice of the f_i. Again, it is routine to check that the assignment of a homomorphism $K[Y] \to K[X]$ to a morphism $X \to Y$ and vice versa provides a pair of inverse bijections $\mathrm{Mor}(X,Y) \leftrightarrow \mathrm{Hom}_K(K[Y], K[X])$.

Finally, we remark that a bijective morphism $X \to Y$ is not necessarily an isomorphism. For example, if $X \subseteq K^2$ is the union of the hyperbola $\{(\xi_1, \xi_2) \in K^2 \mid \xi_1\xi_2 = 1\}$ and the singleton $\{(0,1)\}$, and $f: X \to K^1$ is the first projection, then f is a bijective morphism, but not an isomorphism. This is shown in Fig. 3.1.

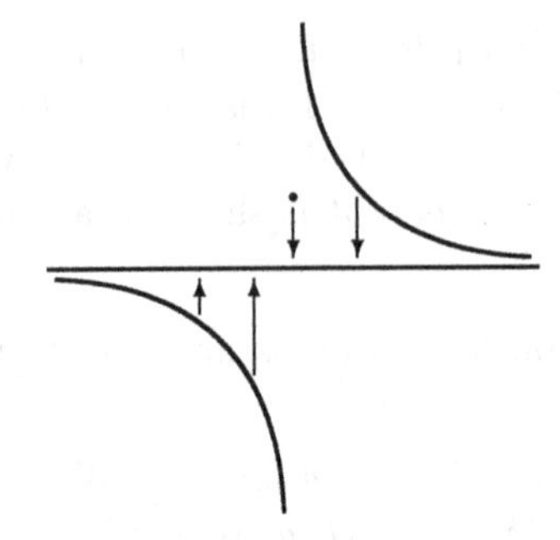

Fig. 3.1. A bijective morphism that is not an isomorphism

3.2 Spectra

In Theorem 1.23 we have seen a one-to-one correspondence between maximal ideals of $K[X]$ and points of X. This suggests that prime ideals can be seen as some sort of "generalized points." Following this idea, we will define a topology on $\operatorname{Spec}(R)$ for any ring R.

Definition 3.5. *Let R be a ring. For a subset $S \subseteq R$ we write*

$$\mathcal{V}_{\operatorname{Spec}(R)}(S) := \{P \in \operatorname{Spec}(R) \mid S \subseteq P\}.$$

For a subset $X \subseteq \operatorname{Spec}(R)$ we write

$$\mathcal{I}_R(X) := \bigcap_{P \in X} P \subseteq R \quad \text{if} \quad X \neq \emptyset, \quad \text{and} \quad \mathcal{I}_R(\emptyset) := R.$$

The **Zariski topology** *on $\operatorname{Spec}(R)$ is defined by saying that all sets of the form $\mathcal{V}_{\operatorname{Spec}(R)}(S)$ with $S \subseteq R$ are closed. By (a) and (b) of the following Proposition 3.6, and since $\emptyset = \mathcal{V}_{\operatorname{Spec}(R)}(\{1\})$ and $\operatorname{Spec}(R) = \mathcal{V}_{\operatorname{Spec}(R)}(\emptyset)$, this indeed defines a topology.*

A subset of $\operatorname{Spec}(R)$ is equipped with the subspace topology induced from the Zariski topology on $\operatorname{Spec}(R)$.

The following proposition contains all the important general facts about the maps $\mathcal{V}_{\operatorname{Spec}(R)}$ and $\mathcal{I}_R$ defined above. In particular, part (e) is an analogy to the ideal–variety correspondence in Corollary 1.19.

Proposition 3.6 (Properties of $\mathcal{V}_{\operatorname{Spec}(R)}$ and $\mathcal{I}_R$). *Let R be a ring.*

(a) Let $S, T \subseteq R$ be subsets. Then

$$\mathcal{V}_{\operatorname{Spec}(R)}(S) \cup \mathcal{V}_{\operatorname{Spec}(R)}(T) = \mathcal{V}_{\operatorname{Spec}(R)}\Big((S)_R \cap (T)_R\Big).$$

(b) Let $\mathcal{M}$ be a nonempty set of subsets of R. Then

$$\bigcap_{S \in \mathcal{M}} \mathcal{V}_{\mathrm{Spec}(R)}(S) = \mathcal{V}_{\mathrm{Spec}(R)}\Big(\bigcup_{S \in \mathcal{M}} S\Big).$$

(c) Let $X \subseteq \mathrm{Spec}(R)$ be a subset. Then $\mathcal{I}_R(X)$ is a radical ideal of R.
(d) Let $I \subseteq R$ be an ideal. Then

$$\mathcal{I}_R\left(\mathcal{V}_{\mathrm{Spec}(R)}(I)\right) = \sqrt{I}.$$

(e) We have a pair of inverse bijections between the set of radical ideals of R and the set of closed subsets of $\mathrm{Spec}(R)$, given by $\mathcal{V}_{\mathrm{Spec}(R)}$ and $\mathcal{I}_R$. Both bijections are inclusion-reversing.

Proof. (a) If $P \in \mathcal{V}_{\mathrm{Spec}(R)}(S)$, then $S \subseteq P$, so also $(S)_R \subseteq P$ and $(S)_R \cap (T)_R \subseteq P$. The same follows if $P \in \mathcal{V}_{\mathrm{Spec}(R)}(T)$, so in both cases $P \in \mathcal{V}_{\mathrm{Spec}(R)}\left((S)_R \cap (T)_R\right)$. Conversely, let $P \in \mathcal{V}_{\mathrm{Spec}(R)}\left((S)_R \cap (T)_R\right)$ and assume $S \not\subseteq P$. So there exists $f \in S \setminus P$. Let $g \in T$. Then $fg \in (S)_R \cap (T)_R$, so $fg \in P$. Since P is a prime ideal, $g \in P$ follows, so $P \in \mathcal{V}_{\mathrm{Spec}(R)}(T)$.

(b) is clear.

(c) This follows since prime ideals are always radical ideals, and intersections of radical ideals are again radical ideals.

(d) is a restatement of Corollary 1.12.

(e) In the light of (c) and (d), we need to show only that $\mathcal{V}_{\mathrm{Spec}(R)}\left(\mathcal{I}_R(X)\right) = X$ for $X \subseteq \mathrm{Spec}(R)$ a closed subset. We have $X = \mathcal{V}_{\mathrm{Spec}(R)}(S)$ with $S \subseteq R$, so $S \subseteq \mathcal{I}_R(X)$. Since the map $\mathcal{V}_{\mathrm{Spec}(R)}$ is inclusion-reversing, we obtain

$$\mathcal{V}_{\mathrm{Spec}(R)}\left(\mathcal{I}_R(X)\right) \subseteq \mathcal{V}_{\mathrm{Spec}(R)}(S) = X \subseteq \mathcal{V}_{\mathrm{Spec}(R)}\left(\mathcal{I}_R(X)\right).$$

This completes the proof. □

In Theorem 1.23 we have exhibited a bijection between points from an affine variety and maximal ideals of its coordinate ring. In Exercise 3.3 it is shown that this map is actually a homeomorphism. This emphasizes our point that prime ideals can be seen as generalized points. It may also be interesting to note that a ring R is a Jacobson ring if and only if for every closed subset $Y \subseteq \mathrm{Spec}(R)$ we have that $\mathrm{Spec}_{\mathrm{max}}(R) \cap Y$ is dense in Y. (Recall that a subset in a topological space is called *dense* if its closure is the whole space.) In fact, this is nothing but a translation of the Jacobson property.

To every ring R we have assigned a topological space $\mathrm{Spec}(R)$. We will make this assignment into a contravariant functor as follows. Let R and S be rings and let $\varphi: R \to S$ be a homomorphism. For every $P \in \mathrm{Spec}(S)$, the preimage $\varphi^{-1}(P)$ is obviously a prime ideal of R, so we obtain a map

$$\varphi^*\colon \mathrm{Spec}(S) \to \mathrm{Spec}(R),\ P \mapsto \varphi^{-1}(P).$$

We will often say that φ^* is **induced** from φ. For $I \subseteq R$ a subset, we have $(\varphi^*)^{-1}\left(\mathcal{V}_{\text{Spec}(R)}(I)\right) = \mathcal{V}_{\text{Spec}(S)}\left(\varphi(I)\right)$, so φ^* is continuous. Maps between spectra of rings that are induced from ring homomorphisms are called *morphisms*.

Going from φ to φ^* is compatible with, and a generalization of, the process of obtaining a morphism $X \to Y$ of affine varieties from a homomorphism $K[Y] \to K[X]$ described on page 35. However, there is no return path from φ^* to φ. In fact, different ring homomorphisms can yield the same induced map even if the rings are reduced. Consider, for example, complex conjugation $\mathbb{C} \to \mathbb{C}$ on the one hand, and the identical map $\mathbb{C} \to \mathbb{C}$ on the other. This behavior is a little unsatisfactory, and it is due to the fact that we were too naive when we assigned φ^* to φ. In algebraic geometry, a morphism between two spectra (more generally, between two ringed spaces) consists of a continuous map between the spectra as topological spaces, together with a morphism of sheaves (see Hartshorne [26, page 72]). This richer concept of a morphism does allow going back and forth between ring homomorphisms and morphisms of spectra.

3.3 Noetherian and Irreducible Spaces

Motivated by the correspondence between ideals and Zariski-closed subsets, we can transport the definition of the Noether property to topological spaces in general.

Definition 3.7. *Let X be a topological space.*

(a) X is called **Noetherian** *if the closed subsets of X satisfy the* descending chain condition, *i.e., for closed subsets $Y_1, Y_2, Y_3, \ldots \subseteq X$ with $Y_{i+1} \subseteq Y_i$ for all positive integers i, there exists an integer n such that $Y_i = Y_n$ for all $i \geq n$. An equivalent condition is that the open subsets satisfy the ascending chain condition.*

(b) X is called **irreducible** *if X is not the union of two proper, closed subsets, and $X \neq \emptyset$. An equivalent condition is that any two nonempty open subsets of X have a nonempty intersection, and $X \neq \emptyset$.*

Example 3.8. (1) $\mathbb{R}$ and $\mathbb{C}$ with the usual Euclidean topology are neither Noetherian nor irreducible.

(2) Every finite space is Noetherian.

(3) Every singleton is irreducible.

(4) If K is an infinite field, then $X = K^1$ with the Zariski topology is irreducible, since the closed subsets are X and its finite subsets. More generally, we will see that K^n is irreducible. (This is a consequence of Theorem 3.10). ◁

The topological spaces that we normally deal with in analysis are almost never Noetherian or irreducible. For example, a Hausdorff space can be irreducible only if it is a singleton (this is obvious), and it can be Noetherian only if it is finite (see Exercise 3.7). However, the following two theorems show that the situation is much better when we consider spaces with the Zariski topology.

Theorem 3.9 (Noether property of the Zariski topology).

(a) Let K be a field and $X \subseteq K^n$ a set of points, equipped with the Zariski topology. Then X is Noetherian.
(b) Let R be a Noetherian ring and $X \subseteq \operatorname{Spec}(R)$ a set of prime ideals, equipped with the Zariski topology. Then X is Noetherian.

Proof. First observe that if X is any Noetherian topological space and $Y \subseteq X$ is a subset equipped with the subset topology, then Y is also Noetherian. So we may assume $X = K^n$ in part (a), and $X = \operatorname{Spec}(R)$ in part (b). To prove (a), let $Y_1, Y_2, Y_3, \ldots \subseteq K^n$ be a descending chain of closed subsets. Then $I_i := \mathcal{I}_{K[x_1,\ldots,x_n]}(Y_i)$ yields an ascending chain of ideals, so by Hilbert's basis theorem (Corollary 2.13), there exists n with $I_i = I_n$ for $i \geq n$. By Lemma 1.18, $Y_i = \mathcal{V}_{K^n}(I_i)$, so $Y_i = Y_n$ for $i \geq n$.

Part (b) follows directly from Proposition 3.6(e). □

In particular, $\operatorname{Spec}(R)$ is a Noetherian space if R is a Noetherian ring. Exercise 3.5 deals with the question whether the converse also holds.

Theorem 3.10 (Irreducible subsets of K^n and $\operatorname{Spec}(R)$).

(a) Let K be a field and $X \subseteq K^n$ a set of points, equipped with the Zariski topology. Then X is irreducible if and only if $\mathcal{I}_{K[x_1,\ldots,x_n]}(X)$ is a prime ideal.
(b) Let R be a ring and $X \subseteq \operatorname{Spec}(R)$ a set of prime ideals, equipped with the Zariski topology. Then X is irreducible if and only if $\mathcal{I}_R(X)$ is a prime ideal.

Proof. (a) First assume that X is irreducible. Then $I := \mathcal{I}_{K[x_1,\ldots,x_n]}(X) \subsetneqq K[x_1,\ldots,x_n]$, since $X \neq \emptyset$. To show that I is a prime ideal, let $f_1, f_2 \in K[x_1,\ldots,x_n]$ with $f_1 f_2 \in I$. Then

$$X = (X \cap \mathcal{V}_{K^n}(f_1)) \cup (X \cap \mathcal{V}_{K^n}(f_2)),$$

so by the irreducibility of X there exists $i \in \{1,2\}$ with $X \subseteq \mathcal{V}_{K^n}(f_i)$. This implies $f_i \in I$. So indeed I is a prime ideal.

Conversely, assume that I is a prime ideal. Then $X \neq \emptyset$, since $\mathcal{I}_{K[x_1,\ldots,x_n]}(\emptyset) = K[x_1,\ldots,x_n]$. To show that X is irreducible, let $X = X_1 \cup X_2$ with X_i closed in X, so $X_i = X \cap \mathcal{V}_{K^n}(I_i)$ with $I_i \subseteq K[x_1,\ldots,x_n]$ ideals. Then

$$X \subseteq \mathcal{V}_{K^n}(I_1) \cup \mathcal{V}_{K^n}(I_2) = \mathcal{V}_{K^n}(I_1 \cap I_2),$$

where we used Proposition 3.1(a) for the equality. This implies

$$I_1 \cap I_2 \subseteq \mathcal{I}_{K[x_1,\ldots,x_n]}\left(\mathcal{V}_{K^n}(I_1 \cap I_2)\right) \subseteq \mathcal{I}_{K[x_1,\ldots,x_n]}(X) = I.$$

Since I is a prime ideal, there exists i with $I_i \subseteq I$, so

$$X \subseteq \mathcal{V}_{K^n}(I) \subseteq \mathcal{V}_{K^n}(I_i).$$

This implies $X_i = X$. Therefore X is irreducible.

(b) The proof of this part is obtained from the proof of part (a) by changing $K[x_1,\ldots,x_n]$ to R, K^n to $\operatorname{Spec}(R)$, and "Proposition 3.1(a)" to "Proposition 3.6(a)." □

The following theorem allows us to view irreducible spaces as the "atoms" of a Noetherian space.

Theorem 3.11 (Decomposition into irreducibles). *Let X be a Noetherian topological space.*

(a) There exist a nonnegative integer n and closed, irreducible subsets $Z_1,\ldots,Z_n \subseteq X$ such that

$$X = Z_1 \cup \cdots \cup Z_n \quad \text{and} \quad Z_i \not\subseteq Z_j \quad \text{for} \quad i \neq j. \tag{3.1}$$

(b) If $Z_1,\ldots,Z_n \subseteq X$ are closed, irreducible subsets satisfying (3.1), *then every irreducible subset $Z \subseteq X$ is contained in some Z_i. (Observe that we do not assume Z to be closed.)*

(c) If $Z_1,\ldots,Z_n \subseteq X$ are closed, irreducible subsets satisfying (3.1), *then they are precisely the maximal irreducible subsets of X. In particular, the Z_i are uniquely determined up to order.*

Proof. First observe that every nonempty set of closed subsets of X has a minimal element, since otherwise it would contain an infinite strictly descending chain. Assume that there exists a nonempty closed subset $Y \subseteq X$ that is not a finite union of closed, irreducible subsets. Then we may assume Y to be minimal with this property. By assumption, Y itself is not irreducible, so $Y = Y_1 \cup Y_2$ with $Y_1, Y_2 \subsetneqq Y$ closed subsets. By the minimality of Y, the Y_i are finite unions of closed irreducible subsets, so the same is true for Y. This is a contradiction. Hence in particular,

$$X = Y_1 \cup \cdots \cup Y_m$$

with $Y_i \subseteq X$ closed and irreducible (where $m = 0$ if $X = \emptyset$). We may assume the Y_i to be pairwise distinct. By deleting those Y_i for which there exists a $j \neq i$ with $Y_i \subseteq Y_j$, we obtain a decomposition as in (3.1).

Now assume that (3.1) is satisfied, and let $Z \subseteq X$ be an irreducible subset. Then

$$Z = (Z \cap Z_1) \cup \cdots \cup (Z \cap Z_n),$$

so $Z = Z \cap Z_i$ for some i, which implies $Z \subseteq Z_i$. This proves (b). Moreover, if Z is maximal among the irreducible subsets of X, then $Z = Z_i$. So all maximal irreducible subsets of X occur among the Z_i. To complete the proof of (c), we need to show that every Z_i is maximal among the irreducible subsets. Indeed, if $Z_i \subseteq Z$ with $Z \subseteq X$ irreducible, then by (b) there exists j with $Z \subseteq Z_j$, so $Z_i \subseteq Z \subseteq Z_j$, which by (3.1) implies $i = j$ and $Z = Z_i$. □

Remark. As we see from the proof, it is only for part (a) of the theorem that we need to assume that X is Noetherian. So if X is not Noetherian but a decomposition as in (a) does exist, then (b) and (c) hold. ◁

Theorem 3.11 has statements on existence and uniqueness, which justifies the following definition.

Definition 3.12. *Let X be a Noetherian topological space. Then the Z_i from Theorem 3.11 are called the* **irreducible components** *of X.*

Example 3.13. Let K be an algebraically closed field and $g \in K[x_1, \ldots, x_n]$ a nonzero polynomial. Let $p_1, \ldots, p_r$ be the distinct prime factors of g. Then

$$\mathcal{V}_{K^n}(g) = \bigcup_{i=1}^{r} \mathcal{V}_{K^n}(p_i).$$

The $(p_i) \subseteq K[x_1, \ldots, x_n]$ are prime ideals, so by the Nullstellensatz (Theorem 1.17), $\mathcal{I}_{K[x_1,\ldots,x_n]}(\mathcal{V}_{K^n}(p_i)) = (p_i)$. By Theorem 3.10(a), the $\mathcal{V}_{K^n}(p_i)$ are the irreducible components of the hypersurface $\mathcal{V}_{K^n}(g)$. ◁

We make the obvious convention of calling a prime ideal $P \in \operatorname{Spec}(R)$ **minimal** if for all $Q \in \operatorname{Spec}(R)$ the inclusion $Q \subseteq P$ implies $Q = P$. So an integral domain has precisely one minimal prime ideal, namely $\{0\}$. By Proposition 3.6(e) and by Theorem 3.10(b), the minimal prime ideals correspond to the maximal closed, irreducible subsets of $X := \operatorname{Spec}(R)$, i.e., to the irreducible components of X (if X is Noetherian).

Corollary 3.14 (Minimal prime ideals). *Let R be a Noetherian ring.*

(a) There exist only finitely many minimal prime ideals $P_1, \ldots, P_n$ of R.
(b) Every prime ideal of R contains at least one of the P_i.
(c) The nilradical is the intersection of the P_i:

$$\operatorname{nil}(R) = \bigcap_{i=1}^{n} P_i.$$

(d) Let $I \subseteq R$ be an ideal. Then the set $\mathcal{V}_{\operatorname{Spec}(R)}(I)$ has finitely many minimal elements $Q_1, \ldots, Q_r$, and

$$\sqrt{I} = \bigcap_{i=1}^{r} Q_i.$$

Proof. By Proposition 3.6(e) and by Theorem 3.10(b), the (maximal) closed, irreducible subsets of $X := \operatorname{Spec}(R)$ correspond to (minimal) prime ideals of R. So for (a) and (b), we need to show that X has only finitely many maximal closed, irreducible subsets, and that every closed, irreducible subset is contained in a maximal one. By Theorem 3.9(b), X is Noetherian, so by Theorem 3.11(c), X has finitely many maximal irreducible subsets $Z_1, \dots, Z_n$, which are all closed. By Theorem 3.11(b), every closed, irreducible subset is contained in a Z_i.

Part (c) follows from (b) and Corollary 1.12, and part (d) follows from applying (a) and (c) to R/I and using the correspondence given by Lemma 1.22. □

Let us remark here that part (b) of Corollary 3.14 generalizes to non-Noetherian rings: It is always true that a prime ideal contains a minimal prime ideal (see Exercise 3.6).

Part (d) of Corollary 3.14 is sometimes expressed by saying that there are only finitely many prime ideals that are minimal over I.

All parts of Corollary 3.14 will be applied in many places throughout this book. As a first application, we complete the proof of Theorem 2.8 from page 27. The implication "(a) ⇒ (b)" of that theorem was already proved in Chapter 2.

Proof of the implication "(b) ⇒ (a)" *in Theorem 2.8.* We assume that R is Noetherian and $\operatorname{Spec}(R) = \operatorname{Spec}_{\max}(R)$, and we need to show that R is Artinian. By Corollary 3.14(c), there exist finitely many prime ideals whose intersection is the nilradical. So

$$\operatorname{nil}(R) = \bigcap_{i=1}^{n} \mathfrak{m}_i \quad \text{with} \quad \mathfrak{m}_i \in \operatorname{Spec}_{\max}(R).$$

This implies $I := \mathfrak{m}_1 \cdots \mathfrak{m}_n \subseteq \operatorname{nil}(R) = \sqrt{\{0\}}$, so by Lemma 2.6 there exists k with $I^k = \{0\}$. Therefore we can apply Lemma 2.7 and conclude that R is Artinian. □

Exercises for Chapter 3

3.1 (Properties of maps). Let $X = \{(\xi_1, \xi_2) \in \mathbb{C}^2 \mid \xi_1\xi_2 = 1\}$. Which of the following maps $\varphi_i \colon X \to X$ are morphisms, isomorphisms, or Zariski continuous?

(a) $\varphi_1(\xi_1, \xi_2) = (\xi_1^{-1}, \xi_2^{-1})$.
(b) $\varphi_2(\xi_1, \xi_2) = (\overline{\xi_1^2}, \overline{\xi_2^2})$.
(c) $\varphi_3(\xi_1, \xi_2) = (\overline{\xi_1}, \overline{\xi_2})$ (complex conjugation).

3.2 (Separating sets by polynomials). Let K be an algebraically closed field and let $X, Y \subseteq K^n$ be two subsets. Show that the following statements are equivalent:

(a) There exists a polynomial $f \in K[x_1, \ldots, x_n]$ such that $f(x) = 0$ for all $x \in X$, and $f(y) = 1$ for all $y \in Y$.
(b) The Zariski closures of X and Y do not meet: $\overline{X} \cap \overline{Y} = \emptyset$.

3.3 (A homeomorphism). Let K be an algebraically closed field and X an affine variety. Show that the bijection $\Phi: X \to \mathrm{Spec}_{\mathrm{max}}(K[X])$ from Theorem 1.23 is a homeomorphism. (Here $\mathrm{Spec}_{\mathrm{max}}(K[X])$ is equipped with the subset topology induced from the Zariski topology on $\mathrm{Spec}(K[X])$.)

3.4 (Another homeomorphism). Let R be a ring and $I \subseteq R$ an ideal. Lemma 1.22 gives a bijection $\mathcal{V}_{\mathrm{Spec}(R)}(I) \to \mathrm{Spec}(R/I)$. Show that this bijection is actually a homeomorphism. (Here $\mathcal{V}_{\mathrm{Spec}(R)}(I)$ is equipped with the subset topology induced from the Zariski topology on $\mathrm{Spec}(R)$.)

***3.5 (A converse of Theorem 3.9?).** Let R be a ring. If $\mathrm{Spec}(R)$ is a Noetherian space, does this imply that R is a Noetherian ring? Give a proof or a counterexample.

The following exercise is due to Martin Kohls.

3.6 (Minimal prime ideals). Let R be a (not necessarily Noetherian) ring and $Q \in \mathrm{Spec}(R)$ a prime ideal. Show that there exists a minimal prime ideal $P \in \mathrm{Spec}(R)$ with $P \subseteq Q$. In particular, if $R \neq \{0\}$, there exist minimal prime ideals in R.
Hint: Use Zorn's lemma with an unusual ordering. *(Solution on page 218)*

3.7 (Hausdorff spaces). Let X be a Noetherian topological space. Show that the following two statements are equivalent:

(a) X is a Hausdorff space.
(b) X is finite and has the discrete topology.

In particular, no infinite subset $Y \subseteq K^n$ with the Zariski topology is Hausdorff.

3.8 (Quasi-compact spaces). Recall that a topological space X is called quasi-compact if for every set $\mathcal{M}$ of open subsets with $X = \bigcup_{U \in \mathcal{M}} U$, there exist $U_1, \ldots, U_n \in \mathcal{M}$ with $X = \bigcup_{i=1}^{n} U_i$.

(a) Show that a topological space X is Noetherian if and only if every subset of X is quasi-compact.
(b) Let R be a ring and $X = \mathrm{Spec}(R)$. Then X is quasi-compact (even if it is not Noetherian).

3.9 (Products of irreducible varieties). Let $X \subseteq K^m$ and $Y \subseteq K^n$ be two irreducible affine varieties over a field. Show that the product variety $X \times Y \subseteq K^{m+n}$ is also irreducible.
Hint: This may be done as follows: For $X \times Y = Z_1 \cup Z_2$ with Z_i closed, consider the sets $X_i := \{x \in X \mid \{x\} \times Y \subseteq Z_i\}$. Show that $X = X_1 \cup X_2$ and that the X_i are closed.

3.10 (Diagonalizable matrices form a dense subset). Let K be an algebraically closed field, and let $D \subset K^{n \times n}$ be the set of all diagonalizable $n \times n$ matrices. Show that the Zariski closure $\overline{D}$ of D is $K^{n \times n}$. (Here $K^{n \times n}$ is identified with affine n^2-space K^{n^2}.) Is D open in $K^{n \times n}$?

Chapter 4
A Summary of the Lexicon

In this chapter we give a brief summary of the algebra–geometry lexicon. All the statements we make here have been proved in Chapter 1 or 3, but for the sake of brevity we will not give any references. This lexicon really comes in two parts. The first links algebraic objects (such as affine algebras) to truly geometric objects (such as affine varieties). The second part is more general and links algebraic objects (such as rings) to objects that are geometric in a more abstract sense (such as spectra of rings).

4.1 True Geometry: Affine Varieties

In this section, K is assumed to be an algebraically closed field. We have the following correspondences between algebraic and geometric objects:

(1) Hilbert's Nullstellensatz gives rise to a bijective correspondence

$$\text{affine varieties in } K^n \quad \longleftrightarrow \quad \text{radical ideals in } K[x_1,\ldots,x_n]. \qquad (4.1)$$

In fact, assigning to any set $X \subseteq K^n$ of points the vanishing ideal $\mathcal{I}(X)$ yields a map from the power set of K^n to the set of radical ideals in $K[x_1,\ldots,x_n]$, and assigning to a set $S \subseteq K[x_1,\ldots,x_n]$ of polynomials the affine variety $\mathcal{V}(S)$ yields a map from the power set of $K[x_1,\ldots,x_n]$ to the set of affine varieties in K^n. Restricting both maps gives the correspondence (4.1). The correspondence is *inclusion-reversing*. An affine variety is irreducible if and only if its vanishing ideal is a prime ideal. So there is a sub-correspondence

$$\text{irreducible affine varieties in } K^n \quad \longleftrightarrow \quad \text{prime ideals in } K[x_1,\ldots,x_n].$$

(2) Every affine K-variety X has a coordinate ring $K[X]$, whose elements give rise to regular functions $X \to K$. This leads us to identify $K[X]$ with the ring of regular functions on X. Assigning to an affine K-variety

G. Kemper, *A Course in Commutative Algebra*, Graduate Texts in Mathematics 256, DOI 10.1007/978-3-642-03545-6_5,

X its coordinate ring yields a map

$$\text{affine } K\text{-varieties} \longrightarrow \text{reduced affine } K\text{-algebras.}$$

Conversely, every reduced affine K-algebra is isomorphic to the coordinate ring of an affine K-variety, which is unique up to isomorphism. For X an affine K-variety, we have the equivalence

$$X \text{ is irreducible} \iff K[X] \text{ is an affine domain.}$$

(3) Let X be an affine K-variety with coordinate ring $K[X]$. Then there is an inclusion-reversing, bijective correspondence

$$\text{Zariski-closed subsets of } X \longleftrightarrow \text{radical ideals in } K[X]. \tag{4.2}$$

A closed subset of X is irreducible if and only if the corresponding ideal in $K[X]$ is a prime ideal. So chains of closed, irreducible subsets of X correspond to chains of prime ideals in $K[X]$, but with all inclusions reversed. The above correspondence has sub-correspondences

$$\text{irreducible components of } X \longleftrightarrow \text{minimal prime ideals in } K[X],$$

and

$$X \longleftrightarrow \operatorname{Spec}_{\max}(K[X]). \tag{4.3}$$

(4) Given two affine K-varieties X and Y, there is a bijective correspondence

$$\text{morphisms } X \to Y \text{ of varieties} \longleftrightarrow \text{homomorphisms } K[Y] \to K[X] \text{ of } K\text{-algebras.}$$

This correspondence translates isomorphisms into isomorphisms, but behaves less well with respect to injectivity (see Exercise 4.1). The composition of two morphisms of varieties corresponds to the composition of the homomorphisms of the coordinate rings, but in reversed order.

We should mention that some parts of the lexicon stay intact if we drop the hypothesis that K is algebraically closed.

4.2 Abstract Geometry: Spectra

There is no variety associated to a general ring R. However, we always have the spectrum $\operatorname{Spec}(R)$, which is an abstract substitute for an affine variety. By (4.3), affine varieties over algebraically closed fields are embedded into the spectrum of the coordinate ring, so, taking a somewhat generous view, we can regard the spectrum as a generalization of an affine variety. In particular, statements about spectra of rings almost always imply statements about affine varieties as special cases. As in the previous section, we will summarize some algebra–geometry correspondences. We will also see that they are generalizations of the correspondences from Section 4.1. In the sequel, R stands for a ring.

(1) There is an inclusion-reversing bijective correspondence

$$\text{Zariski-closed subsets of } \operatorname{Spec}(R) \quad \longleftrightarrow \quad \text{radical ideals in } R. \tag{4.4}$$

In fact, assigning to any subset $X \subseteq \operatorname{Spec}(R)$ the intersection of all prime ideals in X yields a map from the power set of $\operatorname{Spec}(R)$ to the set of radical ideals in R, and assigning to a set $S \subseteq R$ of ring elements the set of all prime ideals that contain S yields a map from the power set of R to the set of Zariski-closed subsets of $\operatorname{Spec}(R)$. Restricting both maps gives the correspondence (4.4).

A closed subset of $\operatorname{Spec}(R)$ is irreducible if and only if the corresponding ideal in R is a prime ideal. So chains of closed, irreducible subsets of $\operatorname{Spec}(R)$ correspond to chains of prime ideals in R, but with all inclusions reversed.

In the special case that $R = K[X]$ is the coordinate ring of an affine variety over an algebraically closed field, we can compose (4.4) with the correspondence (4.2), and get a correspondence

$$\text{Zariski-closed subsets of } \operatorname{Spec}(K[X]) \quad \longleftrightarrow \quad \text{Zariski-closed subsets of } X,$$

given by intersecting a closed subset of $\operatorname{Spec}(K[X])$ with $\operatorname{Spec}_{\max}(K[X])$ and then applying (4.3) to the points. Via the above correspondence, (4.4) can be viewed as a generalization of (4.2).

(2) A ring homomorphism $\varphi\colon R \to S$ induces a morphism

$$\varphi^*\colon \operatorname{Spec}(S) \to \operatorname{Spec}(R),\ Q \mapsto \varphi^{-1}(Q)$$

of spectra. In the special case that $R \subseteq S$ and φ is the inclusion, we have $\varphi^*(Q) = R \cap Q$. Notice that the correspondence between ring homomorphisms and morphisms of spectra is not bijective. If $\psi\colon S \to T$ is a further ring homomorphism, then $(\psi \circ \varphi)^* = \varphi^* \circ \psi^*$.

In general, φ^* does not restrict to a map $\operatorname{Spec}_{\max}(S) \to \operatorname{Spec}_{\max}(R)$; but if φ is a homomorphism of affine K-algebras, it does. If in addition $R = K[Y]$ and $S = K[X]$ are coordinate rings of affine varieties over an algebraically closed field, then (4.3) translates this restriction of φ^* into a map $X \to Y$, which is exactly the morphism from (4) in Section 4.1 corresponding to φ.

Exercises for Chapter 4

4.1 (Dominant and injective morphisms). Let X and Y be affine varieties over a field K, and let $f: X \to Y$ be a morphism with induced homomorphism $\varphi: K[Y] \to K[X]$. We say that f is **dominant** if the image $f(X)$ is dense in Y, i.e., $\overline{f(X)} = Y$.

(a) Show that f is dominant if and only if φ is injective.
(b) Show that if φ is surjective, then f is injective.
(c) Give examples in which f is dominant but not surjective, and in which the converse of part (b) does not hold.

4.2 (When is φ^* dominant?). Let $\varphi: R \to S$ be a homomorphism of rings. Show that the following two statements are equivalent:

(a) The map φ^*: $\operatorname{Spec}(S) \to \operatorname{Spec}(R)$ is dominant.
(b) The kernel $\ker(\varphi)$ is contained in the nilradical $\operatorname{nil}(R)$ of R.

4.3 (The coproduct of spectra and affine varieties). Let $R_1, \ldots, R_n$ be rings. Recall that the direct sum $R := R_1 \oplus \cdots \oplus R_n$ is defined as the Cartesian product of the R_i with componentwise addition and multiplication. The projections $\pi_i: R \to R_i$ induce morphisms f_i: $\operatorname{Spec}(R_i) \to \operatorname{Spec}(R)$. Prove the following:

(a) If S is a ring with morphisms g_i: $\operatorname{Spec}(R_i) \to \operatorname{Spec}(S)$, then there exists a unique morphism g: $\operatorname{Spec}(R) \to \operatorname{Spec}(S)$ with $g \circ f_i = g_i$ for all i. This is expressed by saying that $\operatorname{Spec}(R)$ together with the f_i is a *coproduct* in the category of spectra of rings.
(b) $\operatorname{Spec}(R)$ is the disjoint union of the images of the f_i. For each i, the image of f_i is closed in $\operatorname{Spec}(R)$.
(c) If $X_1, \ldots, X_n$ are affine varieties over an algebraically closed field K, then there exists a coproduct X in the category of affine K-varieties. The analogue of part (b) also holds in this case, and for every i the image of X_i in X is isomorphic to X_i. The universal property of the coproduct is shown in the following diagram:

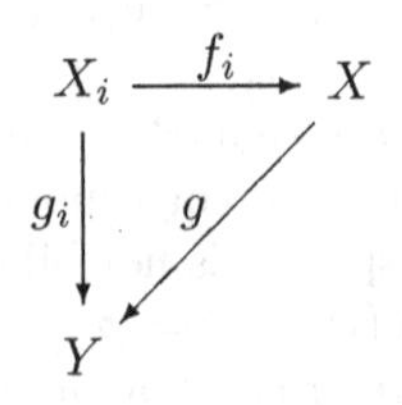

Part II
Dimension

Chapter 5
Krull Dimension and Transcendence Degree

In this chapter, we introduce the Krull dimension, which is the "correct" concept of dimension in algebraic geometry and commutative algebra. Then we prove that the dimension of an affine algebra is equal to its transcendence degree. This makes the dimension more accessible both to computation and to interpretation.

We start by introducing the following ad hoc notation. Let $\mathcal{M}$ be a set whose elements are sets. By a **chain** in $\mathcal{M}$ we mean a subset $\mathcal{C} \subseteq \mathcal{M}$ that is totally ordered by inclusion "$\subseteq$". The length of $\mathcal{C}$ is defined to be $\text{length}(\mathcal{C}) := |\mathcal{C}| - 1 \in \mathbb{N}_0 \cup \{-1, \infty\}$. A finite chain of length n is usually written as

$$X_0 \subsetneqq X_1 \subsetneqq \cdots \subsetneqq X_n.$$

We write

$$\text{length}(\mathcal{M}) := \sup \{\text{length}(\mathcal{C}) \mid \mathcal{C} \text{ is a chain in } \mathcal{M}\} \in \mathbb{N}_0 \cup \{-1, \infty\}$$

(the length -1 occurs if $\mathcal{M} = \emptyset$).

Observe that the dimension of a vector space V is the maximal length of a chain of subspaces, i.e.,

$$\dim(V) = \text{length}\Big(\{U \subseteq V \mid U \text{ subspace}\}\Big).$$

With this in mind, the following definition does not appear too far-fetched.

Definition 5.1 (Krull dimension).

(a) Let X be a topological space. Set $\mathcal{M}$ to be the set of all closed, irreducible subsets of X. Then the **dimension** *of X (also called the* **Krull dimension***) is defined as*

$$\dim(X) := \text{length}(\mathcal{M}).$$

(b) Let R be a ring. Then the **dimension** *of R (also called the* **Krull dimension***) is defined as*

G. Kemper, *A Course in Commutative Algebra*, Graduate Texts in Mathematics 256, DOI 10.1007/978-3-642-03545-6_6,

$$\dim(R) := \dim(\operatorname{Spec}(R)).$$

So $\dim(R) = \operatorname{length}(\operatorname{Spec}(R))$, *since the closed, irreducible subsets of* $\operatorname{Spec}(R)$ *correspond to prime ideals of* R *by Proposition 3.6*(e) *and Theorem 3.10*(b). *In other words, the dimension of* R *is the maximal length of a chain of prime ideals of* R.

(c) *Let* K *be a field. The dimension of a subset* $X \subseteq K^n$ *is the dimension of* X *with the Zariski topology. So if* K *is algebraically closed and* X *is an affine variety, then*

$$\dim(X) = \dim(K[X]),$$

since the closed, irreducible subsets of X *correspond to prime ideals in the coordinate ring* $K[X]$ *by Theorem 1.23 and Theorem 3.10*(a).

Example 5.2. (1) If $X = \{P\}$ is a singleton or (more generally) a nonempty, finite, discrete topological space, then $\dim(X) = 0$. Moreover, $\dim(\emptyset) = -1$.

(2) If K is an infinite field and $X = K^1$, then the closed, irreducible subsets are the singletons and all of X, so $\dim(K^1) = 1$.

(3) Let $X \subseteq \mathbb{R}^3$ be the union of a plane P and a line L that is not contained in the plane. We can see two types of nonrefinable chains of closed, irreducible subsets:

   (a) A point of L not lying in P, followed by all of L
   (b) A point of P, followed by a line in P that contains the point, followed by all of P

   From this, we see that $\dim(X) \geq 2$. Intuition tells us that the dimension should be equal to 2, but we cannot verify this yet.

(4) Every field has Krull dimension 0.

(5) The ring $\mathbb{Z}$ of integers has Krull dimension 1, with all maximal chains of prime ideals of the form $\{0\} \subsetneqq (p)$ with p a prime number.
   More generally, every principal ideal domain that is not a field has Krull dimension 1.

(6) In particular, a polynomial ring $K[x]$ over a field has $\dim(K[x]) = 1$.

(7) Let K be a field and $R = K[x_1, x_2, \ldots]$ a polynomial ring in countably many indeterminates x_i, $i \in \mathbb{N}$. Then $P_i = (x_1, \ldots, x_i)$ provides an infinite chain of prime ideals, so $\dim(R) = \infty$. ◁

Remark. The ring in Example 5.2(7) is not Noetherian. It is tempting to hope that Noetherian rings are always finite-dimensional. However, Exercise 7.7 dashes this hope. The converse is also not true: Combining Example 2.3 and Exercise 5.3 yields a non-Noetherian integral domain of Krull dimension 2. ◁

Remark. If X is a Noetherian topological space with irreducible components $Z_1, \ldots, Z_n$, then

$$\dim(X) = \max\{\dim(Z_1), \ldots, \dim(Z_n), -1\}.$$

This follows from Theorem 3.11(b). We will call X **equidimensional** if all Z_i have the same dimension. Likewise, a Noetherian ring R is called **equidimensional** if $\operatorname{Spec}(R)$ is equidimensional. ◁

As we see from Example 5.2, it is very difficult to apply Definition 5.1 directly for determining the dimension of a variety. At this point we are not even able to determine the dimension of K^n (or of the polynomial ring $K[x_1, \ldots, x_n]$), although we easily get n as a lower bound. Another disadvantage is that at this point it is far from clear that the Krull dimension of an affine variety coincides with what we intuitively understand by dimension. The main result of this chapter is an "alternative definition" of the dimension of an affine algebra, which is much more accessible and more intuitive (see Remark 5.4). Another, less well-known, alternative definition, which holds for general rings, is given in Exercise 6.8.

Recall that a subset $\{a_1, \ldots, a_n\} \subseteq A$ of size n of an algebra A over a field K is called algebraically independent if for all nonzero polynomials $f \in K[x_1, \ldots, x_n]$ we have $f(a_1, \ldots, a_n) \neq 0$.

Definition 5.3. *Let A be an algebra over a field K. Then the* **transcendence degree** *of A is defined as*

$$\operatorname{trdeg}(A) := \sup\{|T| \mid T \subseteq A \text{ is finite and algebraically independent}\}.$$

So $\operatorname{trdeg}(A) \in \mathbb{N}_0 \cup \{-1, \infty\}$, *where* -1 *occurs if* $A = \{0\}$ *is the zero ring. (We set* $\sup \emptyset := -1$.*)*

Our next goal is to show that the dimension and the transcendence degree of an affine algebra coincide. The following remark is intended to convince the reader that this is a worthy goal.

Remark 5.4. (a) Let $A = K[X]$ be the coordinate ring of an affine variety over an infinite field. Finding an algebraically independent subset of size n of A is equivalent to finding an injective homomorphism $K[x_1, \ldots, x_n] \to A$. By Exercise 4.1(a), this is the same as giving a dominant morphism $X \to K^n$. So $\operatorname{trdeg}(A)$ is the largest number n such that there exists a dominant morphism $X \to K^n$. This already links the transcendence degree to an intuitive concept of dimension. In fact, we will be able to do even better: In Chapter 8, we will see that such a morphism can be chosen to be surjective, and such that every point in K^n has only finitely many preimages (see after Remark 8.20 on page 105).

(b) If $A = K[x_1, \ldots, x_n]/I$ is an affine algebra given by generators of an ideal $I \subseteq K[x_1, \ldots, x_n]$, then $\operatorname{trdeg}(A)$ can be computed algorithmically by Gröbner basis methods. We will see this in Chapter 9 (see on page 128). So equating dimension and transcendence degree brings the dimension into the realm of computability. ◁

Theorem 5.5 (Dimension of algebras, upper bound). *Let A be a (not necessarily finitely generated) algebra over a field K. Then*

$$\dim(A) \leq \operatorname{trdeg}(A).$$

Proof. This is the special case $S = A$ of the following lemma. □

Lemma 5.6. *Let A be an algebra over a field K, and let $S \subseteq A$ be a subset that generates A as an algebra. Then*

$$\dim(A) \leq \sup\{|T| \mid T \subseteq S \textit{ is finite and algebraically independent}\}.$$

Proof. Let n be the supremum on the right-hand side of the claimed inequality. There is nothing to show if $n = \infty$, and the lemma is correct if $n = -1$. So assume $n \in \mathbb{N}_0$. We need to show that $\dim(A/P) \leq n$ for all $P \in \operatorname{Spec}(A)$. If we substitute A by A/P and S by $\{a + P \mid a \in S\}$, then n cannot increase. Therefore we may assume that A is an integral domain.

First consider the case $n = 0$. Then all elements from S are algebraic, so the field of fractions $\operatorname{Quot}(A)$ is generated as a field extension of K by algebraic elements. It follows that $\operatorname{Quot}(A)$ is algebraic, so A is algebraic, too. By Lemma 1.1(a), this implies that A is a field, so $\dim(A) = 0$.

Now assume $n > 0$, and let

$$P_0 \subsetneqq P_1 \subsetneqq \cdots \subsetneqq P_m$$

be a chain in $\operatorname{Spec}(A)$ of length $m > 0$. Factoring by P_1 yields a chain in $\operatorname{Spec}(A/P_1)$ of length $m - 1$ (see Lemma 1.22). If we can show that all algebraically independent subsets $T \subseteq \{a + P_1 \mid a \in S\} \subseteq A/P_1$ have size $|T| < n$, then we can use induction on n and conclude that $\dim(A/P_1) < n$, so $m - 1 < n$, which yields the lemma.

By way of contradiction, assume that there exist $a_1, \ldots, a_n \in S$ such that $\{a_1 + P_1, \ldots, a_n + P_1\} \subseteq A/P_1$ is algebraically independent of size n. Then also $\{a_1, \ldots, a_n\} \subseteq S$ is algebraically independent. By the definition of n, all $a \in S$ are algebraic over $L := \operatorname{Quot}(K[a_1, \ldots, a_n])$, so $\operatorname{Quot}(A)$ is algebraic over L, too. There exists a nonzero element $a \in P_1$. We have a nonzero polynomial $G = \sum_{i=0}^{k} g_i x^i \in L[x]$ with $G(a) = 0$. Since $a \neq 0$, we may assume $g_0 \neq 0$. Furthermore, we may assume $g_i \in K[a_1, \ldots, a_n]$. Then

$$g_0 = -\sum_{i=1}^{k} g_i a^i \in P_1,$$

so viewing g_0 as a polynomial in n indeterminates over K, we obtain $g_0(a_1 + P_1, \ldots, a_n + P_1) = 0$, contradicting the algebraic independence of the $a_i + P_1 \in A/P_1$. This completes the proof. □

We can now determine the dimension of polynomial rings over fields and of affine n-space K^n.

Corollary 5.7 (Dimension of a polynomial ring). *If K is a field, then*

$$\dim (K[x_1, \ldots, x_n]) = n.$$

Moreover,

$$\dim (K^n) = \begin{cases} n & \textit{if } K \textit{ is infinite,} \\ 0 & \textit{if } K \textit{ is finite.} \end{cases}$$

Proof. With $S := \{x_1, \ldots, x_n\}$, Lemma 5.6 yields $\dim (K[x_1, \ldots, x_n]) \leq n$. Since we have the chain

$$\{0\} \subsetneqq (x_1) \subsetneqq (x_1, x_2) \subsetneqq \cdots \subsetneqq (x_1, \ldots, x_n) \tag{5.1}$$

of length n in $\operatorname{Spec}(K[x_1, \ldots, x_n])$, equality holds.

Moreover, a chain of length m of closed, irreducible subsets $X_i \subseteq K^n$ gives rise to a chain of length m of ideals $\mathcal{I}(X_i) \subset K[x_1, \ldots, x_n]$, which are prime by Theorem 3.10(a), so $m \leq n$ by the above. On the other hand, if K is infinite, the affine varieties corresponding to the ideals in (5.1) are irreducible and provide a chain of length n, so $\dim (K^n) = n$. If K is a finite field, then $\dim (K^n) = 0$ by Example 5.2(1). □

Example 5.8. The bound from Theorem 5.5 is not always sharp. Indeed, consider the rational function field $A = K(x_1, \ldots, x_n) := \operatorname{Quot}(K[x_1, \ldots, x_n])$. We have

$$\dim(A) = 0 < n = \operatorname{trdeg}(A).$$

◁

In Chapter 7 we will prove that if $R \neq \{0\}$ is a Noetherian ring, then

$$\dim (R[x]) = \dim(R) + 1$$

(Corollary 7.13), generalizing Corollary 5.7. In Exercise 7.10, the analogous result will be proved for the formal power series ring $R[[x]]$. For the formal power series ring in n indeterminates over a field K, this implies

$$\dim (K[[x_1, \ldots, x_n]]) = n.$$

Theorem 5.9 (Dimension and transcendence degree). *Let A be an affine algebra. Then*

$$\dim(A) = \operatorname{trdeg}(A).$$

We will prove the theorem together with the following proposition, which often facilitates the computation of the transcendence degree since the set S can be taken to be finite.

Proposition 5.10 (Calculating the transcendence degree). *Let A be an affine algebra, and let $S \subseteq A$ be a generating set. Then*

$$\operatorname{trdeg}(A) = \sup \{|T| \mid T \subseteq S \textit{ is finite and algebraically independent}\}.$$

Proof of Theorem 5.9 and Proposition 5.10. By Lemma 5.6 we have

$$\dim(A) \leq \sup\{|T| \mid T \subseteq S \text{ is finite and algebraically independent}\},$$

and this supremum is clearly less than or equal to $\operatorname{trdeg}(A)$. So we need to show only that $\operatorname{trdeg}(A) \leq \dim(A)$. Using induction on n, we will show that if $\operatorname{trdeg}(A) \geq n$, then $\dim(A) \geq n$. We may assume $n > 0$. So let $a_1, \ldots, a_n \in A$ be algebraically independent. By Corollary 2.12, A is Noetherian, so by Corollary 3.14(a), there exist only finitely many minimal prime ideals $M_1, \ldots, M_r$ of A. Assume that for all $i \in \{1, \ldots, r\}$ we have that $a_1 + M_i, \ldots, a_n + M_i \in A/M_i$ are algebraically dependent. Then there exist polynomials $f_i \in K[x_1, \ldots, x_n] \setminus \{0\}$ such that $f_i(a_1, \ldots, a_n) \in M_i$, so

$$a := \prod_{i=1}^{r} f_i(a_1, \ldots, a_n) \in \bigcap_{i=1}^{r} M_i = \operatorname{nil}(A),$$

where the last equality follows from Corollary 3.14(c). So there exists a k with $a^k = 0$, so with $f := \prod_{i=1}^{r} f_i^k \neq 0$ we have $f(a_1, \ldots, a_n) = 0$, contradicting the algebraic independence of the a_i. Hence for some M_i the elements $a_1 + M_i, \ldots, a_n + M_i \in A/M_i$ are algebraically independent. It suffices to show that $\dim(A/M_i) \geq n$, so by replacing A by A/M_i, we may assume that A is an affine domain.

Consider the field $L := \operatorname{Quot}(K[a_1])$, which is a subfield of $\operatorname{Quot}(A)$, and the subalgebra $A' := L \cdot A \subseteq \operatorname{Quot}(A)$. Clearly A' is an affine L-domain, and $a_2, \ldots, a_n \in A'$ are algebraically independent over L. By induction, $\dim(A') \geq n - 1$, so there exists a chain

$$P'_0 \subsetneqq P'_1 \subsetneqq \cdots \subsetneqq P'_{n-1}$$

in $\operatorname{Spec}(A')$. Set $P_i := A \cap P'_i \in \operatorname{Spec}(A)$. Then $P_{i-1} \subseteq P_i$ for $i = 1, \ldots, n-1$. These inclusions are strict since clearly $L \cdot P_i = P'_i$ for all i. Moreover, $L \cap P_{n-1} = \{0\}$, since otherwise P'_{n-1} would contain an invertible element from L, leading to $P'_{n-1} = A'$. It follows that $a_1 + P_{n-1} \in A/P_{n-1}$ is not algebraic over K. By Lemma 1.1(b), A/P_{n-1} is not a field, so P_{n-1} is not a maximal ideal. Let $P_n \subset A$ be a maximal ideal containing P_{n-1}. Then we have a chain

$$P_0 \subsetneqq P_1 \subsetneqq \cdots \subsetneqq P_{n-1} \subsetneqq P_n$$

in $\operatorname{Spec}(A)$, and $\dim(A) \geq n$ follows. □

In Exercise 5.3, the scope of Theorem 5.9 will be extended to all subalgebras of affine algebras. In Chapter 8, we will learn more about chains of prime ideals in affine domains (see Theorem 8.22).

We will now use Theorem 5.9 in order to characterize 0-dimensional affine algebras. To avoid ambiguities, we write $\dim_K(V)$ for the dimension (= size of a basis) of a vector space V over a field K.

Theorem 5.11 (0-dimensional affine algebras). *Let $A \neq \{0\}$ be an affine K-algebra. Then the following statements are equivalent:*

(a) $\dim(A) = 0$.
(b) A is algebraic over K.
(c) $\dim_K(A) < \infty$.
(d) A is Artinian.
(e) $|\mathrm{Spec}_{\max}(A)| < \infty$.

Proof. If $\dim(A) = 0$, then A is algebraic by Theorem 5.9. Assume that A is algebraic. We can write $A = K[a_1, \ldots, a_n]$, so there exist nonzero polynomials $g_i \in K[x]$ with $g_i(a_i) = 0$. It is easy to see that the set

$$\left\{\prod_{i=1}^{n} a_i^{e_i} \mid 0 \leq e_i < \deg(g_i) \text{ for all } i\right\}$$

generates A as a K-vector space. Now assume that A is finite-dimensional as a K-vector space. Then the linear subspaces satisfy the descending chain condition. Therefore so do the ideals, so A is Artinian.

Assume that A is Artinian. By Corollary 3.14(a) and (b), every maximal ideal of A contains one of the minimal prime ideals $P_1, \ldots, P_n$. But by Theorem 2.8, the P_i themselves are maximal. This implies (e).

Finally, assume that there exist only finitely many maximal ideals, and let $P \in \mathrm{Spec}(A)$. By Theorem 1.13, P is the intersection of all maximal ideals of A containing P, so P is a finite intersection of maximal ideals. Since P is a prime ideal, it follows that P itself is maximal. Therefore $\dim(A) = 0$. □

Exercise 5.4 gives an interpretation of $\dim_K(A)$ in the case that $A = K[X]$ is the coordinate ring of a finite set X. The following proposition describes 0-dimensional subsets of K^n. For K algebraically closed, this is just a reformulation of Theorem 5.11(e).

Proposition 5.12 (0-dimensional sets). *Let K be a field and $X \subseteq K^n$ nonempty. Then* $\dim(X) = 0$ *if and only if X is finite.*

Proof. Assume $\dim(X) = 0$. Since X is a subset of a Noetherian space, X is Noetherian, too. By Theorem 3.11(a), X is a finite union of closed, irreducible subsets Z_i. Choose $x_i \in Z_i$. Then $\{x_i\} \subseteq Z_i$ is a chain of closed, irreducible subsets, so $Z_i = \{x_i\}$. It follows that X is finite.

Conversely, if X is finite, then the irreducible subsets are precisely the subsets of size 1, so $\dim(X) = 0$. □

The following theorem deals with a situation that is, in a sense, opposite to the one from Theorem 5.11: equidimensional algebras whose dimension is only 1 less than the number of generators. These correspond to equidimensional affine varieties in K^n of dimension $n - 1$. Such varieties are usually called *hypersurfaces.*

Theorem 5.13 (Hypersurfaces). *Let $I \subseteq K[x_1, \ldots, x_n]$ be an ideal in a polynomial ring over a field, and $A := K[x_1, \ldots, x_n]/I$. Then the following statements are equivalent:*

(a) A is equidimensional of dimension $n-1$.
(b) $I \neq K[x_1, \ldots, x_n]$, and every prime ideal in $K[x_1, \ldots, x_n]$ that is minimal over I is minimal among all nonzero prime ideals. (According to Definition 6.10, this means that I has height 1.)
(c) $\sqrt{I} = (g)$ with $g \in K[x_1, \ldots, x_n]$ a nonconstant polynomial.

Proof. In the proof we will make frequent use of the bijection between $\operatorname{Spec}(A)$ and $\mathcal{V}_{\operatorname{Spec}(K[x_1,\ldots,x_n])}(I)$ given by Lemma 1.22. Let $\mathcal{M} \subseteq \operatorname{Spec}(K[x_1, \ldots, x_n])$ be the set of all prime ideals that are minimal over I. Then $\mathcal{M}$ is finite by Corollary 3.14(d), and the minimal prime ideals of A are the P/I, $P \in \mathcal{M}$.

First assume that A is equidimensional of dimension $n-1$, so for all $P \in \mathcal{M}$ we have

$$\dim\left(K[x_1, \ldots, x_n]/P\right) = \dim\left(A/(P/I)\right) = n - 1. \tag{5.2}$$

It follows from Corollary 5.7 that $P \neq \{0\}$. If P were not minimal among all nonzero primes, we could build a chain of prime ideals in $\operatorname{Spec}(K[x_1, \ldots, x_n])$ by going two steps down from P, and, using (5.2), going $n-1$ steps up from P. This chain would have length $n+1$, contradicting Corollary 5.7. Since (a) also implies that $I \neq K[x_1, \ldots, x_n]$, (b) follows.

Now assume (b), and again take $P \in \mathcal{M}$. By Lemma 5.14, which we prove below, there exists an irreducible polynomial g_P such that $P = (g_P)$. With Corollary 3.14(d), it follows that

$$\sqrt{I} = \bigcap_{P \in \mathcal{M}} P = \bigcap_{P \in \mathcal{M}} (g_P) = (g),$$

where we set $g := \prod_{P \in \mathcal{M}} g_P$. Since $I \neq K[x_1, \ldots, x_n]$, g is nonconstant, so (c) holds.

Finally, assume (c), and let $g = g_1 \cdots g_r$ be a decomposition into irreducible polynomials. For $i \neq j$, g_i does not divide g_j since (g) is a radical ideal. We obtain prime ideals $P_i := (g_i) \in \operatorname{Spec}(K[x_1, \ldots, x_n])$, and $\bigcap_{i=1}^r P_i = \sqrt{I}$. It follows that $\mathcal{M} = \{P_1, \ldots, P_r\}$. Since $\dim\left(A/(P_i/I)\right) = \dim\left(K[x_1, \ldots, x_n]/P_i\right)$ we need to show that $\dim\left(K[x_1, \ldots, x_n]/(g_i)\right) = n-1$ for all i. But it is clear that by excluding an indeterminate x_j that occurs in g_i from the set $\{x_1, \ldots, x_n\}$, we obtain a maximal subset of $\{x_1, \ldots, x_n\}$ that is algebraically independent modulo g_i. So the claim follows by Proposition 5.10 and Theorem 5.9. □

The following lemma was used in the proof. Recall that a *factorial ring* is the same as a *unique factorization domain.*

Lemma 5.14 (Height-one prime ideals in a factorial ring). *Let R be a factorial ring and let $P \in \operatorname{Spec}(R)$ be prime ideal that is minimal among all* *nonzero prime ideals. (According to Definition 6.10, this means that P has height 1.) Then $P = (a)$ with $a \in R$ a prime element.*

Proof. Let $a \in P \setminus \{0\}$. Since P is a prime ideal, at least one factor of a factorization of a into prime elements also lies in P, so we may assume a to be a prime element. Then (a) is a prime ideal and $\{0\} \subsetneqq (a) \subseteq P$, so $(a) = P$ by the minimality hypothesis. □

Part (c) of Theorem 5.13 talks about principal ideals. This should be compared to Theorem 8.25, which talks about ideals generated by n polynomials. Readers may also take a look at Theorem 7.4, where the implication (c) $\Rightarrow$ (b) of Theorem 5.13 is generalized from $K[x_1, \ldots, x_n]$ to arbitrary Noetherian rings.

As a further application of Theorem 5.9 and Proposition 5.10, we determine the dimension of a product of affine varieties.

Theorem 5.15 (Dimension of a product variety). *Let $X \subseteq K^n$ and $Y \subseteq K^m$ be nonempty affine varieties over an algebraically closed field K. Then the product variety $X \times Y \subseteq K^{n+m}$ satisfies*

$$\dim(X \times Y) = \dim(X) + \dim(Y).$$

Proof. The proof is very easy and straightforward, even if it takes some space to write it down.

Write $d = \dim(X) = \dim(K[X])$ and $e = \dim(Y) = \dim(K[Y])$. By Theorem 5.9 and Proposition 5.10, d is the largest nonnegative integer m such that there exist pairwise distinct indeterminates $x_{i_1}, \ldots, x_{i_m} \in \{x_1, \ldots, x_n\}$ such that

$$\{f \in K[x_{i_1}, \ldots, x_{i_m}] \mid f \in \mathcal{I}(X)\} = \{0\}. \tag{5.3}$$

So we have $x_{i_1}, \ldots, x_{i_d} \in \{x_1, \ldots, x_n\}$ and $y_{j_1}, \ldots, y_{j_e} \in \{y_1, \ldots, y_m\}$ (with $y_1, \ldots, y_m$ a new set of indeterminates) satisfying (5.3) for X and Y, respectively. To show that the union of these satisfy (5.3) for $X \times Y$, let $f \in K[x_{i_1}, \ldots, x_{i_d}, y_{j_1}, \ldots, y_{j_e}]$ be a polynomial that vanishes on $X \times Y$. Write $f = \sum_{k=1}^{r} g_k t_k$ with $g_k \in K[x_{i_1}, \ldots, x_{i_d}]$ and t_k pairwise distinct products of powers of the y_{j_ν}. Let $(\xi_1, \ldots, \xi_n) \in X$. Then the polynomial $\sum_{k=1}^{r} g_k(\xi_1, \ldots, \xi_n) t_k \in K[y_{j_1}, \ldots, y_{j_e}]$ lies in $\mathcal{I}(Y)$, so it is zero. Since the t_k are linearly independent over K, this implies $g_k(\xi_1, \ldots, \xi_n) = 0$ for all k. Since this holds for all points in X, we conclude $g_k = 0$, so $f = 0$. This shows that $\dim(X \times Y) \geq d + e$.

To see that $\dim(X \times Y)$ is not greater than $d + e$, let $T \subseteq \{x_1, \ldots, x_n, y_1, \ldots, y_m\}$ be a subset with $|T| > d + e$. Then $|T \cap \{x_1, \ldots, x_n\}| > d$ or $|T \cap \{y_1, \ldots, y_m\}| > e$. By symmetry, we may assume the first case, so there exist pairwise distinct $x_{i_1}, \ldots, x_{i_m} \in T$ with $m > e$. Therefore we have $f \in K[x_{i_1}, \ldots, x_{i_m}] \setminus \{0\}$ which vanishes on X. So f, viewed as a polynomial in the indeterminates from T, vanishes on $X \times Y$. This completes the proof. □

Exercises for Chapter 5

5.1 (The dimension of a subset). Let X be a topological space and let $Y \subseteq X$ be a subset equipped with the subset topology.

(a) Show that Y is irreducible if and only if the closure $\overline{Y}$ is irreducible.
(b) Show that $\dim(Y) \leq \dim(X)$.

5.2 (Dimension of the power series ring). Let K be a field and $R = K[[x]]$ the formal power series ring over K. Show that $\dim(R) = 1$.

***5.3 (Subalgebras of affine algebras).** Let A be a (not necessarily finitely generated) subalgebra of an affine algebra. Show that Theorem 5.9 and Proposition 5.10 hold for A. *(Solution on page 218)*

5.4 (Coordinate rings of finite sets of points). Let K be a field and $X \subseteq K^n$ a finite set of points. Show that

$$\dim_K(K[X]) = |X|.$$

5.5 (The ring of Laurent polynomials). Let K be a field, $K(x)$ the rational function field, and $R = K[x, x^{-1}] \subset K(x)$ the ring of Laurent polynomials. Determine the Krull dimension of R.

5.6 (Right or wrong?). Decide whether each of the following statements is true or false. Give reasons for your answers.

(a) If $R \subseteq S$ is a subring, then $\dim(R) \leq \dim(S)$.
(b) If A is an affine algebra and $B \subseteq A$ a subalgebra, then $\dim(B) \leq \dim(A)$.
(c) If R is a ring and $I \subseteq R$ an ideal, then $\dim(R/I) \leq \dim(R)$.
(d) If A is an affine K-algebra, then the transcendence degree of A is the size of a maximal algebraically independent subset of A.
(e) If A is an affine K-domain, then the transcendence degree of A is the size of a maximal algebraically independent subset of A.
(f) Let A be a zero-dimensional algebra over a field K. Then $\dim_K(A) < \infty$.

5.7 (Matrices of small rank). Let K be an infinite field and $K^{n\times m}$ the set of all $n \times m$ matrices with entries in K, which we identify with affine $n \cdot m$-space $K^{n \cdot m}$. For an integer k with $0 \leq k \leq \min\{n, m\}$, let

$$X_k := \{A \in K^{n\times m} \mid \operatorname{rank}(A) \leq k\} \subseteq K^{n\times m}.$$

(a) Show that X_k is closed and irreducible.
Hint: Pick a matrix $M \in K^{n\times m}$ of rank k and consider the map $f\colon K^{n\times n} \times K^{m\times m} \to K^{n\times m}$, $(A, B) \mapsto AMB$.

(b) Show that

$$\dim(X_k) = k \cdot (n + m - k).$$

Hint: Determine the transcendence degree of $K[X_k]$ using (a), Exercise 5.6(e), and Remark 5.4(a).

5.8 (Images of morphisms). Let X and Y be affine varieties over an algebraically closed field K, $f\colon X \to Y$ a morphism, and $\overline{\operatorname{im}(f)}$ the Zariski closure of its image. Show that

$$\dim\left(\overline{\operatorname{im}(f)}\right) \leq \dim(X). \tag{5.4}$$

Does (5.4) extend to the case that f is a morphism of spectra?
Remark: By Exercise 5.1, the inequality (5.4) implies $\dim(\operatorname{im}(f)) \leq \dim(X)$.

5.9 (The polynomial ring over a principal ideal domain). Let R be a principal ideal domain that is not a field. Show that the polynomial ring $R[x]$ has dimension 2.
Hint: For a chain of prime ideals P_i in $R[x]$, consider the ideals in $\operatorname{Quot}(R)[x]$ generated by the P_i. Show that $R \cap P_2 \neq \{0\}$.
Remark: This result is a special case of Corollary 7.13 on page 84, which requires much more work.

Chapter 6
Localization

In commutative algebra, localization is a construction that is almost as important as the formation of quotient structures. In this chapter we define localization and give the basic properties. In particular, we will see what localization does to the spectrum of a ring. Localization naturally leads to the topics of local rings and the height of an ideal, which will be dealt with here.

The construction of $\mathbb{Q}$ from $\mathbb{Z}$ or, more generally, of $\operatorname{Quot}(R)$ from an integral domain R is a model for the following definition of localization. However, localization is more general in two ways: It allows one to make only a selection of ring elements invertible, which may include zero divisors, and we extend the definition to modules.

Definition 6.1. *Let R be a ring, M an R-module (where $M = R$ is an important special case), and $U \subseteq R$ a submonoid of the multiplicative monoid of R (i.e., $1 \in U$, and with $a, b \in U$ the product $a \cdot b$ also lies in U; we do* not *assume that $0 \notin U$). Such a set U is called a* **multiplicative subset** *of R. Define a relation $\sim$ on the Cartesian product $U \times M$ by*

$$(u_1, m_1) \sim (u_2, m_2) \quad :\Longleftrightarrow \quad \text{there exists } u \in U \text{ such that } uu_2m_1 = uu_1m_2.$$

(It is routine to check that this is indeed an equivalence relation.) We will write the equivalence class of $(u, m) \in U \times M$ as a fraction:

$$[(u, m)]_\sim =: \frac{m}{u}.$$

(This notation makes it clear that the intention of the equivalence relation is to allow the reduction of fractions, as well as the reverse process.) The **localization of M with respect to U**, *written as $U^{-1}M$, is the set of equivalence classes:*

$$U^{-1}M := (U \times M)/\sim \; = \left\{\frac{m}{u} \mid m \in M,\ u \in U\right\}.$$

There is a **canonical map** *given by*

G. Kemper, *A Course in Commutative Algebra*, Graduate Texts in Mathematics 256, DOI 10.1007/978-3-642-03545-6_7,

$$\varepsilon\colon M \to U^{-1}M,\ m \mapsto \frac{m}{1}.$$

$U^{-1}M$ is made into an R-module by

$$\frac{m_1}{u_1} + \frac{m_2}{u_2} := \frac{u_2 m_1 + u_1 m_2}{u_1 u_2} \quad \textit{for} \quad m_i \in M \textit{ and } u_i \in U$$

and

$$a \cdot \frac{m}{u} := \frac{am}{u} \quad \textit{for} \quad a \in R,\ m \in M,\ \textit{and } u \in U.$$

(Again, it is routine to check that these operations are well defined and the axioms of a module are satisfied.)

In the special case that $U = R \setminus P$ with $P \in \operatorname{Spec}(R)$, we write

$$U^{-1}M =: M_P$$

and call this the **localization of M at P**.

The generality and flexibility of localization are best demonstrated by examples.

Example 6.2. (1) Let R be an integral domain and $U = R \setminus \{0\}$. Then $U^{-1}R = \operatorname{Quot}(R)$, the field of fractions. $\operatorname{Quot}(R)$ is the localization of R at the prime ideal $\{0\}$.

(2) More generally, let R be any ring and

$$U := \{a \in R \mid a \text{ is not a zero divisor}\},$$

which is a multiplicative subset. $U^{-1}R$ is called the **total ring of fractions** of R. The canonical map $R \to U^{-1}R$ is injective, and U is maximal with this property. In fact, if $S \subseteq R$ is any multiplicative subset, then the canonical map $R \to S^{-1}R$ is injective if and only if $S \subseteq U$.

(3) Consider $R = \mathbb{Z}$ and $P = (2) \in \operatorname{Spec}(\mathbb{Z})$. Then R_P is (isomorphic to) the ring of all rational numbers with odd denominator. We have mentioned this ring on page 15 as an example of a non-Jacobson ring.

(4) Let $R = \mathbb{Z}/(6)$ and $P = (2) \in \operatorname{Spec}(R)$. In R_P, we have $\frac{2}{1} = \frac{0}{1}$, since $3 \notin P$ and

$$3 \cdot 1 \cdot 2 = 3 \cdot 1 \cdot 0.$$

With this, it is easy to see that $R_P \cong \mathbb{Z}/(2)$. So a localization can be "smaller" than the original ring.

Also notice that the total ring of fractions of $\mathbb{Z}/(6)$ is isomorphic to $\mathbb{Z}/(6)$.

(5) Let K be a field and $X \subseteq K^n$ an affine variety with coordinate ring $K[X]$. For $x \in X$, consider the maximal ideal $\mathfrak{m}_x \in \operatorname{Spec}_{\max}(K[X])$ of all regular functions vanishing at x. Then

$$K[X]_x := K[X]_{\mathfrak{m}_x}$$

consists of all fractions of regular functions on X whose denominator does not vanish at x. This example gives a first hint that localization has something to do with locality. A second hint is contained in Exercise 6.6. $K[X]_x$ is also called the **localization of $K[X]$ at x**.

(6) Let R be a ring and $a \in R$. Then $U = \{1, a, a^2, \ldots\} = \{a^k \mid k \in \mathbb{N}_0\} \subseteq R$ is a multiplicative subset. It is customary to write

$$M_a := U^{-1}M,$$

although this may sometimes lead to confusion. For example, with this notation $\mathbb{Z}_2$ is (isomorphic to) the ring of all rational numbers with a power of 2 as denominator.

(7) If $0 \in U$, then $U^{-1}M = \{0\}$ for every R-module M, including $M = R$. This follows from Definition 6.1.

(8) Let $(G, +)$ be a finite abelian group, which becomes a $\mathbb{Z}$-module by defining $a \cdot \sigma := \sum_{i=1}^{a} \sigma$ and $(-a) \cdot \sigma := -(a \cdot \sigma)$ for $\sigma \in G$ and $a \in \mathbb{Z}$ nonnegative. Let $U = \mathbb{Z} \setminus \{0\}$. Then $U^{-1}G$ is the zero module, since each $\sigma \in G$ has positive order $\operatorname{ord}(\sigma) \in U$, and $\operatorname{ord}(\sigma) \cdot \sigma = 0$. $\triangleleft$

The following proposition is a collection of basic properties of localization. The proofs of all parts are straightforward but sometimes a little tedious. We leave them as an exercise for the reader, who should be prepared to spend a small pile of paper on them.

Proposition 6.3 (Properties of localization). *Let R be a ring, $U \subseteq R$ a multiplicative subset, and M an R-module.*

(a) $U^{-1}R$ becomes a ring with the addition defined as in $U^{-1}M$, and multiplication defined as multiplying numerators and denominators.

(b) The canonical map $\varepsilon\colon R \to U^{-1}R$ is a homomorphism of rings. So $U^{-1}R$ becomes an R-algebra.

(c) $U^{-1}M$ becomes a $U^{-1}R$-module, with multiplication of an element of $U^{-1}R$ and an element of $U^{-1}M$ defined as multiplying numerators and denominators.

(d) All $\varepsilon(u)$ with $u \in U$ are invertible in $U^{-1}R$.

(e) Let $\varphi\colon R \to S$ be a ring homomorphism such that all $\varphi(u)$ with $u \in U$ are invertible in S. Then there exists a unique homomorphism $U^{-1}R \to S$ of R-algebras. This universal property tells us that $S^{-1}R$ is the "smallest" and "freest" R-algebra in which the elements from U become invertible.

(f) If R is an integral domain and $0 \notin U$, then $U^{-1}R$ is embedded in $\operatorname{Quot}(R)$ in the obvious way. Therefore we may (and often will) identify $U^{-1}R$ with a subalgebra of $\operatorname{Quot}(R)$.

(g) If $V \subseteq R$ is a multiplicative subset containing U, then

$$V^{-1}(U^{-1}M) \cong \varepsilon(V)^{-1}(U^{-1}M) \cong V^{-1}M$$

(isomorphisms of R-modules). So "step-by-step" localization is the same as "all-at-once" localization. For $M = R$, the second isomorphism is also

a ring isomorphism

$$\varepsilon(V)^{-1}(U^{-1}R) \cong V^{-1}R.$$

(h) *Let $N \subseteq M$ be a submodule. Then $U^{-1}N$ is isomorphic to a submodule of $U^{-1}M$. In fact, with $\varepsilon_M \colon M \to U^{-1}M$ the canonical map, the map*

$$U^{-1}N \to \Big(\varepsilon_M(N)\Big)_{U^{-1}R} = U^{-1}R \cdot \varepsilon_M(N), \ \frac{n}{u} \mapsto \frac{1}{u} \cdot \varepsilon_M(n),$$

is an isomorphism of $U^{-1}R$-modules. Therefore we may (and will) identify $U^{-1}N$ with $(\varepsilon_M(N))_{U^{-1}R} \subseteq U^{-1}M$. In particular, for an ideal $I \subseteq R$ we identify $U^{-1}I$ with the ideal $(\varepsilon(I))_{U^{-1}R} \subseteq U^{-1}R$.

(i) *Let $\mathfrak{N} \subseteq U^{-1}M$ be a $U^{-1}R$-submodule. With $\varepsilon_M \colon M \to U^{-1}M$ the canonical map, the preimage $N := \varepsilon_M^{-1}(\mathfrak{N}) \subseteq M$ is a submodule, and*

$$U^{-1}N = \mathfrak{N}.$$

In particular, if $\mathfrak{I} \subseteq U^{-1}R$ is an ideal, then

$$U^{-1}\varepsilon^{-1}(\mathfrak{I}) = \mathfrak{I}.$$

The above properties of localization will often be used without explicit reference to Proposition 6.3. As an immediate consequence of part (i), we get the following:

Corollary 6.4 (Localization preserves the Noether property). *Let R be a ring, $U \subseteq R$ a multiplicative subset, and M an R-module. If M is Noetherian, then so is $U^{-1}M$ (as a $U^{-1}R$-module). In particular, if R is Noetherian, then so is $U^{-1}R$.*

The following result gives a description of the spectrum of a localized ring $U^{-1}R$. It is a counterpart of Lemma 1.22, which deals with quotient rings.

Theorem 6.5 (The spectrum of a localized ring). *Let R be a ring and $U \subseteq R$ a multiplicative subset. Let $\varepsilon \colon R \to U^{-1}R$ be the canonical map and*

$$\mathcal{A} := \{Q \in \operatorname{Spec}(R) \mid U \cap Q = \emptyset\}.$$

Then the map

$$\operatorname{Spec}\left(U^{-1}R\right) \to \mathcal{A}, \ \mathfrak{Q} \mapsto \varepsilon^{-1}(\mathfrak{Q})$$

is an inclusion-preserving bijection with inverse map

$$\mathcal{A} \to \operatorname{Spec}\left(U^{-1}R\right), \ Q \mapsto U^{-1}Q.$$

In particular, for a prime ideal $P \in \operatorname{Spec}(R)$, the prime ideals of R_P correspond to prime ideals $Q \in \operatorname{Spec}(R)$ with $Q \subseteq P$.

Proof. Since preimages of prime ideals under ring homomorphisms are always prime ideals, $\varepsilon^{-1}(\mathfrak{Q}) \in \mathrm{Spec}(R)$ for $\mathfrak{Q} \in \mathrm{Spec}\left(U^{-1}R\right)$. Moreover, $U \cap \varepsilon^{-1}(\mathfrak{Q}) = \emptyset$, since otherwise $\mathfrak{Q}$ would contain an invertible element from $U^{-1}R$. So $\varepsilon^{-1}(\mathfrak{Q}) \in \mathcal{A}$. By Proposition 6.3(i) we also have

$$U^{-1}\varepsilon^{-1}(\mathfrak{Q}) = \mathfrak{Q}.$$

Now let $Q \in \mathcal{A}$. We claim that

$$\varepsilon^{-1}\left(U^{-1}Q\right) = Q. \tag{6.1}$$

It is clear that $Q \subseteq \varepsilon^{-1}\left(U^{-1}Q\right)$. For the reverse inclusion, take $a \in \varepsilon^{-1}\left(U^{-1}Q\right)$. Then there exist $q \in Q$ and $u \in U$ with

$$\frac{a}{1} = \frac{q}{u},$$

so $u'ua = u'q$ with $u' \in U$. With the definition of $\mathcal{A}$, this implies $a \in Q$, proving (6.1).

We still need to show that $U^{-1}Q$ is a prime ideal. We see that $U^{-1}Q$ is an ideal by Proposition 6.3(h), and it follows from (6.1) that $U^{-1}Q \neq U^{-1}R$. Take $a_1, a_2 \in R$ and $u_1, u_2 \in U$ with

$$\frac{a_1}{u_1} \cdot \frac{a_2}{u_2} \in U^{-1}Q.$$

Then $\varepsilon(a_1a_2) \in U^{-1}Q$, so $a_1a_2 \in Q$ by (6.1). This implies that at least one of the a_i lies in Q, so $\frac{a_i}{u_i} \in U^{-1}Q$, and $U^{-1}Q$ is a prime ideal indeed.

It is immediately clear that our maps preserve inclusions. This completes the proof. □

In Exercise 6.5 it is shown that the bijections from Theorem 6.5 are actually homeomorphisms. Theorem 6.5 has two immediate consequences, Corollaries 6.6 and 6.8.

Corollary 6.6 (Dimension of a localized ring). *Let R be a ring and $U \subseteq R$ a multiplicative set. Then*

$$\dim\left(U^{-1}R\right) \leq \dim\left(R\right).$$

Definition 6.7. *A ring R is called a* **local ring** *if it has precisely one maximal ideal.*

Corollary 6.8 (Localizing at a prime ideal gives a local ring). *Let R be a ring and $P \subset R$ a prime ideal. Then the localization R_P is a local ring with P_P as unique maximal ideal.*

Example 6.9. The rings in Example 6.2(1), (3), and (5) are examples of local rings. We give a few more.

(1) Every field is a local ring.
(2) Let $K[x]$ be a polynomial ring over a field. Then $K[x]/(x^2)$ is a local ring with $(x)/(x^2)$ as unique maximal ideal.
(3) The formal power series ring $K[[x]]$ over a field is a local ring with (x) as unique maximal ideal (see Exercise 1.2).
(4) The zero ring $R = \{0\}$ is *not* a local ring. ◁

Definition 6.10. *Let R be a ring.*

(a) Let $P \subset R$ be a prime ideal. Then the **height** *of P is defined as*

$$\operatorname{ht}(P) := \dim(R_P) \in \mathbb{N}_0 \cup \{\infty\}.$$

So by Theorem 6.5, $\operatorname{ht}(P)$ is the maximal length n of a chain

$$P_0 \subsetneqq P_1 \subsetneqq \cdots \subsetneqq P_n = P$$

of prime ideals $P_i \in \operatorname{Spec}(R)$ ending with P.

(b) Let $I \subseteq R$ be an ideal. If $I \neq R$, the **height** *of I is defined as*

$$\operatorname{ht}(I) := \min\left\{\operatorname{ht}(P) \mid P \in \mathcal{V}_{\operatorname{Spec}(R)}(I)\right\}.$$

If $I = R$, we set

$$\operatorname{ht}(I) := \dim(R) + 1.$$

Since the height, as defined in (a), *gets smaller when we pass to a sub–prime ideal, the definitions in* (a) *and* (b) *are consistent.*

Remark 6.11. (a) If $P \subset R$ is a prime ideal, then Lemma 1.22 tells us that $\dim(R/P)$ is the maximal length of a chain of prime ideals in $\operatorname{Spec}(R)$ *starting* with P. Therefore

$$\operatorname{ht}(P) + \dim(R/P) \leq \dim(R). \tag{6.2}$$

This is often an equality, for example in the case that $R = K[X]$ with X an equidimensional affine variety (see Corollary 8.23). For this reason, some authors use the term *codimension* for the height. Example 6.12(3) shows that the inequality (6.2) can also be strict.

(b) It is not hard to give a geometric interpretation of height. If X is an affine variety over an algebraically closed field, then the prime ideals in the coordinate ring $K[X]$ correspond to the closed, irreducible subsets of X (see Theorem 1.23 and Theorem 3.10(a)). So if $P \in \operatorname{Spec}(K[X])$ corresponds to $Y \subseteq X$, i.e., $Y = \mathcal{V}_X(P)$, then $\operatorname{ht}(P)$ is the maximal length k of a chain

$$Y = Y_0 \subsetneqq Y_1 \subsetneqq \cdots \subsetneqq Y_k$$

of closed, irreducible subsets of X starting with Y. On the other hand, $\dim(K[X]/P)$ is the maximal length of a chain *ending* with Y. So $\operatorname{ht}(P) + \dim(K[X]/P)$ is the maximal length of a chain passing through Y. ◁

Example 6.12. (1) Every minimal prime ideal has height 0. If R is a Noetherian ring, then $\mathrm{ht}(\{0\}) = 0$. By Exercise 3.6, this is also true for R not Noetherian.

(2) Let K be a field, $(\xi_1, \ldots, \xi_n) \in K^n$, and $P = \mathcal{I}_{K[x_1,\ldots,x_n]}(\{(\xi_1, \ldots, \xi_n)\})$. Then $\mathrm{ht}(P) = n$, since we have a chain of prime ideals

$$\{0\} \subsetneqq (x_1 - \xi_1) \subsetneqq (x_1 - \xi_1, x_2 - \xi_2) \subsetneqq \cdots \subsetneqq (x_1 - \xi_1, \ldots, x_n - \xi_n) = P,$$

and on the other hand $\mathrm{ht}(P) \leq \dim(K[x_1, \ldots, x_n]) = n$ by (6.2).

(3) Let $X = Z_1 \cup Z_2$ be an affine variety over an algebraically closed field with irreducible components Z_1 and Z_2 such that $\dim(Z_1) < \dim(Z_2)$. Let x be a point of Z_1 not lying in Z_2. Then a chain of closed, irreducible subsets of X that starts with $\{x\}$ lies completely in Z_1, so for $P := \mathcal{I}_{K[X]}(x)$ we have

$$\mathrm{ht}(P) \leq \dim(Z_1).$$

(In fact, equality holds as a consequence of Corollary 8.24.) Since $\dim(K[X]/P) = 0$, we have

$$\mathrm{ht}(P) + \dim(K[X]/P) < \dim(K[X]),$$

so the inequality (6.2) is strict here. ◁

We conclude this chapter with a definition.

Definition 6.13. *Let R be a ring and M an R-module.*

(a) For an element $m \in M$, the **annihilator** *of m is*

$$\mathrm{Ann}(m) := \{a \in R \mid a \cdot m = 0\}.$$

This is an ideal in R.

(b) The **annihilator** *of M is*

$$\mathrm{Ann}(M) := \bigcap_{m \in M} \mathrm{Ann}(m) \subseteq R.$$

Clearly one can restrict this intersection to the elements m of a generating set of M.

(c) The **(Krull) dimension** *of M is*

$$\dim(M) := \dim(R/\mathrm{Ann}(M)),$$

where the dimension on the right-hand side denotes the Krull dimension of the ring. Readers should notice that for R a field and M a nonzero vector space, $\dim(M)$ is always 0, so this has nothing to do with dimension as a vector space.

(d) The **support** *of M is*

$$\mathrm{Supp}(M) := \{P \in \mathrm{Spec}(R) \mid M_P \neq \{0\}\} \subseteq \mathrm{Spec}(R).$$

So a $P \in \operatorname{Spec}(R)$ *lies in the support if and only if there exists* $m \in M$ *with* $\operatorname{Ann}(m) \subseteq P$.

Example 6.14. Let $I \subseteq R$ be an ideal in a ring, and consider the quotient ring $M := R/I$ as an R-module. Then it is easy to see that $\operatorname{Ann}(M) = I$ and $\operatorname{Supp}(M) = \mathcal{V}_{\operatorname{Spec}(R)}(I)$. A generalization can be found in Exercise 6.10. ◁

Exercises for Chapter 6

6.1 (Properties of localization). Check all assertions made in Definition 6.1 and Proposition 6.3.

6.2 (Some examples of localization). In each of the following examples, we give a ring R, a multiplicative subset $U \subseteq R$, and an R-module M. Give a description of the localization $U^{-1}M$. The letter K always stands for a field and x for an indeterminate.

(a) $M = R = K[x]$, $U = \{x^k \mid k \in \mathbb{N}_0\}$.
(b) $M = R = \mathbb{Z}$, $U = \{1\} \cup \{12n \mid n \in \mathbb{Z}, n > 0\}$.
(c) $R = \mathbb{Z}$, $U = \mathbb{Z} \setminus \{0\}$, $M = \mathbb{Z}[x]$.
(d) $R = K[x]$, $M = K[x]/(x^2)$, $U = \{x^k \mid k \in \mathbb{N}_0\}$.
(e) $R = K[x]$, $M = K[x]/(x^2)$, $U = K[x] \setminus (x)$.

6.3 (Localization is an exact functor). Let R be a ring and $U \subseteq R$ a multiplicative set. Let $\varphi\colon M \to N$ be a homomorphism of R-modules. Show that the map

$$U^{-1}\varphi\colon U^{-1}M \to U^{-1}N,\ \frac{m}{u} \mapsto \frac{\varphi(m)}{u},$$

is a homomorphism of $U^{-1}R$-modules. (Since passing from φ to $U^{-1}\varphi$ is compatible with composition of homomorphisms, this makes localization with respect to U into a functor from the category of R-modules to the category of $U^{-1}R$-modules.)

By an *exact sequence of R-modules*, we mean a sequence

$$\cdots \xrightarrow{\varphi_{-2}} M_{-1} \xrightarrow{\varphi_{-1}} M_0 \xrightarrow{\varphi_0} M_1 \xrightarrow{\varphi_1} M_2 \xrightarrow{\varphi_2} M_3 \xrightarrow{\varphi_3} \cdots \tag{6.3}$$

with M_i modules over R and $\varphi_i\colon M_i \to M_{i+1}$ module homomorphisms such that $\operatorname{im}(\varphi_i) = \ker(\varphi_{i+1})$ for all $i \in \mathbb{Z}$. More formally, a sequence is a direct sum $M = \bigoplus_{i\in\mathbb{Z}} M_i$ of R-modules together with a homomorphism $\varphi\colon M \to M$ such that $\varphi(M_i) \subseteq M_{i+1}$ for all i, and exactness means that $\operatorname{im}(\varphi) = \ker(\varphi)$. Assume that the sequence (6.3) is exact and show that the localized sequence

$$\cdots \longrightarrow U^{-1}M_0 \xrightarrow{U^{-1}\varphi_0} U^{-1}M_1 \xrightarrow{U^{-1}\varphi_1} U^{-1}M_2 \longrightarrow \cdots \tag{6.4}$$

is also exact. We express this by saying that localization is an exact functor. *Remark:* Most exact sequences appearing in the real life of a mathematician are finite, meaning that only finitely many M_i are nonzero. The most frequent example is a "short exact sequence," in which all M_i except M_1, M_2, and M_3 are zero. In that case, exactness implies $M_3 \cong M_2/M_1$.

A consequence of this exercise is that injective (surjective) homomorphisms localize to injective (surjective) maps.

6.4 (Local–global principle). A *local-global principle* is a theorem that states that some property holds "globally" if and only if it holds everywhere "locally." Here are two examples.

(a) Let M be a module over a ring R with submodules $L, N \subseteq M$. Prove the equivalence

$$L \subseteq N \quad \Longleftrightarrow \quad L_{\mathfrak{m}} \subseteq N_{\mathfrak{m}} \quad \text{for all} \quad \mathfrak{m} \in \operatorname{Spec}_{\max}(R).$$

(b) Let $\varphi\colon M \to N$ be a homomorphism of modules over a ring R. For $\mathfrak{m} \in \operatorname{Spec}_{\max}(R)$ there is a homomorphism $\varphi_{\mathfrak{m}}\colon M_{\mathfrak{m}} \to N_{\mathfrak{m}}$ as defined in Exercise 6.3. Show that φ is injective or surjective if and only if the same property holds for every $\varphi_{\mathfrak{m}}$ with $\mathfrak{m} \in \operatorname{Spec}_{\max}(R)$.

6.5 (Homeomorphisms). Show that the bijections from Theorem 6.5 are homeomorphisms. (Here $\mathcal{A}$ is equipped with the subset topology induced from the Zariski topology on $\operatorname{Spec}(R)$.)

6.6 (Localization hides components). Let $X = Y_1 \cup Y_2$ be an affine variety over a field K, decomposed as a union of two closed subsets. Let $x \in Y_1 \setminus Y_2$ be a point. Show that the restriction homomorphism $\varphi\colon K[X] \to K[Y_1]$ induces an isomorphism (of K-algebras)

$$\varphi_x\colon K[X]_x \to K[Y_1]_x, \quad \frac{f}{u} \mapsto \frac{\varphi(f)}{\varphi(u)},$$

of the coordinate rings localized at x.
Remark: This result may be expressed thus: "Localization at x sees only those components in which x lies" – a further hint that localization has something to do with locality.

6.7 (A characterization of local rings). Let R be a ring. Prove the following.

(a) R is local if and only if the set of noninvertible elements of R is an ideal. Then this is the unique maximal ideal.
(b) Let $\mathfrak{m} \subsetneq R$ be a proper ideal. Then R is local with $\mathfrak{m}$ as maximal ideal if and only if all elements from $R \setminus \mathfrak{m}$ are invertible.

6.8 (An alternative definition of Krull dimension). In this exercise we develop a prime-ideal-free definition of the Krull dimension of a ring. It is based on an article by Coquand and Lombardi [11], which was brought to my attention by Peter Heinig.

Let R be a ring. For $a \in R$, define the multiplicative set

$$U_a := \{a^m(1 + xa) \mid m \in \mathbb{N}_0,\ x \in R\}.$$

(a) For $a \in R$ and $P \in \operatorname{Spec}(R)$ a prime ideal, prove the following equivalence:

$$U_a \cap P = \emptyset \quad \Longleftrightarrow \quad a \notin P \quad \text{and} \quad P + (a)_R \neq R.$$

(b) For $n \in \mathbb{N}_0$ a nonnegative integer, prove the following equivalence:

$$\dim(R) \leq n \quad \Longleftrightarrow \quad \dim\left(U_a^{-1}R\right) \leq n - 1 \quad \text{for all} \quad a \in R.$$

(c) For $n \in \mathbb{N}_0$ a nonnegative integer, show that $\dim(R) \leq n$ holds if and only if for every $a_0, \ldots, a_n \in R$ there exist $m_0, \ldots, m_n \in \mathbb{N}_0$ such that

$$\prod_{i=0}^{n} a_i^{m_i} \in \left(a_j \cdot \prod_{i=0}^{j} a_i^{m_i} \,\middle|\, j = 0, \ldots, n\right)_R. \tag{6.5}$$

Remark: Part (c) provides the desired alternative definition of the Krull dimension. The condition (6.5) looks a bit messy at first glance, but it is easy to understand and to remember in terms of the lexicographic monomial ordering, which we will introduce in Example 9.2(1) on page 119. In fact, (c) says that $\dim(R) \leq n$ if and only if for every $a_0, \ldots, a_n \in R$ there exists a monomial in the a_i that can be written as an R-linear combination of lexicographically *larger* monomials in the a_i. As a nice application, it is easy to derive Theorem 5.5 and the first part of Corollary 5.7 from (c) using the lexicographic ordering (see Exercise 9.5). *(Solution on page 219)*

6.9 (Localizing an affine domain). Let A be an affine domain and $a \in A \setminus \{0\}$. Show that the localization A_a has the same dimension as A. Does this remain true for A an affine algebra or A an integral domain? Does it remain true if one localizes with respect to an arbitrary multiplicative subset $U \subseteq A \setminus \{0\}$?

6.10 (Support of modules). Let R be a ring and M an R-module.

(a) Assume that M is finitely generated and show that

$$\operatorname{Supp}(M) = \mathcal{V}_{\operatorname{Spec}(R)}\left(\operatorname{Ann}(M)\right).$$

In particular, $\operatorname{Supp}(M)$ is Zariski closed in $\operatorname{Spec}(R)$.

*(b) Give an example in which $\operatorname{Supp}(M)$ is not Zariski closed.

6.11 (Associated primes). Let R be a Noetherian ring and M an R-module. A prime ideal $P \in \operatorname{Spec}(R)$ is called an **associated prime** of M if there exists $m \in M$ with $P = \operatorname{Ann}(m)$. (But notice that not all annihilators of elements of M are prime ideals!) We write the set of all associated primes as $\operatorname{Ass}(M)$.

(a) Let I be an ideal that is maximal among all $\operatorname{Ann}(m)$ with $m \in M \setminus \{0\}$. Show that $I \in \operatorname{Ass}(M)$. So in particular $\operatorname{Ass}(M) \neq \emptyset$ if $M \neq \{0\}$.
(b) Let $U \subseteq R$ be a multiplicative subset and consider the R-module $U^{-1}M$. Show that

$$\operatorname{Ass}(U^{-1}M) = \{P \in \operatorname{Ass}(M) \mid U \cap P = \emptyset\}.$$

(c) Consider the special case $M = R/I$ with I a radical ideal. Show that $\operatorname{Ass}(M)$ is the set of all prime ideals that are minimal over I.
(d) Let $R = K[x_1, x_2]$ be a polynomial ring over a field and $M = R/I$ with $I := (x_1^2, x_1x_2)$. Determine $\operatorname{Ass}(M)$. Does the conclusion of part (c) hold?

Remark: Part (c) suggests that associated primes may be seen as a generalization of irreducible components. The theory of associated primes and primary decomposition is treated in most textbooks on commutative algebra, but not in this one.

Chapter 7
The Principal Ideal Theorem

This chapter has very few definitions, but many results. In the first section we prove Krull's principal ideal theorem, which says, roughly speaking, that an ideal generated by n elements has height at most n. This theorem is one of the workhorses of commutative algebra. As corollaries, we obtain the existence of systems of parameters of Noetherian local rings, and the fact that every Noetherian local ring has finite dimension. Along the way, two important lemmas are proved: Nakayama's lemma and the prime avoidance lemma.

The second section of this chapter deals with the dimension of fibers of a morphism of spectra of rings. The principal ideal theorem leads to a lower bound for the fiber dimension. From this, we obtain a formula for the dimension of a polynomial ring over an arbitrary Noetherian ring. More work is required to show that under suitable hypotheses, the lower bound is exact "almost everywhere." This project will be completed in Chapter 10. Large parts of the second section may be skipped by readers who are not interested in fiber dimension. Details are given at the beginning of the section.

7.1 Nakayama's Lemma and the Principal Ideal Theorem

For a square matrix $A \in R^{n \times n}$ with entries in a ring R, the determinant $\det(A)$ is defined by the Leibniz formula.

Lemma 7.1 (Adjugate matrix over rings). *Let $A = (a_{i,j})_{1 \le i,j \le n} \in R^{n \times n}$ be a square matrix with entries in a ring R. For $i, k \in \{1, \dots, n\}$, let $c_{i,k} \in R$ be the determinant of the matrix obtained from A by deleting the ith row and the kth column. Then for $j, k \in \{1, \dots, n\}$:*

$$\sum_{i=1}^{n} (-1)^{i+k} c_{i,k} a_{i,j} = \delta_{j,k} \cdot \det(A) \quad \textit{with} \quad \delta_{j,k} := \begin{cases} 1 & \textit{if } j = k, \\ 0 & \textit{if } j \neq k. \end{cases}$$

G. Kemper, *A Course in Commutative Algebra*, Graduate Texts in Mathematics 256, DOI 10.1007/978-3-642-03545-6_8,

Proof. If determinant theory is developed over a ring, this is the standard result on the adjugate matrix. For readers who are familiar with determinants only over a field, we present a proof by reduction to the field case.

For $i, j \in \{1, \ldots, n\}$, let $x_{i,j}$ be an indeterminate over $\mathbb{Q}$, and consider the matrix $\widehat{A} := (x_{i,j}) \in \mathbb{Q}(x_{1,1}, x_{1,2}, \ldots, x_{n,n})^{n \times n}$ with coefficients in the rational function field in n^2 indeterminates. Let $\widehat{c}_{i,k} \in \mathbb{Q}(x_{1,1}, x_{1,2}, \ldots, x_{n,n})$ be the minors of $\widehat{A}$, formed as the $c_{i,k}$ from A. The rule of the adjugate matrix states that

$$\sum_{i=1}^{n} (-1)^{i+k} \widehat{c}_{i,k} x_{i,j} = \delta_{j,k} \cdot \det(\widehat{A}). \tag{7.1}$$

Both sides of (7.1) lie in the polynomial ring $\mathbb{Z}[x_{1,1}, x_{1,2}, \ldots, x_{n,n}]$. There exists a (unique) ring homomorphism

$$\varphi\colon \mathbb{Z}[x_{1,1}, x_{1,2}, \ldots, x_{n,n}] \to R \quad \text{with} \quad \varphi(x_{i,j}) = a_{i,j}.$$

Applying φ to both sides of (7.1) yields the lemma. □

The following lemma will be used in the proof of Nakayama's lemma, but also in the development of the theory of integral ring extensions (see Lemma 8.3).

Lemma 7.2. *Let R be a ring, $M = (m_1, \ldots, m_n)_R$ a finitely generated R-module, and $a_{i,j} \in R$ ring elements ($i, j \in \{1, \ldots, n\}$) with*

$$\sum_{j=1}^{n} a_{i,j} m_j = 0 \quad \textit{for} \quad i \in \{1, \ldots, n\}.$$

Then

$$\det(a_{i,j})_{1 \le i,j \le n} \in \operatorname{Ann}(M).$$

Proof. Let $A := (a_{i,j}) \in R^{n \times n}$, and let $c_{i,k} \in R$ be as in Lemma 7.1. For every $k \in \{1, \ldots, n\}$, it follows from Lemma 7.1 that

$$\det(A) \cdot m_k = \sum_{j=1}^{n} \delta_{j,k} \det(A) m_j = \sum_{j=1}^{n} \sum_{i=1}^{n} (-1)^{i+k} c_{i,k} a_{i,j} m_j = 0,$$

so indeed $\det(A) \in \operatorname{Ann}(M)$. □

Nakayama's lemma, which we prove now, is one of the key tools in commutative algebra. It is one of those results that seldom arouse spontaneous enthusiasm, but then develop a habit of appearing at crucial steps in many proofs. (Readers may look up the index entry for "Nakayama's lemma" at the end of this book to locate some examples.) If R is a ring, then the intersection

$$J := \bigcap_{\mathfrak{m} \in \operatorname{Spec}_{\max}(R)} \mathfrak{m}$$

is called the **Jacobson radical** of R. (For $R = \{0\}$, we define the Jacobson radical to be $J = R$.) For example, if R is a local ring, then J is the unique maximal ideal. Because of its importance, we give Nakayama's lemma the status of a theorem.

Theorem 7.3 (Nakayama's lemma). *Let R be a ring with Jacobson radical J, and let M be a finitely generated R-module. If*

$$J \cdot M = M,$$

then $M = \{0\}$.

Proof. Write $M = (m_1, \dots, m_n)_R$. By hypothesis, $m_i = \sum_{j=1}^n a_{i,j} m_j$ with $a_{i,j} \in J$. By Lemma 7.2,

$$d := \det(\delta_{i,j} - a_{i,j})_{1 \leq i,j \leq n} \in \operatorname{Ann}(M).$$

But $d \equiv 1 \mod J$, so d lies in no maximal ideal of R. This implies that d is invertible, so $M = \{0\}$. □

We can now prove the first version of Krull's principal ideal theorem. This is a generalization of the implication (c) ⇒ (b) of Theorem 5.13 (which is about polynomial rings) to arbitrary Noetherian rings. Recall that a prime ideal $P \in \operatorname{Spec}(R)$ is said to be minimal over an ideal $I \subseteq R$ if P is a minimal element of $\mathcal{V}_{\operatorname{Spec}(R)}(I)$, i.e., $I \subseteq P$, but no proper sub–prime ideal of P contains I.

Theorem 7.4 (Principal ideal theorem, first version). *Let R be a Noetherian ring and $P \in \operatorname{Spec}(R)$ a prime ideal that is minimal over a principal ideal $(a) \subsetneq R$. Then*

$$\operatorname{ht}(P) \leq 1.$$

In particular, a proper principal ideal of R has height at most 1.

Proof. Let R_P be the localization at P. Using Theorem 6.5, we see that P_P is a prime ideal that is minimal over $\left(\frac{a}{1}\right)_{R_P}$, and $\operatorname{ht}(P_P) = \operatorname{ht}(P)$. So by replacing R with R_P, we may assume that R is local with maximal ideal P. Then the quotient ring $R/(a)$ has the unique prime ideal $P/(a)$, so $R/(a)$ is Artinian by Theorem 2.8. We will transport the Artin property from $R/(a)$ to the localization R_a using the canonical maps ε and φ shown in Fig. 7.1.

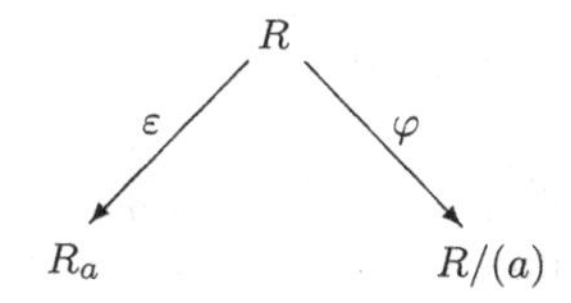

Fig. 7.1. The canonical maps ε and φ

Let $I, J \subseteq R_a$ be ideals with $I \supseteq J$ and $\varphi\left(\varepsilon^{-1}(I)\right) = \varphi\left(\varepsilon^{-1}(J)\right)$. If we can prove that this implies $I = J$, it follows that a descending chain of ideals I_n in R_a stops at the same point or earlier than the chain of ideals $\varphi\left(\varepsilon^{-1}(I_n)\right) \subseteq R/(a)$. To prove the claim, let $x \in \varepsilon^{-1}(I)$. Then there exist $y \in \varepsilon^{-1}(J)$ and $z \in R$ with $x = y + az$. This implies $az \in \varepsilon^{-1}(I)$. Since $\varepsilon(a)$ is invertible in R_a, we obtain $z \in \varepsilon^{-1}(I)$, so $x \in \varepsilon^{-1}(J) + (a) \cdot \varepsilon^{-1}(I)$. Therefore the R-module $M := \varepsilon^{-1}(I)/\varepsilon^{-1}(J)$ satisfies

$$M \subseteq (a)M \subseteq PM \subseteq M.$$

Nakayama's lemma (Theorem 7.3) yields $M = \{0\}$, so $\varepsilon^{-1}(I) = \varepsilon^{-1}(J)$. By Proposition 6.3(i), this implies $I = J$. So indeed R_a is Artinian. Applying Theorem 2.8 again, we obtain $\dim(R_a) \in \{0, -1\}$, with 0 occurring if $R_a \neq \{0\}$.

Let $Q \in \operatorname{Spec}(R)$ be a prime ideal with $Q \subsetneqq P$. The minimality of P implies $a \notin Q$, so by Proposition 6.3(g), R_Q is a localization of R_a (which implies $R_a \neq \{0\}$). By Corollary 6.6, this implies $\dim(R_Q) \leq \dim(R_a) = 0$, so $\operatorname{ht}(Q) = 0$. It follows that $\operatorname{ht}(P) \leq 1$. □

The principal ideal theorem can fail badly for non-Noetherian rings. Examples are given in Exercise 7.4.

Theorem 7.4 will be used in the proof of a second version of the principal ideal theorem, which generalizes from principal ideals to ideals generated by n elements.

Theorem 7.5 (Principal ideal theorem, generalized version). *Let R be a Noetherian ring and $P \in \operatorname{Spec}(R)$ a prime ideal that is minimal over an ideal $(a_1, \ldots, a_n) \subseteq R$ generated by n elements. Then*

$$\operatorname{ht}(P) \leq n.$$

Proof. We use induction on n. The result is correct for $n = 0$, so assume $n > 0$. As in the proof of Theorem 7.4, we may assume that R is a local ring with maximal ideal P. Let $Q \subsetneqq P$ be a prime ideal such that no other prime ideals lie between Q and P. We need to show that $\operatorname{ht}(Q) \leq n - 1$. By assumption, Q does not lie over $(a_1, \ldots, a_n)$, so by relabeling we may assume that $a_1 \notin Q$. So P is a prime ideal that is minimal over $Q + (a_1)$, and since P is the unique maximal ideal of R, it is the only prime ideal over $Q + (a_1)$. By Corollary 1.12, we obtain $P = \sqrt{Q + (a_1)}$, so for $i \in \{2, \ldots, n\}$ we have

$$a_i^{k_i} = b_i + x_i a_1 \quad \text{with} \quad k_i > 0,\ b_i \in Q, \quad \text{and} \quad x_i \in R.$$

This implies $(a_1, a_2^{k_2}, \ldots, a_n^{k_n}) = (a_1, b_2, \ldots, b_n)$. Therefore P is minimal over $(a_1, b_2, \ldots, b_n)$, so by Lemma 1.22, $P/(b_2, \ldots, b_n) \in \operatorname{Spec}\left(R/(b_2, \ldots, b_n)\right)$ is minimal over $(a_1, b_2, \ldots, b_n)/(b_2, \ldots, b_n)$. But this last ideal is a principal ideal in $R/(b_2, \ldots, b_n)$, so

$$\operatorname{ht}(P/(b_2,\ldots,b_n)) \leq 1$$

by Theorem 7.4. This implies $\operatorname{ht}(Q/(b_2,\ldots,b_n)) = 0$, so Q is a prime ideal that is minimal over $(b_2,\ldots,b_n)$. By induction, this implies $\operatorname{ht}(Q) \leq n-1$, which completes the proof. □

In the case that R is an affine domain, Theorem 7.5 translates into a statement on dimension, which is given in Theorem 8.25. Readers may be interested in taking a look at that theorem now, along with its geometric interpretation that is discussed after Theorem 8.25.

Theorem 7.5 has some important consequences. Here is a first corollary.

Corollary 7.6 (Finiteness of height). *Let R be a Noetherian ring and $P \in \operatorname{Spec}(R)$. If P is generated by n elements, then*

$$\operatorname{ht}(P) \leq n.$$

In particular, every Noetherian local ring has finite Krull dimension, bounded above by the number of generators of the maximal ideal.

Exercise 7.7 shows that it is not always true that a Noetherian ring has finite Krull dimension.

Just like Nakayama's lemma, the following result is rather technical and inconspicuous, but often used.

Lemma 7.7 (Prime avoidance). *Let R be a ring and $I, P_1, \ldots, P_n \subseteq R$ ideals, with n a positive integer. Assume that P_i is a prime ideal for $i > 2$. Then*

$$I \subseteq \bigcup_{i=1}^{n} P_i$$

implies that there exists an i with $I \subseteq P_i$.

Proof. The proof is by induction on n. There is nothing to show for $n = 1$, so assume $n > 1$. By way of contradiction, assume that for each $i \in \{1,\ldots,n\}$ there exists

$$x_i \in I \setminus \bigcup_{j \neq i} P_j.$$

So by assumption, $x_i \in P_i$. It follows that $x_1 + x_2$ lies in I but neither in P_1 nor in P_2, so $n > 2$. But then $x_1 \cdot x_2 \cdots x_{n-1} + x_n$ lies in I but in none of the P_i, a contradiction. Therefore there exists i with

$$I \subseteq \bigcup_{j \neq i} P_j,$$

and the result follows by induction. □

The prime avoidance lemma is about interchanging quantifiers: If for each $x \in I$ there exists $i \in \{1,\ldots,n\}$ such that $x \in P_i$, then there exists i such that for all $x \in I$ one has $x \in P_i$. The name "prime avoidance" comes from reading the lemma backwards: If I is contained in none of the P_i, then there exists an element x of I that lies in none of the P_i, i.e., x *avoids* all P_i.

Using prime avoidance, we will now see that Theorem 7.5 has its own converse as a consequence.

Theorem 7.8 (A converse of the principal ideal theorem). *Let R be a Noetherian ring and $P \in \operatorname{Spec}(R)$ a prime ideal of height n. Then there exist $a_1, \ldots, a_n \in R$ such that P is minimal over $(a_1, \ldots, a_n)$.*

Proof. We will show that there exist $a_1, \ldots, a_n \in P$ with

$$\operatorname{ht}((a_1, \ldots, a_k)) = k \quad \text{for all} \quad k \leq n.$$

Assume that $a_1, \ldots, a_{k-1}$ have been found. Let $\mathcal{M} \subseteq \operatorname{Spec}(R)$ be the set of all prime ideals that are minimal over $(a_1, \ldots, a_{k-1})$. By Definition 6.10(b), every $Q \in \mathcal{M}$ has height at least $k-1$, so by Theorem 7.5 we must have $\operatorname{ht}(Q) = k-1$ (and this is also true in the case $k = 1$). This implies $P \not\subseteq Q$. Since $\mathcal{M}$ is finite by Corollary 3.14(d), Lemma 7.7 yields the existence of $a_k \in P$ with $a_k \notin Q$ for all $Q \in \mathcal{M}$. Every prime ideal Q' lying over $(a_1, \ldots, a_k)$ also lies over $(a_1, \ldots, a_{k-1})$, so Q' contains a $Q \in \mathcal{M}$. Since $a_k \notin Q$, the containment is proper, so $\operatorname{ht}(Q') > \operatorname{ht}(Q) = k - 1$. This implies $\operatorname{ht}((a_1, \ldots, a_k)) \geq k$, and Theorem 7.5 yields equality.

For $k = n$, we obtain that P lies minimally over $(a_1, \ldots, a_n)$, since otherwise $\operatorname{ht}(P) > n$. □

Corollary 7.9 (Systems of parameters). *Let R be a Noetherian local ring with maximal ideal $\mathfrak{m}$. Then $\dim(R)$ is the least number n such that there exist $a_1, \ldots, a_n \in \mathfrak{m}$ with*

$$\mathfrak{m} = \sqrt{(a_1, \ldots, a_n)}. \tag{7.2}$$

A sequence $a_1, \ldots, a_n \in \mathfrak{m}$ satisfying (7.2) with $n = \dim(R)$ will be called a **system of parameters** *of R.*

Proof. Using Corollary 1.12, we see that (7.2) is equivalent to the condition that $\mathfrak{m}$ is minimal over $(a_1, \ldots, a_n)$. The existence of $a_1, \ldots, a_n$ with $n = \operatorname{ht}(\mathfrak{m}) = \dim(R)$ is guaranteed by Theorem 7.8. By Theorem 7.5, n cannot be chosen smaller than $\dim(R)$. □

In Exercise 8.11, a connection between systems of parameters and Noether normalization is given. This gives rise to a (rather rough) geometric interpretation of a system of parameters as a "good local coordinate system."

7.2 The Dimension of Fibers

In this section we study images of morphisms and the dimensions of fibers, a project that will be completed in Chapter 10. The results of Chapter 10 will not be used anywhere else in this book, and of this section, only Proposition 7.11 and the term *going down* will be used outside of Chapter 10. Therefore it is possible to restrict the reading of this section to Proposition 7.11 and the discussion of going down on page 85, and to skip Chapter 10 altogether.

In algebraic geometry, a fiber is the preimage $f^{-1}(\{x\})$ of a point x under a morphism f. In this section we study the dimension of fibers. In the case that f is a surjective linear map between vector spaces V and W, we know from linear algebra that all fibers have the dimension $\dim\left(f^{-1}(x)\right) = \dim(V) - \dim(W)$, so it is reasonable to expect a similar formula in more general cases. It may be instructive to take a look at Example 7.14 now to get an impression of what can happen in the affine variety case. As we will see, it requires a good deal of work to transport the above formula to situations beyond linear maps. We start by considering the algebraic counterpart of fibers. The following lemma, which deals with the local case, is a consequence of Corollary 7.9. After giving the proof, we will discuss why we refer to the lemma under the keyword fiber dimension, and in particular we will see that the quotient ring S/I in the lemma belongs to the fiber (see the discussion after Proposition 7.11).

Lemma 7.10 (Fiber dimension, lower bound, local case)**.** *Let R and S be Noetherian local rings with maximal ideals $\mathfrak{m}$ and $\mathfrak{n}$, respectively. Let $\varphi\colon R \to S$ be a homomorphism with $\varphi(\mathfrak{m}) \subseteq \mathfrak{n}$, and let $I := (\varphi(\mathfrak{m}))_S$ be the ideal in S generated by the image of $\mathfrak{m}$. Then*

$$\dim(S/I) \geq \dim(S) - \dim(R).$$

Proof. Let $a_1, \ldots, a_m \in \mathfrak{m}$ be a system of parameters of R, so $m = \dim(R)$ by Corollary 7.9. By Lemma 2.6, there exists a nonnegative integer k with $\mathfrak{m}^k \subseteq (a_1, \ldots, a_m)_R$. It is easy to check that this implies

$$I^k \subseteq (\varphi(a_1), \ldots, \varphi(a_m))_S\,. \tag{7.3}$$

Since S/I is local and Noetherian by Lemma 1.22 and Proposition 2.4, there exists a system of parameters $b_1 + I, \ldots, b_n + I \in \mathfrak{n}/I$ (with $b_i \in \mathfrak{n}$) of S/I, so $n = \dim(S/I)$ by Corollary 7.9. We claim that

$$\mathfrak{n} = \sqrt{(\varphi(a_1), \ldots, \varphi(a_m), b_1, \ldots, b_n)_S}. \tag{7.4}$$

It is clear that the right-hand side of (7.4) is contained in the left-hand side. Conversely, let $x \in \mathfrak{n}$. There exists a positive integer l with $(x + I)^l \in (b_1 + I, \ldots, b_n + I)_{S/I}$, so $x^l \in (b_1, \ldots, b_n)_S + I$. With (7.3), this yields

$$x^{kl} \in (b_1, \ldots, b_n)_S + I^k \subseteq (\varphi(a_1), \ldots, \varphi(a_m), b_1, \ldots, b_n)_S .$$

So (7.4) is proved. By Corollary 7.9, this implies $\dim(S) \leq m + n$. □

Given a homomorphism $\varphi\colon R \to S$ of rings, we have an induced map

$$f\colon \operatorname{Spec}(S) \to \operatorname{Spec}(R),\ Q \mapsto \varphi^{-1}(Q) \tag{7.5}$$

(see on page 37; if R and S are coordinate rings of affine varieties, f corresponds to a morphism of varieties). For $P \in \operatorname{Spec}(R)$, the set $f^{-1}(\{P\})$ is called the **fiber** of f over P. We now wish to give an algebraic counterpart of the fiber. The answer is as follows. Consider the ideal $I = (\varphi(P))_S$ in S generated by the image of P, and the multiplicative subset

$$U := \{\varphi(a) + I \mid a \in R \setminus P\} \subseteq S/I.$$

Then form the ring

$$S_{[P]} := U^{-1}(S/I).$$

With the canonical homomorphisms $\pi\colon S \to S/I$ and $\varepsilon\colon S/I \to S_{[P]}$, we get the following result.

Proposition 7.11. *In the above situation, the map*

$$\Phi\colon \operatorname{Spec}\left(S_{[P]}\right) \to f^{-1}(\{P\}),\ \mathfrak{Q} \mapsto \pi^{-1}\left(\varepsilon^{-1}(\mathfrak{Q})\right),$$

is an inclusion-preserving bijection.

Proof. By Lemma 1.22 and Theorem 6.5, Φ is an inclusion-preserving injection $\operatorname{Spec}\left(S_{[P]}\right) \to \operatorname{Spec}(S)$, and its image is

$$\operatorname{im}(\Phi) = \{Q \in \operatorname{Spec}(S) \mid I \subseteq Q \text{ and } U \cap (Q/I) = \emptyset\}.$$

It is easy to verify that the above conditions on Q are equivalent to $P = \varphi^{-1}(Q)$, i.e., $Q \in f^{-1}(\{P\})$. □

In fact, the bijection Φ from Proposition 7.11 can be shown to be a homeomorphism (this follows from Exercises 3.4 and 6.5). So $S_{[P]}$ is the desired algebraic counterpart of the fiber, and the **fiber dimension** is equal to the Krull dimension of $S_{[P]}$. Motivated by Proposition 7.11, we call $S_{[P]}$ the **fiber ring** of φ over P. Notice that $S_{[P]} \cong S/I$ if P is a maximal ideal, since in that case all $\varphi(a) + I$ with $a \in R \setminus P$ are already invertible in S/I. The reader should be warned that the symbol $S_{[P]}$ for the fiber ring is not standard notation. In Exercise 7.9 we will study a more abstract way of defining the fiber ring, and introduce an alternative notation that is more standard.

Theorem 7.12 (Fiber dimension, lower bound). *Let $\varphi\colon R \to S$ be a homomorphism of Noetherian rings. Moreover, let $Q \in \operatorname{Spec}(S)$ and $P := \varphi^{-1}(Q)$. Then*

$$\dim\left(S_{[P]}\right) \geq \operatorname{ht}(Q) - \operatorname{ht}(P). \tag{7.6}$$

In fact, if $\mathfrak{Q} \in \operatorname{Spec}\left(S_{[P]}\right)$ *is the image of* Q *in* $S_{[P]}$ *(i.e.,* $\mathfrak{Q} = U^{-1}(Q/I)$ *with the above notation), then*

$$\operatorname{ht}(\mathfrak{Q}) \geq \operatorname{ht}(Q) - \operatorname{ht}(P). \tag{7.7}$$

Proof. The second inequality (7.7) implies the first, so we only need to prove (7.7). We do this by reduction to the local case. We have a (well-defined) homomorphism

$$\psi\colon R_P \to S_Q,\ \frac{a}{b} \mapsto \frac{\varphi(a)}{\varphi(b)},$$

mapping P_P into Q_Q. Setting $J := (\psi(P_P))_{S_Q}$ and applying Lemma 7.10, we obtain

$$\dim\left(S_Q/J\right) \geq \dim(S_Q) - \dim(R_P) = \operatorname{ht}(Q) - \operatorname{ht}(P). \tag{7.8}$$

We claim that $\dim\left(S_Q/J\right) = \operatorname{ht}(\mathfrak{Q})$. To prove this, we study the spectrum of S_Q/J. With the canonical homomorphisms $\varepsilon\colon S \to S_Q$ and $\pi\colon S_Q \to S_Q/J$, we get a map

$$\operatorname{Spec}\left(S_Q/J\right) \to \operatorname{Spec}(S),\ \mathfrak{q} \mapsto \varepsilon^{-1}\left(\pi^{-1}(\mathfrak{q})\right).$$

By Lemma 1.22 and Theorem 6.5, this map is inclusion-preserving and injective, and its image is

$$\begin{aligned}\mathcal{M} &:= \{Q' \in \operatorname{Spec}(S) \mid \varphi(P) \subseteq Q' \text{ and } Q' \subseteq Q\} \\ &= \left\{Q' \in \operatorname{Spec}(S) \mid \varphi^{-1}(Q') = P \text{ and } Q' \subseteq Q\right\}.\end{aligned}$$

So $\dim\left(S_Q/J\right)$ is the maximal length of a chain in $\mathcal{M}$. On the other hand, by Proposition 7.11, $\mathcal{M}$ is in an inclusion-preserving, bijective correspondence with

$$\left\{\mathfrak{Q}' \in \operatorname{Spec}\left(S_{[P]}\right) \mid \mathfrak{Q}' \subseteq \mathfrak{Q}\right\}.$$

So $\operatorname{ht}(\mathfrak{Q})$ is also the maximal length of a chain in $\mathcal{M}$, and we conclude that $\operatorname{ht}(\mathfrak{Q}) = \dim\left(S_Q/J\right)$. □

The most important special case of Theorem 7.12 is the "truly geometric" case in which φ comes from a morphism of affine varieties. Readers may already take a look at the inequality (10.8) in Corollary 10.6 on page 142, which gives a translation of Theorem 7.12 into geometric terms.

We will continue to investigate whether (or when) the inequalities in Theorem 7.12 are actually equalities (see Theorem 10.5 for the final result). Before we start doing that, we draw an important conclusion, which requires only the lower bound from Theorem 7.12.

Corollary 7.13 (Dimension of polynomial rings). *Let $R \neq \{0\}$ be a Noetherian ring and $R[x]$ the polynomial ring in one indeterminate. Then*

$$\dim(R[x]) = \dim(R) + 1.$$

Proof. Let $\varphi\colon R \to R[x] =: S$ be the natural embedding. For an ideal $I \subseteq R$ we have $S/(\varphi(I))_S \cong (R/I)[x]$ and $\varphi^{-1}((\varphi(I))_S) = I$. Let $P_0 \subsetneqq P_1 \subsetneqq \cdots \subsetneqq P_n$ be a chain of prime ideals in $\operatorname{Spec}(R)$. By the above, $Q_i := (\varphi(P_i))_S$ yields a strictly ascending chain of prime ideals in S. Since $S/Q_n \cong (R/P_n)[x]$ is not a field, Q_n is not maximal. Therefore $\dim(S) \geq n+1$, and we conclude that

$$\dim(S) \geq \dim(R) + 1.$$

For showing the reverse inequality, let $Q \in \operatorname{Spec}(S)$ be a prime ideal and set $P := \varphi^{-1}(Q) \in \operatorname{Spec}(R)$. We are done if we can show that $\operatorname{ht}(Q) \leq \operatorname{ht}(P)+1$. By Theorem 2.11, S is Noetherian, so we may apply Theorem 7.12. The inequality (7.6) yields $\operatorname{ht}(Q) \leq \operatorname{ht}(P) + \dim(S_{[P]})$, so it remains to show that $\dim(S_{[P]}) \leq 1$. We have $S/(\varphi(P))_S \cong (R/P)[x]$, and under this isomorphism the set $U = \{\varphi(a) + (\varphi(P))_S \mid a \in R \setminus P\}$ maps to $(R/P) \setminus \{0\}$, so the fiber ring is

$$S_{[P]} \cong ((R/P) \setminus \{0\})^{-1} (R/P)[x] = \operatorname{Quot}(R/P)[x].$$

So by Example 5.2(6) we get $\dim(S_{[P]}) = 1$. □

By repeated application of Corollary 7.13 we obtain a new proof of the first part of Corollary 5.7. In Exercise 7.10, the analogous result is shown for power series rings.

When are the inequalities of Theorem 7.12 actually equalities? We first look at two examples.

Example 7.14. We assume that K is an algebraically closed field.

(1) Let $X = \mathcal{V}_{K^2}(x_1 \cdot x_2)$ be the "coordinate cross," $Y = K^1$ and

$$f\colon X \to Y,\ (\xi_1, \xi_2) \mapsto \xi_1$$

the first projection. This is the morphism of varieties induced by

$$\varphi\colon K[x] \to K[x_1, x_2]/(x_1 \cdot x_2),\ x \mapsto x_1 + (x_1 \cdot x_2).$$

Every maximal ideal in $K[x]$ corresponds to a point $\xi \in K^1$, and has height 1 by the geometric interpretation of height (Remark 6.11(b)). Likewise, every maximal ideal in $K[x_1, x_2]/(x_1 \cdot x_2)$ has height 1. For $\xi \in K^1 \setminus \{0\}$, the fiber $f^{-1}(\xi)$ consists of one point, so both inequalities from Theorem 7.12 are equalities. But for $\xi = 0$, the fiber is the entire x_2-axis, so here the inequalities are strict.

(2) The variety X in the previous example is reducible. Now consider the irreducible variety

$$X = \mathcal{V}_{K^3}(x_1^2 + x_2^2 - x_3^2) = \mathcal{V}_{K^3}\left(x_1^2 + (x_2 + x_3)(x_2 - x_3)\right),$$

which in $\mathbb{R}^3$ visualizes as a circular cone, and the morphism

$$f\colon X \to Y := K^2,\ (\xi_1, \xi_2, \xi_3) \mapsto (\xi_1, \xi_2 + \xi_3).$$

We assume that K is not of characteristic 2. As above, we see that all maximal ideals in the coordinate rings $K[X]$ and $K[Y]$ have height 2. If $(\alpha, \beta) \in K^2$ with $\beta \neq 0$, then

$$f^{-1}(\alpha, \beta) = \left\{\left(\alpha, \frac{\beta^2 - \alpha^2}{2\beta}, \frac{\beta^2 + \alpha^2}{2\beta}\right)\right\},$$

so the fiber dimension is 0 and equality holds in Theorem 7.12. But if $\beta = 0$ and $\alpha \neq 0$, the fiber is empty (so Theorem 7.12 does not apply), and if $\alpha = \beta = 0$, then

$$f^{-1}(0,0) = \{(0, \eta, -\eta) \mid \eta \in K\},$$

which is one-dimensional. So here the inequalities are strict. ◁

In both of the above examples, the inequalities from Theorem 7.12 are equalities on an open, dense subset of Y. But we could easily destroy equality by substituting Y with a larger affine variety of higher dimension. This shows that for Y an irreducible affine variety, equality can hold only if the morphism is dominant. So Exercise 4.2 tells us that a reasonable hypothesis for which we can expect equality almost everywhere is that $\varphi\colon R \to S$ is injective. This explains why Theorem 10.5, our final result on fiber dimension, has this hypothesis.

To push the theory further, we need a few lemmas. Before stating the first, we introduce the following terminology. We will say that **going down** holds for a homomorphism $\varphi\colon R \to S$ of rings if for every $P \in \operatorname{Spec}(R)$ and every $Q' \in \operatorname{Spec}(S)$ with $\varphi(P) \subseteq Q'$, there exists $Q \in \operatorname{Spec}(S)$ with $Q \subseteq Q'$ and $\varphi^{-1}(Q) = P$. This is illustrated in Fig. 7.2.

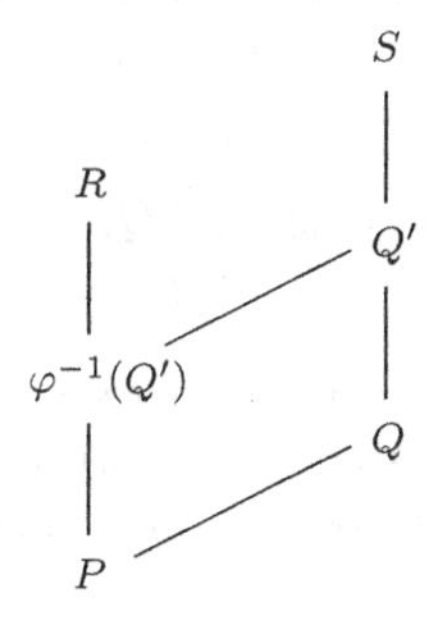

Fig. 7.2. Going down

The term "going down" refers to the descent from Q' to Q. Exercise 8.9 contains an example in which going down fails. It is easy to see that if $U \subseteq R$ is a multiplicative subset and going down holds for the homomorphism $U^{-1}R \to \varphi(U)^{-1}S$ induced by φ, then it also holds for the original homomorphism $\varphi: R \to S$ with the additional hypothesis that $\varphi(U) \cap Q' = \emptyset$.

Lemma 7.15 (Going down and fiber dimension). *In the situation of Theorem 7.12, let* $U := \varphi(R \setminus P)$. *If going down holds for the homomorphism* $R_P \to U^{-1}S$ *induced by* φ, *then equality holds in* (7.7).

Proof. By Proposition 7.11, $\operatorname{ht}(\mathfrak{Q})$ is the maximal length of a chain

$$Q_0 \subsetneqq Q_1 \subsetneqq \cdots \subsetneqq Q_m = Q \tag{7.9}$$

of prime ideals $Q_i \in \operatorname{Spec}(S)$ with $\varphi^{-1}(Q_i) = P$. We have a chain $P_0 \subsetneqq \cdots \subsetneqq P_n = P$ in $\operatorname{Spec}(R)$ of length $n = \operatorname{ht}(P)$. Since $U \cap Q_0 = \emptyset$ and $\varphi(P_{n-1}) \subseteq \varphi(P) \subseteq Q_0$, we can use the remark preceding Lemma 7.15 to work downwards along the chain of P_i and find $Q_{-1}, \ldots, Q_{-n} \in \operatorname{Spec}(S)$ extending the chain (7.9) downwards. This yields $\operatorname{ht}(Q) \geq n + m = \operatorname{ht}(P) + \operatorname{ht}(\mathfrak{Q})$, so (7.7) is an equality. □

A ring homomorphism $R \to S$ makes S into an R-module. Recall that modules over a ring do not always have a basis (= a linearly independent generating set). If a module does have a basis, it is called free.

Lemma 7.16 (Freeness implies going down). *Let* $\varphi: R \to S$ *be a ring homomorphism with* S *Noetherian.*

(a) If S *is free as an* R*-module, then going down holds for* φ.
(b) If additionally there exists a basis B *of* S *over* R *with* $1 \in B$, *then the induced map* $\varphi^*: \operatorname{Spec}(S) \to \operatorname{Spec}(R)$ *is surjective.*

Proof. For the proof of (a), let $P \in \operatorname{Spec}(R)$ and $Q' \in \operatorname{Spec}(S)$ with $\varphi(P) \subseteq Q'$. Set $I := (\varphi(P))_S$, and let $Q \in \operatorname{Spec}(S)$ be minimal among the prime ideals that lie between I and Q'. In particular, Q is a minimal prime ideal over I. Let $Q_1, \ldots, Q_n$ be the other minimal prime ideals over I. (There are finitely many of them by Corollary 3.14(d).) We have $\bigcap_{i=1}^n Q_i \nsubseteq Q$, so we may choose $y \in \bigcap_{i=1}^n Q_i \setminus Q$. We claim that $\varphi^{-1}(Q) \subseteq P$, which together with $I \subseteq Q$ yields $\varphi^{-1}(Q) = P$, proving (a). So take $a \in \varphi^{-1}(Q)$. Then

$$\varphi(a)y \in Q \cap \bigcap_{i=1}^n Q_i = \sqrt{I},$$

where Corollary 3.14(d) was used. So there exists a positive integer k with $\varphi(a)^k y^k \in I$. We have $y^k \notin I$, since the contrary would imply $y \in Q$. So the smallest j with $\varphi(a)^j y^k \in I$ is positive. Set $z := \varphi(a)^{j-1} y^k$. Then $z \notin I$ but $\varphi(a) z \in I$, so

$$\varphi(a)z = \sum_{i=1}^{m} x_i \varphi(a_i) \tag{7.10}$$

with $x_i \in S$ and $a_i \in P$. If B is a basis of S as an R-module, we can write $z = \sum_{b \in B} \varphi(z_b) \cdot b$ and $x_i = \sum_{b \in B} \varphi(x_{i,b}) \cdot b$ with $z_b, x_{i,b} \in R$ (where only finitely many coefficients in the sums are nonzero). From (7.10) and the linear independence of B we get

$$a \cdot z_b = \sum_{i=1}^{m} x_{i,b} a_i \in P$$

for all $b \in B$. But there exists $b \in B$ with $z_b \notin P$ (otherwise, $z \in I$), so $a \in P$. This completes the proof of (a).

The hypothesis of (b) implies that there exists a homomorphism $\psi: S \to R$ of R-modules with $\psi \circ \varphi = \mathrm{id}_R$. Let $P \in \mathrm{Spec}(R)$. In view of (a), we have only to show that there exists $Q' \in \mathrm{Spec}(S)$ with $\varphi(P) \subseteq Q'$. Assume the contrary. Then $(\varphi(P))_S = S$, so we have $1 = \sum_{i=1}^{m} s_i \varphi(a_i)$ with $s_i \in S$ and $a_i \in P$. Applying ψ yields $1 = \sum_{i=1}^{m} \psi(s_i) a_i \in P$, a contradiction. This proves (b). □

In Chapter 8, we will obtain another set of conditions under which going down holds (see Theorem 8.17).

This is how far we can push the theory with the present methods. We will continue the investigation of fiber dimensions in Chapter 10. There we will prove the generic freeness lemma (Corollary 10.2), which under rather weak assumptions says that the hypotheses of Lemma 7.16(a) and (b) are satisfied after localization at all prime ideals lying in an open, dense subset of $\mathrm{Spec}(R)$. Putting things together, this yields exact formulas for the fiber dimension, which hold almost everywhere (see Theorem 10.5).

Exercises for Chapter 7

7.1 (The Cayley–Hamilton theorem). Deduce the Cayley–Hamilton theorem ("substituting a square matrix $A \in R^{n \times n}$ over a ring into its own characteristic polynomial f yields $f(A) = 0$") from Lemma 7.2.

7.2 (Hypotheses of Nakayama's lemma). Give an example that shows that the hypothesis on finite generation of M cannot be dropped from Nakayama's lemma (Theorem 7.3).

7.3 (Nakayama's lemma and systems of generators). Let R be a ring with Jacobson radical $J \subseteq R$, and let M be a finitely generated R-module. Write $\pi: M \to M/JM$ for the canonical map. Observe that $M/JM = \pi(M)$

is an R/J-module, and for an R-submodule N of M, $\pi(N) \subseteq \pi(M)$ is an R/J-submodule.

(a) Let $N \subseteq M$ be a submodule. Prove the equivalence

$$N = M \iff \pi(N) = \pi(M).$$

(b) Let $x_1, \dots, x_n \in M$. Prove the equivalence

$$M = (x_1, \dots, x_n)_R \iff \pi(M) = (\pi(x_1), \dots, \pi(x_n))_{R/J}.$$

(c) Assume that R is local with maximal ideal $\mathfrak{m}$, and write $K := R/\mathfrak{m}$. Show that all minimal systems of generators of M have the same number n of elements, namely $n = \dim_K(M/\mathfrak{m}M)$.
(d) Give an example of a ring R and a finitely generated module M such that not all minimal systems of generators have the same size.

***7.4 (Hypotheses of the principal ideal theorem).** In this exercise we learn that Krull's principal ideal theorem (Theorem 7.4) may fail for non-Noetherian rings. The example is adapted from Gilmer [20, page 321, Exercise 21]. Let $K[x, y]$ be a polynomial ring over a field, and consider the subalgebra $R := K[x, xy, xy^2, xy^3, \dots] \subset K[x, y]$, which we have already seen as an example of a non-Noetherian domain (Example 2.3).

(a) Show that there exists precisely one prime ideal $P \in \operatorname{Spec}(R)$ lying over the principal ideal $(x)_R$.
(b) Show that $\operatorname{ht}(P) = 2$.
(c) By generalizing this, construct a ring R_n for each $n \in \mathbb{N} \cup \{\infty\}$ that has a proper principal ideal of height n.

(Solution on page 220)

***7.5 (Can the spectrum be just one chain?).** This exercise originated from the question whether there exists a Noetherian ring with just three prime ideals $P_0 \subsetneqq P_1 \subsetneqq P_2$. When I posed this question to Viet-Trung Ngo, he immediately answered it in the negative. His answer led to the following statement, which should be proved in this exercise: If $P \subseteq Q$ are two prime ideals in a Noetherian ring R that have at most finitely many prime ideals in between, then in fact there exists no prime ideal that properly lies between P and Q.

7.6 (Semilocal rings). A ring R is called **semilocal** if it has (at most) finitely many maximal ideals. For example, semilocal rings occur as coordinate rings of affine varieties consisting of finitely many points. Here is how semilocal rings can be constructed by localization. Let $P_1, \dots, P_n \in \operatorname{Spec}(R)$ be finitely many prime ideals in a ring such that $P_i \not\subseteq P_j$ for $i \neq j$. Show that

$$U := R \setminus \bigcup_{i=1}^{n} P_i \subseteq R$$

is a multiplicative subset. Furthermore, show that the localization $U^{-1}R$ is semilocal with maximal ideals $U^{-1}P_i$.

***7.7 (An infinite-dimensional Noetherian ring).** In this exercise we study an example of a Noetherian ring that has infinite Krull dimension. The example is due to Nagata [41, Appendix, Example E1], and the hints given for the proof are adapted from Eisenbud [17, Exercise 9.6]. Let K be a field and $R = K[x_1, x_2, \dots]$ a polynomial ring in countably many indeterminates x_i, $i \in \mathbb{N}$. Consider the prime ideals

$$P_i := \left(x_{i^2+1}, x_{i^2+2}, \dots, x_{(i+1)^2}\right) \subset R \quad (i \in \mathbb{N}_0)$$

and set $U := R \setminus \bigcup_{i \in \mathbb{N}_0} P_i$. Show that $S := U^{-1}R$ is Noetherian but has infinite Krull dimension.
Hint: The hard part is to show that S is Noetherian. For this, consider a nonzero ideal $I \subseteq R$, take $f \in I \setminus \{0\}$, and choose $n \in \mathbb{N}_0$ such that all indeterminates x_j occurring in f satisfy $j \le (n+1)^2$. Show that there exist $f_1, \dots, f_m \in I$ such that $(I)_{R_{P_i}} = (f_1, \dots, f_m)_{R_{P_i}}$ for $i \le n$. Now take $g \in I$ and set

$$J := \{h \in R \mid h \cdot g \in (f_1, \dots, f_m, f)_R\}.$$

Use Lemma 7.7 to show that there exists $h \in J \setminus \bigcup_{i=0}^{n} P_i$. Assume that $J \subseteq \bigcup_{i \in \mathbb{N}_0} P_i$ and derive a contradiction. From this, conclude that $g \in (f_1, \dots, f_m, f)_S$, and that S is Noetherian. *(Solution on page 221)*

7.8 (Systems of parameters). Parts (a)–(c) of this exercise give examples of affine varieties $X \subseteq K^n$ over an algebraically closed field. Consider the localization R_x of the coordinate ring $R = K[X]$ at the point $x = (0, \dots, 0) \in X$ and find a system of parameters of R_x. Does there exist a system of parameters that generates the maximal ideal?

(a) $X = \{(\xi_1, \xi_2) \in K^2 \mid \xi_1\xi_2 = 0\}$ (see Example 7.14(1)).
(b) $X = \{(\xi_1, \xi_2, \xi_3) \in KL^3 \mid \xi_1^2 + \xi_2^2 - \xi_3^2 = 0\}$ (see Example 7.14(2)).
(c) $X = \{(\xi_1, \xi_2) \in K^2 \mid \xi_2^2 - \xi_1(\xi_1^2 + 1) = 0\}$ (an elliptic curve, shown as X_3 in Fig. 12.1 on page 179).

Hint: You may use Exercise 7.3(c) for answering the second question.
Remark: A local ring whose maximal ideal is generated by a system of parameters is called *regular* (see Definition 13.2).

7.9 (The fiber ring as a tensor product). This exercise gives a more abstract description of the fiber ring. Let $\varphi\colon R \to S$ be a homomorphism of rings and $P \in \operatorname{Spec}(R)$. Let $K := \operatorname{Quot}(R/P)$ and $\psi\colon R \to K$, $a \mapsto \frac{a+P}{1+P}$, the canonical map.

(a) Show that the fiber ring $S_{[P]}$ is the pushout of φ and ψ. Here is the definition of the pushout.
For two R-algebras A and B (given by homomorphisms α and β), the **pushout** of α and β is defined to be a ring C together with homomorphisms $\gamma\colon A \to C$ and $\delta\colon B \to C$ making the square in the below diagram commutative (i.e., $\gamma \circ \alpha = \delta \circ \beta$) such that the following universal property holds: For a ring T with homomorphisms $\Gamma\colon A \to T$ and $\Delta\colon B \to T$ with $\Gamma \circ \alpha = \Delta \circ \beta$, there exists a unique homomorphism $\Theta\colon C \to T$ such that the diagram

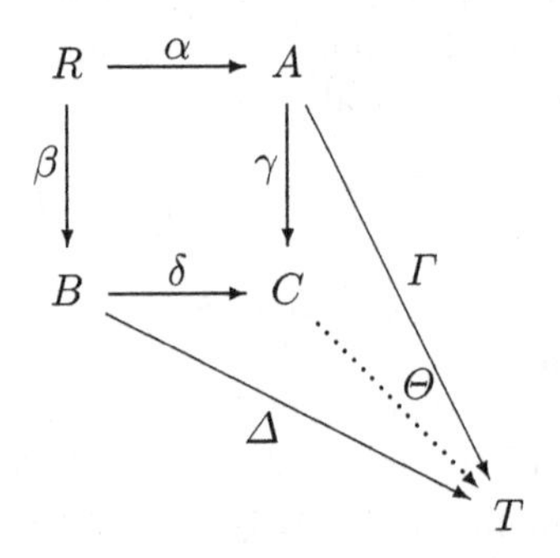

commutes (i.e., $\Theta \circ \gamma = \Gamma$ and $\Theta \circ \delta = \Delta$). As usual with universal properties, this implies that the pushout (if it exists) is unique up to isomorphism. More precisely, between two pushouts there exists a unique map that is simultaneously an isomorphism of A-algebras and of B-algebras.

(b) Conclude that $\operatorname{Spec}\left(S_{[P]}\right)$ is the pullback of the maps $f\colon \operatorname{Spec}(S) \to \operatorname{Spec}(R)$ and $g\colon \operatorname{Spec}(K) \to \operatorname{Spec}(R)$ induced by φ and ψ. (The pullback is defined as the pushout but with all arrows reversed, and the maps considered are morphisms of spectra of rings.)

(c) Describe the map $g\colon \operatorname{Spec}(K) \to \operatorname{Spec}(R)$ explicitly, and show that the pullback of f and g is the fiber $f^{-1}(\{P\})$. So $\operatorname{Spec}\left(S_{[P]}\right) \cong f^{-1}(\{P\})$ by the uniqueness of pullbacks, which re-proves Proposition 7.11.

Remark: The pushout of two homomorphisms $\alpha\colon R \to A$ and $\beta\colon R \to B$ is isomorphic to the tensor product $A \otimes_R B$, which is equipped with a natural structure as a ring (see Lang [33, Chapter XVI, Proposition 6.1]). So by (a) we have $S_{[P]} \cong K \otimes_R S$. In fact, a notation for the fiber ring over P more commonly found in the literature is $\kappa(P) \otimes_R S$, where $\kappa(P) := \operatorname{Quot}(R/P) = K$ stands for the residue class field at P.

7.10 (Dimension of the formal power series ring). Let $R \neq \{0\}$ be a Noetherian ring and $R[[x]]$ the formal power series ring over R. Show that

$$\dim\left(R[[x]]\right) = \dim(R) + 1.$$

The proof may be broken up into the following steps.

(a) If $I \subseteq R$ is an ideal, then the kernel of the epimorphism $S := R[[x]] \to (R/I)\,[[x]]$, obtained from applying the canonical map $R \to R/I$ coefficientwise, is $(I)_S$. Conclude that $\dim(S) \geq \dim(R) = 1$.
(b) Show that $1 - xf$ is invertible in S for every $f \in S$. Let $\mathfrak{m} \in \operatorname{Spec}_{\max}(S)$ be a maximal ideal. Conclude that $x \in \mathfrak{m}$. Show that $\mathfrak{n} := R \cap \mathfrak{m}$ is a maximal ideal in R.
(c) Show that

$$\operatorname{ht}(\mathfrak{m}) \leq \operatorname{ht}(\mathfrak{n}) + 1.$$

This finishes the proof.

Remark: By repeatedly using this result, we obtain

$$\dim\left(K[[x_1, \ldots, x_n]]\right) = n$$

for the formal power series ring in n indeterminates over a field.

7.11 (Free modules and the locus of freeness). Parts (a)–(c) of this exercise give examples of ring extensions $R \subseteq S$. Decide whether S is free as an R-module. Determine the "locus of freeness," i.e., the set $X_{\text{free}} \subseteq \operatorname{Spec}(R)$ of all $P \in \operatorname{Spec}(R)$ such that $(R \setminus P)^{-1}S$ is free as an R_P-module. Draw conclusions on the fibers of the induced morphism $\operatorname{Spec}(S) \to \operatorname{Spec}(R)$. Here $K[x_1, x_2, \ldots]$ is always a polynomial ring over a field.

(a) $R = \mathbb{Z}$ and $S = \mathbb{Z}[1/2]$.
(b) $S = K[x_1, x_2]/(x_2^2 - x_1^2(x_1 + 1))$ and $R = K[\overline{x}_1]$, where $\overline{x}_i \in S$ denotes the residue class of x_i. See Example 8.9(4) for the variety belonging to S.
(c) $S = K[x_1, x_2, x_3]/(x_1^2 + x_2^2 - x_3^2)$ and $R - K[\overline{x}_1, \overline{x}_2 - \overline{x}_3]$ (see Example 7.14(2)). We assume $\operatorname{char}(K) \neq 2$.

Chapter 8
Integral Extensions

The concept of an integral ring extension is a generalization of the concept of an algebraic field extension. In the first section of this chapter, we develop the algebraic theory of integral extensions, and introduce the concept of a normal ring. Section 8.2 studies the morphism $\operatorname{Spec}(S) \to \operatorname{Spec}(R)$ induced from an integral extension $R \subseteq S$. In Section 8.3, we turn our attention to affine algebras again. We prove the Noether normalization theorem, and use it to prove, among other results, that all maximal ideals of an affine domain have equal height.

8.1 Integral Closure

In the previous section we have considered ring homomorphisms $\varphi\colon R \to S$. We will now assume that φ is injective, so we view R as a subring of S or (equivalently) S as a ring extension of R.

Definition 8.1. *Let S be a ring and $R \subseteq S$ a subring.*

(a) Let $s \in S$. A monic polynomial

$$g = x^n + a_1 x^{n-1} + \cdots + a_{n-1}x + a_n \in R[x]$$

with $g(s) = 0$ is called an **integral equation** *for s over R.*

(b) An element $s \in S$ is called **integral** *over R if there exists an integral equation for s over R. (The difference between this definition and that of "algebraic" is that here we insist that the polynomial equation for s be monic.)*

(c) S is called **integral** *over R if all elements from S are integral over R. In this case we call S an* **integral extension** *of R.*

Example 8.2. (1) $\sqrt{2} \in \mathbb{R}$ is integral over $\mathbb{Z}$. The ring $\mathbb{Z}[\sqrt{2}]$ is an integral extension of $\mathbb{Z}$.

G. Kemper, *A Course in Commutative Algebra*, Graduate Texts in Mathematics 256, DOI 10.1007/978-3-642-03545-6_9,

(2) $1/\sqrt{2} \in \mathbb{R}$ is not integral over $\mathbb{Z}$ (although it is algebraic). To see this, assume

$$\frac{1}{\sqrt{2}^n} + a_1 \frac{1}{\sqrt{2}^{n-1}} + \cdots + a_{n-1} \frac{1}{\sqrt{2}} + a_n = 0$$

with $a_i \in \mathbb{Z}$. Observe that 1 and $\sqrt{2}$ are linearly independent over $\mathbb{Q}$. Multiplying the above equation by $\sqrt{2}^n$ and picking out the summands that lie in $\mathbb{Q}$ yields

$$1 + 2a_2 + 4a_4 + \cdots = 0,$$

a contradiction.

(3) $s = \frac{1+\sqrt{5}}{2} \in \mathbb{R}$ is integral over $\mathbb{Z}$, since $s^2 - s - 1 = 0$. Therefore s is also integral over $R := \mathbb{Z}\left[\sqrt{5}\right] \subset \mathbb{R}$ (the subalgebra generated by $\sqrt{5}$). What is remarkable about this is that there exists an algebraic equation for s over R of degree 1 (so $s \in \operatorname{Quot}(R)$), but the smallest *integral* equation has degree 2. ◁

We wish to prove that products and sums of integral elements are again integral. The proof is quite similar to the standard proof of the analogous result in field theory, and requires the following lemma.

Lemma 8.3 (Integral elements and finite modules). *Let S be a ring, $R \subseteq S$ a subring, and $s \in S$. Then the following statements are equivalent:*

(a) The element s is integral over R.
(b) The subalgebra $R[s] \subseteq S$ generated by s is finitely generated as an R-module.
(c) There exists an $R[s]$-module M with $\operatorname{Ann}(M) = \{0\}$ such that M is finitely generated as an R-module.

Proof. Assume that s is integral over R, so we have an integral equation $x^n + a_1 x^{n-1} + \cdots + a_{n-1}x + a_n \in R[x]$ for s. We claim that $R[s]$ is generated by the s^i, $i \in \{0, \ldots, n-1\}$, i.e.,

$$R[s] = \left(1, s, \ldots, s^{n-1}\right)_R = \sum_{i=0}^{n-1} R s^i =: N.$$

Indeed, for $k \geq n$, we have $s^k = -\left(a_1 s^{k-1} + \cdots + a_n s^{k-n}\right)$, so it follows by induction that all s^k lie in N. So (a) implies (b). Moreover, it is clear that (b) implies (c): Take $M = R[s]$, then $1 \in M$, so $\operatorname{Ann}(M) = \{0\}$.

Now assume (c). We have $M = (m_1, \ldots, m_r)_R$, so for each $i \in \{1, \ldots, r\}$ there exist $a_{i,j} \in R$ with $s \cdot m_i = \sum_{j=1}^r a_{i,j} m_j$. By Lemma 7.2 this implies

$$\det\left(\delta_{i,j} s - a_{i,j}\right)_{1 \leq i,j \leq r} \in \operatorname{Ann}(M),$$

so by hypothesis the determinant is zero. Therefore $\det\left(\delta_{i,j} x - a_{i,j}\right)_{1 \leq i,j \leq r} \in R[x]$ is an integral equation for s. □

The following theorem is in perfect analogy to the result that a finitely generated field extension is finite if and only if it is algebraic. It also implies that sums and products of integral elements are again integral.

Theorem 8.4 (Generated by integral elements implies integral). *Let S be a ring and $R \subseteq S$ a subring such that $S = R[a_1, \dots, a_n]$ is finitely generated as an R-algebra. Then the following statements are equivalent:*

(a) All a_i are integral over R.
(b) S is integral over R.
(c) S is finitely generated as an R-module.

Proof. Clearly (b) implies (a). We use induction on n to show that (a) implies (c). We may assume $n > 0$. By induction, $S' := R[a_1, \dots, a_{n-1}]$ is finitely generated as an R-module, so $S' = (m_1, \dots, m_r)_R = \sum_{i=1}^r Rm_i$ with $m_i \in S'$. We also have that a_n is integral over S', so Lemma 8.3 yields $S'[a_n] = \sum_{j=1}^l S'n_j$ with $n_j \in S$. Putting things together, we obtain

$$S = S'[a_n] = \sum_{j=1}^{l} \sum_{i=1}^{r} Rm_i n_j,$$

so (c) holds.

Finally, (c) implies (b) by Lemma 8.3 (take $M = S$ in Lemma 8.3(c)). □

Corollary 8.5 (Integral elements form a subalgebra). *Let S be a ring and $R \subseteq S$ a subring. Then the set*

$$S' := \{s \in S \mid s \text{ is integral over } R\} \subseteq S$$

is an R-subalgebra.

Proof. Clearly all elements from R lie in S'. So all we need to show is that if $a, b \in S'$, then also $a + b \in S'$ and $a \cdot b \in S'$. But this follows since $R[a, b]$ is integral over R by Theorem 8.4. □

We obtain a further consequence of Lemma 8.3 and Theorem 8.4.

Corollary 8.6 (Towers of integral extensions). *Let T be a ring and $R \subseteq S \subseteq T$ subrings. If T is integral over S and S is integral over R, then T is integral over R.*

Proof. For every $t \in T$ we have an integral equation

$$t^n + s_1 t^{n-1} + \cdots + s_{n-1} t + s_n = 0$$

with $s_i \in S$. So t is integral over $S' := R[s_1, \dots, s_n] \subseteq S$. By Lemma 8.3, $S'[t]$ is finitely generated as an S'-module, and by Theorem 8.4, S' is finitely generated as an R-module. It follows that $S'[t]$ is finitely generated as an R-module, so applying Lemma 8.3 again shows that t is integral over R. □

Corollary 8.5 prompts the following definition.

Definition 8.7.

(a) *Let S be a ring and $R \subseteq S$ a subring. Then the set S' of all elements from S that are integral over R is called the* **integral closure** *of R in S. If $S' = R$, we say that R is* **integrally closed** *in S.*
(b) *An integral domain R is called* **normal** *if it is integrally closed in its field of fractions* $\operatorname{Quot}(R)$. *One can extend this definition to rings that need not be integral domains by calling a ring normal if it is integrally closed in its total ring of fractions. In this book, normality is understood in the above narrower sense.*
(c) *If R is an integral domain, the* **normalization** *of R, often written as $\widetilde{R}$, is the integral closure of R in its field of fractions* $\operatorname{Quot}(R)$. *Observe that $\widetilde{R}$ is normal by Corollary 8.6.*
(d) *An irreducible affine variety X over a field K is called* **normal** *if the coordinate ring $K[X]$ is normal.*

Before giving some examples, we prove an elementary result.

Proposition 8.8. *Every factorial ring is normal.*

Proof. Let R be a factorial ring, and let $a/b \in \operatorname{Quot}(R)$ be integral over R with $a, b \in R$ coprime. So we have

$$\frac{a^n}{b^n} + a_1 \frac{a^{n-1}}{b^{n-1}} + \cdots + a_{n-1}\frac{a}{b} + a_n = 0$$

with $a_i \in R$. Multiplying this by b^n shows that b divides a^n, so every prime factor of b divides a. By the coprimality, b has no prime factors, so it is invertible in R. Therefore $a/b \in R$. □

Example 8.9. (1) By Proposition 8.8, $\mathbb{Z}$ is normal, and so is every polynomial ring $K[x_1, \ldots, x_n]$ over a field.

(2) By Example 8.2(3), $R := \mathbb{Z}\left[\sqrt{5}\right]$ is not normal. In fact, the normalization is

$$\widetilde{R} = \mathbb{Z}\left[\left(1+\sqrt{5}\right)/2\right] =: S.$$

To see this, let $a + b\sqrt{5} \in \mathbb{Q}\left[\sqrt{5}\right] = \operatorname{Quot}(S)$ (with $a, b \in \mathbb{Q}$) be integral over S. Since S is integral over $\mathbb{Z}$ by Theorem 8.4, $a + b\sqrt{5}$ is integral over $\mathbb{Z}$ by Corollary 8.6, and so is $a - b\sqrt{5}$ (satisfying the same integral equation over $\mathbb{Z}$). So the sum $2a$ and the product $a^2 - 5b^2$ of these two elements are also integral over $\mathbb{Z}$. Since $\mathbb{Z}$ is integrally closed, it follows that $2a \in \mathbb{Z}$ and $a^2 - 5b^2 \in \mathbb{Z}$. Now it is easy to see that this implies $a + b\sqrt{5} \in S$.

It may be interesting to note that the ring S is actually factorial.

(3) A rather different case is $R = \mathbb{Z}\left[\sqrt{-5}\right] \subseteq \mathbb{C}$. For an element $a + b\sqrt{-5} \in \mathbb{Q}\left[\sqrt{-5}\right]$, we obtain the conditions $2a \in \mathbb{Z}$ and $a^2 + 5b^2 \in \mathbb{Z}$ for integrality

over $\mathbb{Z}$. It is easy to see that this implies $a, b \in \mathbb{Z}$, so R is normal. However, R is not factorial, as the nonunique factorization

$$6 = 2 \cdot 3 = (1 + \sqrt{-5})(1 - \sqrt{-5}) \tag{8.1}$$

shows. In fact, one needs to show that the factors in (8.1) really are irreducible, and that the factorizations are essentially distinct, i.e., not the same up to the order of the factors and up to invertible elements. For $z = a + b\sqrt{-5} \in R$, write $N(z) := a^2 + 5b^2 = z \cdot \overline{z}$ (z times its complex conjugate) for the so-called **norm** of z. Assume that $2 = z_1 z_2$ with $z_i \in R$. Since the norm is multiplicative, it follows that $4 = N(z_1) \cdot N(z_2)$. But 2 does not occur as a norm of an element of R, so z_1 or z_2 has norm 1. But this means $z_1 = \pm 1$ or $z_2 = \pm 1$, so z_1 or z_2 is invertible. Since every invertible element of R has norm 1, 2 itself is not invertible, so 2 is irreducible in R. Since 3 is not a norm, either, it follows by the same argument that 3 and $1 \pm \sqrt{-5}$ are irreducible, too. Finally, none of the quotients $(1 \pm \sqrt{-5})/2$ and $(1 \pm \sqrt{-5})/3$ lie in R, so the factorizations in (8.1) are essentially different.
This example shows that the converse of Proposition 8.8 does not hold.

(4) Let K be an algebraically closed field. An example from geometry is the singular cubic curve

$$X = \mathcal{V}_{K^2}\left(y^2 - x^2(x+1)\right)$$

over a field K, which is shown in Fig. 8.1, and which has a (visible) singular point at the origin. The coordinate ring of X is

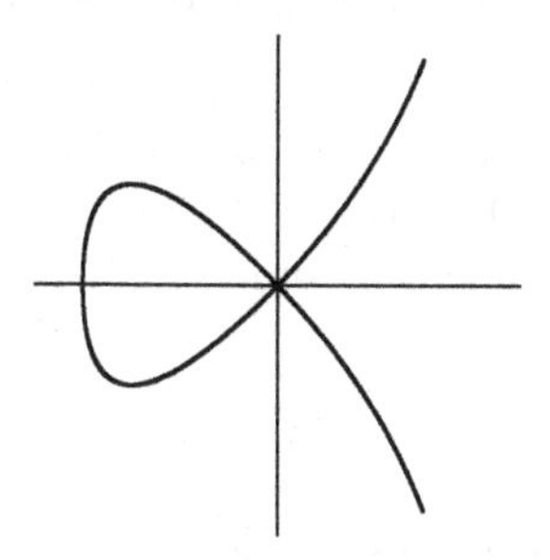

Fig. 8.1. A singular cubic curve

$$A := K[X] = K[x, y] \,/\left(y^2 - x^2(x+1)\right) =: K[\overline{x}, \overline{y}].$$

We have

$$(\overline{y}/\overline{x})^2 - \overline{x} - 1 = 0,$$

so $\overline{y}/\overline{x} \in \operatorname{Quot}(A)$ is integral over A. The above equation also tells us that $\overline{x}$ and $\overline{y} = (\overline{y}/\overline{x}) \cdot \overline{x}$ lie in $K[\overline{y}/\overline{x}]$, so $A \subseteq K[\overline{y}/\overline{x}] \subseteq \widetilde{A}$. Since $K[\overline{y}/\overline{x}]$ is normal by Example 8.9(1), we obtain

$$\widetilde{A} = K[\overline{y}/\overline{x}].$$

It is interesting to consider the morphism of varieties induced by the embedding $A \hookrightarrow \widetilde{A}$. This is given by

$$K^1 \to X, \ \zeta \mapsto (\zeta^2 - 1, \zeta^3 - \zeta).$$

Observe that K^1 has no singular points, and that every nonsingular point of X has precisely one preimage in K^1, whereas the unique singular point of X has two preimages. So the normalization amounts to a desingularization here. As we will see later, these observations are no coincidence. In fact, we will prove in Section 14.1 that normality and nonsingularity coincide in dimension 1. This is one (but not the only) reason why normal rings are interesting. ◁

As the following proposition shows, normality is a *local property*, meaning that it holds globally if and only if it holds locally everywhere.

Proposition 8.10 (Normal rings and localization). *For an integral domain R, the following statements are equivalent:*

(a) R is normal.
(b) For every multiplicative subset $U \subset R$ with $0 \notin U$, the localization $U^{-1}R$ is normal.
(c) For every maximal ideal $\mathfrak{m} \in \operatorname{Spec}_{\max}(R)$, the localization $R_\mathfrak{m}$ is normal.

Proof. Let $K = \operatorname{Quot}(R)$ be the field of fractions. Assume that R is normal, and let $U \subset R$ be a multiplicative subset with $0 \notin U$. We have $U^{-1}R \subseteq K$ and $\operatorname{Quot}(U^{-1}R) = K$. To show that $U^{-1}R$ is normal, let $a \in K$ be integral over $U^{-1}R$. Then there exist $u \in U$ and $a_1, \ldots, a_n \in R$ such that

$$a^n + \frac{a_1}{u} a^{n-1} + \cdots + \frac{a_{n-1}}{u} a + \frac{a_n}{u} = 0.$$

Multiplying this by u^n yields an integral equation for ua over R. So by assumption, $ua \in R$, so $a \in U^{-1}R$. We have shown that the statement (a) implies (b). Clearly (b) implies (c).

Now assume that (c) holds, and let $a \in K$ be integral over R. Consider the ideal $I := \{b \in R \mid ba \in R\} \subseteq R$. For every $\mathfrak{m} \in \operatorname{Spec}_{\max}(R)$, a is integral over $R_\mathfrak{m}$, so $a \in R_\mathfrak{m}$ by assumption. It follows that there exists $b \in I \setminus \mathfrak{m}$. This means that I is not contained in any maximal ideal. But if $I \subsetneqq R$, Zorn's lemma would yield the existence of a maximal ideal containing I. So $1 \in I$, and $a \in R$ follows. So we have shown that (c) implies (a). □

Proposition 8.10 implies that an irreducible affine variety X is normal if and only if for every point $x \in X$ the local ring $K[X]_x$ is normal. Normality also behaves well with respect to passing from R to the polynomial ring $R[x]$, as Exercise 8.7 shows.

We finish the section with a lemma that will be used in Chapter 12. If R is an integral domain, then an element $s \in \mathrm{Quot}(R)$ is said to be **almost integral** (over R) if there exists a nonzero $c \in R$ such that $cs^n \in R$ for all nonnegative integers n.

Lemma 8.11 (Almost integral elements). *In the above setting, if s is integral, then it is almost integral. If R is Noetherian, the converse holds.*

Proof. By Lemma 8.3, s is integral if and only if $R[s] \subseteq \mathrm{Quot}(R)$ is finitely generated as an R-module. In this case there exists $c \in R \setminus \{0\}$ such that $cf \in R$ for all $f \in R[s]$. In particular, $cs^n \in R$ for all n.

Conversely, if s is almost integral, then $R[s]$ is contained in $c^{-1}R \subseteq \mathrm{Quot}(R)$, which is finitely generated (by c^{-1}) as an R-module. If R is Noetherian, it follows with Theorem 2.10 that the same holds for $R[s]$. □

8.2 Lying Over, Going Up, and Going Down

If $R \subseteq S$ is an extension of rings, we have a map $f\colon \mathrm{Spec}(S) \to \mathrm{Spec}(R)$, $Q \mapsto R \cap Q$, induced from the inclusion. We know from Exercise 4.2 that this map is dominant. The following theorem shows that if S is integral over R, then f is, in fact, surjective, and its fibers are finite if S is finitely generated as an R-algebra.

Theorem 8.12 (Lying over and going up). *Let $R \subseteq S$ be an integral extension of rings, $P \in \mathrm{Spec}(R)$ a prime ideal, and $I \subseteq S$ an ideal with $R \cap I \subseteq P$. (Notice that the zero ideal always satisfies the condition on I.) Set*

$$\mathcal{M} := \{Q \in \mathrm{Spec}(S) \mid R \cap Q = P \text{ and } I \subseteq Q\}.$$

Then the following hold:

(a) $\mathcal{M}$ is nonempty.
(b) There exist no $Q, Q' \in \mathcal{M}$ with $Q \subsetneqq Q'$.
(c) If S is finitely generated as an R-algebra, then $\mathcal{M}$ is finite.

The keywords "lying over" and "going up," with which we advertised Theorem 8.12, refer to the following: A prime ideal $Q \in \mathrm{Spec}(S)$ with $R \cap Q = P$ is said to *lie over* P. If additionally I is contained in Q, we say that we are *going up* from I. The situation is illustrated in Fig. 8.2.

Proof of Theorem 8.12. With $S' := S/I$, $R' := R/(R \cap I)$, and $P' := P/(R \cap I)$, we have an integral extension $R' \subseteq S'$, and Lemma 1.22 yields an inclusion-preserving bijection

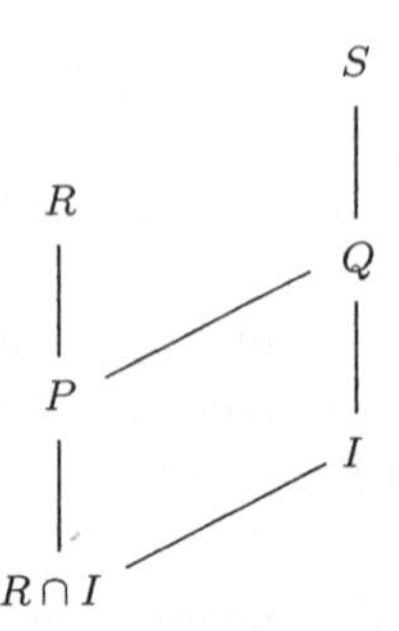

Fig. 8.2. Lying over and going up

$$\mathcal{M} \to \{Q' \in \operatorname{Spec}(S') \mid R' \cap Q' = P'\}.$$

Substituting all objects by their primed versions, we may therefore assume that $I = \{0\}$. By Proposition 7.11, we have to show that the fiber ring $S_{[P]}$ is not the zero ring (implying (a)), has Krull dimension 0 (implying (b)), and has a finite spectrum if S is finitely generated (implying (c)).

By way of contradiction, assume that $S_{[P]} = \{0\}$. By the definition of $S_{[P]}$, this is equivalent to the existence of $u \in R \setminus P$ with $u \in (P)_S$. Forming the localization $S_P := (R \setminus P)^{-1}S$, we obtain $1 \in (P_P)_{S_P}$, so

$$1 = \sum_{i=1}^{n} s_i a_i \quad \text{with} \quad s_i \in S_P,\ a_i \in P_P.$$

Form $\widetilde{S} := R_P[s_1, \ldots, s_n] \subseteq S_P$. Then the above equation implies $(P_P)_{\widetilde{S}} = \widetilde{S}$, which we may write as $P_P\widetilde{S} = \widetilde{S}$. Since $\widetilde{S}$ is an integral extension of R_P, it is finitely generated as an R_P-module by Theorem 8.4. Applying Nakayama's lemma (Theorem 7.3) yields $\widetilde{S} = \{0\}$. Since R_P is embedded into $\widetilde{S}$, this contradicts the fact that local rings are never zero. So we conclude that $S_{[P]}$ is nonzero.

The homomorphism

$$K := \operatorname{Quot}(R/P) \to S_{[P]}, \quad \frac{a+P}{b+P} \mapsto \frac{a+(P)_S}{b+(P)_S},$$

makes $S_{[P]}$ into a K-algebra. The hypothesis that S is integral over R translates into the fact that $S_{[P]}$ is algebraic over K. So if $Q \in \operatorname{Spec}\left(S_{[P]}\right)$, then the quotient ring $S_{[P]}/Q$ is algebraic over K as well, and Lemma 1.1(a) yields that $S_{[P]}/Q$ is a field. This shows that $\dim\left(S_{[P]}\right) = 0$.

Finally, if S is finitely generated as an R-algebra, then $S_{[P]}$ is an affine K-algebra, so Theorem 5.11 yields that $\operatorname{Spec}_{\max}\left(S_{[P]}\right)$ is finite. Since $S_{[P]}$ has dimension 0, $\operatorname{Spec}\left(S_{[P]}\right) = \operatorname{Spec}_{\max}\left(S_{[P]}\right)$, so we are done. □

Let $R \subseteq S$ be an integral extension of rings. If $P_0 \subsetneqq P_1 \subsetneqq \cdots \subsetneqq P_n$ is a chain of prime ideals $P_i \in \operatorname{Spec}(R)$, we can use Theorem 8.12 to construct a chain $Q_0 \subseteq \cdots \subseteq Q_n$ of prime ideals in $\operatorname{Spec}(S)$ with $R \cap Q_i = P_i$. In particular, all inclusions of the Q_i are proper. So $\dim(S) \geq n$, which implies

$$\dim(R) \leq \dim(S). \tag{8.2}$$

On the other hand, if $Q \in \operatorname{Spec}(S)$ is a prime ideal and $Q_0 \subsetneqq Q_1 \subsetneqq \cdots \subsetneqq Q_n \subseteq Q$ is a chain of prime ideals in $\operatorname{Spec}(S)$, then $P_i := R \cap Q_i$ yields a chain in $\operatorname{Spec}(S)$, and it follows from Theorem 8.12(b) that the inclusions of the P_i are proper. So with $P := R \cap Q$ we obtain

$$\operatorname{ht}(Q) \leq \operatorname{ht}(P). \tag{8.3}$$

This implies

$$\dim(S) \leq \dim(R). \tag{8.4}$$

By putting (8.2) and (8.4) together, we obtain the following corollary.

Corollary 8.13. *Let $R \subseteq S$ be an integral extension of rings. Then*

$$\dim(R) = \dim(S).$$

We now pose the question whether the reverse inequality of (8.3) also holds, i.e., whether (8.3) is in fact an equality. For proving this, we need to start with a chain of prime ideals in $\operatorname{Spec}(R)$ that are all contained in P, and construct an equally long chain of prime ideals in $\operatorname{Spec}(S)$ that are all contained in Q. The way to do this is to work our way downwards from Q. But what we need for being able to do this is the going down property, which was discussed in Section 7.2 (see on page 85). We have proved the following:

Corollary 8.14. *Let $R \subseteq S$ be an integral extension of rings such that going down holds for the inclusion $R \hookrightarrow S$. If $Q \in \operatorname{Spec}(S)$ and $P := R \cap Q$, then*

$$\operatorname{ht}(P) = \operatorname{ht}(Q).$$

Unfortunately, going down does not always hold for integral ring extensions, as Exercise 8.9 shows. We have proved that a sufficient condition for going down is freeness (see Lemma 7.16). However, freeness is rarely found for integral extensions. We will exhibit another sufficient condition for going down (see Theorem 8.17). For proving this, we need two lemmas. The effort is worth it, since the reverse inequality of (8.3) is of crucial importance for proving some important results about affine algebras, such as Theorem 8.22 and its corollaries. The first lemma is a result from field theory. The proof uses some standard results from field theory, which we will quote from Lang [33].

Lemma 8.15 (Elements fixed by field automorphisms). *Let N be a field of characteristic $p \geq 0$ and let $K \subseteq N$ be a subfield such that N is finite and* *normal over K (see Lang [33, Chapter VII, Theorem 3.3] for the definition of a normal field extension). Let $G := \mathrm{Aut}_K(N)$ be the group of automorphisms of N fixing K elementwise. Then for every $\alpha \in N^G$ in the fixed field of G, there exists $n \in \mathbb{N}_0$ such that $\alpha^{p^n} \in K$. If N is separable over K, then $n = 0$, so $\alpha \in K$.*

Proof. In the separable case, the lemma follows directly from Galois theory. The proof we give works for the separable case, too.

Let $g = \mathrm{irr}(\alpha, K) \in K[x]$ be the minimal polynomial of α over K. Let $\overline{N}$ be the algebraic closure of N, and let $\beta \in \overline{N}$ be a zero of g. Since $K[\alpha] \cong K[x]/(g) \cong K[\beta]$ with an isomorphism sending α to β, we have a homomorphism $\sigma\colon K[\alpha] \to \overline{N}$ of K-algebras with $\sigma(\alpha) = \beta$. By Lang [33, Chapter VII, Theorem 2.8], this extends to a homomorphism $\sigma\colon N \to \overline{N}$. The normality of N implies $\sigma \in G$ (see Lang [33, Chapter VII, Theorem 3.3]). Since $\sigma(\alpha) = \beta$, the hypothesis of the lemma implies $\beta = \alpha$. So α is the only zero of g, and we obtain $g = (x - \alpha)^m$ with $m \in \mathbb{N}$. Write $m = k \cdot p^n$ with $p \nmid k$. If N is separable over K, then g has to be separable, so $m = 1$ and $n = 0$. We have

$$g = (x - \alpha)^m = (x^{p^n} - \alpha^{p^n})^k = x^{kp^n} - k \cdot \alpha^{p^n} \cdot x^{(k-1)p^n} + (\text{lower terms}),$$

so $g \in K[x]$ implies $a^{p^n} \in K$. □

Lemma 8.16. *Let N be a field and $K \subseteq N$ a subfield such that N is finite and normal over K. Let $R \subseteq K$ be a subring that is integrally closed in K, and let $S \subseteq N$ be the integral closure of R in N. Then for two prime ideals $Q, \widetilde{Q} \in \mathrm{Spec}(S)$ with $R \cap Q = R \cap \widetilde{Q}$, there exists $\sigma \in G := \mathrm{Aut}_K(N)$ with $\widetilde{Q} = \sigma(Q)$.*

Proof. Let $a \in \widetilde{Q}$. Then the product $\prod_{\sigma \in G} \sigma(a)$ lies in N^G, so by Lemma 8.15 there exists $n \in \mathbb{N}_0$ with

$$b := \prod_{\sigma \in G} \sigma(a)^{p^n} \in K, \tag{8.5}$$

where $p = \mathrm{char}(K)$ and $n = 0$ if $p = 0$. Since a is integral over R and all $\sigma \in G$ fix R elementwise, all $\sigma(a)$ are integral over R as well. So b is integral over R, too, and (8.5) implies $b \in R$. Moreover, b is an S-multiple of a, so $b \in R \cap \widetilde{Q} = R \cap Q \subseteq Q$. Since Q is a prime ideal, it follows from (8.5) that there exists $\sigma \in G$ with $\sigma(a) \in Q$. Since this holds for all $a \in \widetilde{Q}$, we conclude that

$$\widetilde{Q} \subseteq \bigcup_{\sigma \in G} \sigma(Q).$$

By the prime avoidance lemma (Lemma 7.7), this implies that there exists $\sigma \in G$ with $\widetilde{Q} \subseteq \sigma(Q)$. Since σ fixes R elementwise, we have $R \cap \sigma(Q) = R \cap Q = R \cap \widetilde{Q}$, so by Theorem 8.12(b), the inclusion $\widetilde{Q} \subseteq \sigma(Q)$ cannot be strict. □

Theorem 8.17 (Going down for integral extensions of normal rings). *Let S be a ring and $R \subseteq S$ a subring such that*

(1) S is an integral domain,
(2) R is normal,
(3) S is integral over R, and
(4) S is finitely generated as an R-algebra.

Then going down holds for the inclusion $R \hookrightarrow S$. In particular, the conclusion of Corollary 8.14 holds.

Proof. The proof is not difficult but a bit involved. Fig. 8.3 shows what is going on. Given prime ideals $P \in \operatorname{Spec}(R)$ and $Q' \in \operatorname{Spec}(S)$ with $P \subseteq Q'$, we need to produce $Q \in \operatorname{Spec}(S)$ with $R \cap Q = P$ and $Q \subseteq Q'$. The field of fractions $L := \operatorname{Quot}(S)$ is a finite field extension of $K := \operatorname{Quot}(R)$. By Lang [33, Chapter VII, Theorem 3.3], there exists a finite normal field extension N of K such that $L \subseteq N$. Let $T \subseteq N$ be the integral closure of R in N, so $S \subseteq T$. By Theorem 8.12, there exist $\widetilde{Z}, Z' \in \operatorname{Spec}(T)$ such that $R \cap \widetilde{Z} = P$ and $S \cap Z' = Q'$. We cannot assume that $\widetilde{Z}$ is contained in Z'. However, applying Theorem 8.12 again, we see that there exists $\widetilde{Z}' \in \operatorname{Spec}(T)$ such that $R \cap \widetilde{Z}' = R \cap Q'$ and $\widetilde{Z} \subseteq \widetilde{Z}'$. We have

$$R \cap Z' = R \cap S \cap Z' = R \cap Q' = R \cap \widetilde{Z}'.$$

So by Lemma 8.16 there exists $\sigma \in \operatorname{Aut}_K(N)$ with $Z' = \sigma(\widetilde{Z}')$. Set $Z := \sigma(\widetilde{Z})$ and $Q := S \cap Z \in \operatorname{Spec}(S)$. Then

$$R \cap Q = R \cap Z = R \cap \sigma(\widetilde{Z}) = R \cap \widetilde{Z} = P$$

and

$$Q = S \cap \sigma(\widetilde{Z}) \subseteq S \cap \sigma(\widetilde{Z}') = S \cap Z' = Q'.$$

This finishes the proof. □

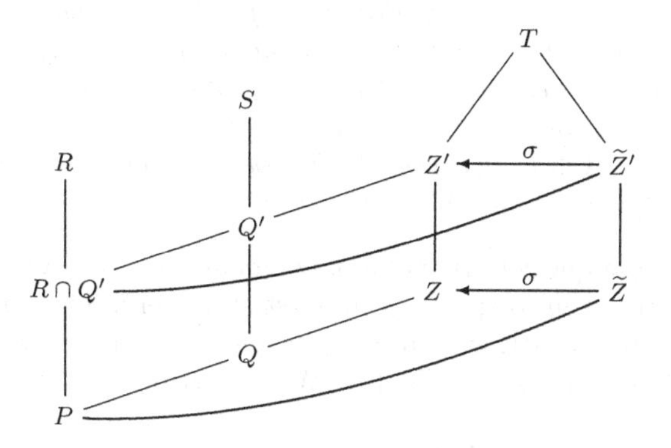

Fig. 8.3. Going down: given P and Q', construct Q

We finish this section by drawing some conclusions about geometric properties of normalization.

Proposition 8.18 (Geometric properties of normalization). *Let R be an integral domain with normalization $\widetilde{R}$, and consider the morphism $f\colon \operatorname{Spec}(\widetilde{R}) \to \operatorname{Spec}(R)$ induced from the inclusion $R \subseteq \widetilde{R}$. Then*

(a) $\dim(\widetilde{R}) = \dim(R)$.
(b) The morphism f is surjective.
(c) Let $P \in \operatorname{Spec}(R)$ be such that R_P is normal. Then the fiber $f^{-1}(\{P\})$ consists of one point.

Proof. Parts (a) and (b) follow from Corollary 8.13 and Theorem 8.12(a).

To prove (c), take $P \in \operatorname{Spec}(R)$ with R_P normal. Both R_P and $\widetilde{R}$ are contained in $\operatorname{Quot}(R)$. With $U := R \setminus P$ we have $U^{-1}\widetilde{R} \subseteq \operatorname{Quot}(R) = \operatorname{Quot}(R_P)$, and $U^{-1}\widetilde{R}$ is integral over R_P, so $U^{-1}\widetilde{R} = R_P$ by the normality of R_P. Let $Q \in \operatorname{Spec}(\widetilde{R})$ be in the fiber of P, so $R \cap Q = P$. By Theorem 6.5 it follows that $U^{-1}Q \in \operatorname{Spec}(U^{-1}\widetilde{R}) = \operatorname{Spec}(R_P)$, and $\widetilde{R} \cap U^{-1}Q = Q$, so $R \cap U^{-1}Q = P$. But Theorem 6.5 also says that P_P is the only prime ideal in R_P whose intersection with R is P, so $U^{-1}Q = P_P$. It follows that $Q = \widetilde{R} \cap P_P$, showing uniqueness. □

8.3 Noether Normalization

We now turn our attention to the special case of affine algebras. Let A be an affine K-algebra with $\dim(A) = n$. By Theorem 5.9, there exist algebraically independent elements $a_1, \ldots, a_n \in A$ such that A is algebraic over the subalgebra $K[a_1, \ldots, a_n]$. As we will see in the following theorem, more can be said.

Theorem 8.19 (Noether normalization). *Let $A \neq \{0\}$ be an affine K-algebra. Then there exist algebraically independent elements $c_1, \ldots, c_n \in A$ (with $n \in \mathbb{N}_0$) such that A is integral over the subalgebra $C := K[c_1, \ldots, c_n]$. In particular, A is finitely generated as a C-module, and C is isomorphic to a polynomial ring (with $C = K$ if $n = 0$).*

If $c_1, \ldots, c_n \in A$ are algebraically independent and A is integral over $K[c_1, \ldots, c_n]$, then $n = \dim(A)$.

Proof. Write A as a quotient ring of a polynomial ring: $A = K[x_1, \ldots, x_m]/I$. We use induction on m for proving the first statement. There is nothing to show for $m = 0$. If $I = \{0\}$, we can set $c_i = x_i + I$, and again there is nothing to show. If $I \neq \{0\}$, choose $f \in I \setminus \{0\}$. We can write f as

$$f = \sum_{(i_1, \ldots, i_m) \in S} \alpha_{i_1, \ldots, i_m} \cdot x_1^{i_1} \cdots x_m^{i_m}$$

with $\emptyset \neq S \subset \mathbb{N}_0^m$ a finite subset and $\alpha_{i_1,\ldots,i_m} \in K \setminus \{0\}$. Choose $d > \deg(f)$ (in fact, it suffices to choose d bigger than all x_i-degrees of f). Then the function $s\colon S \to \mathbb{N}_0$, $(i_1,\ldots,i_m) \mapsto \sum_{j=1}^m i_j \cdot d^{j-1}$ is injective. For $i = 2,\ldots,m$ set $y_i := x_i - x_1^{d^{i-1}}$. Then

$$\begin{aligned} f &= f\left(x_1, y_2 + x_1^d, \ldots, y_m + x_1^{d^{m-1}}\right) \\ &= \sum_{(i_1,\ldots,i_m)\in S} \alpha_{i_1,\ldots,i_m} \left(x_1^{s(i_1,\ldots,i_m)} + g_{i_1,\ldots,i_m}(x_1, y_2, \ldots, y_m)\right) \end{aligned}$$

with $g_{i_1,\ldots,i_m}$ polynomials satisfying $\deg_{x_1}(g_{i_1,\ldots,i_m}) < s(i_1,\ldots,i_m)$. We have exactly one $(i_1,\ldots,i_m) \in S$ such that $k := s(i_1,\ldots,i_m)$ becomes maximal. Since $A \neq \{0\}$, f is not constant, so $k > 0$. We obtain

$$f = \alpha_{i_1,\ldots,i_m} \cdot x_1^k + h(x_1, y_2, \ldots, y_m)$$

with $\deg_{x_1}(h) < k$, so

$$x_1^k + \alpha_{i_1,\ldots,i_m}^{-1} h(x_1, y_2, \ldots, y_m) \in I.$$

Set $B := K[y_2 + I, \ldots, y_m + I] \subseteq A$. Then $A = B[x_1 + I]$, and the above equation and Theorem 8.4 show that A is integral over B. By induction, there exist algebraically independent $c_1,\ldots,c_n \in B$ such that B is integral over $K[c_1,\ldots,c_n]$, and the same follows for A by Corollary 8.6.

The statement $n = \dim(A)$ follows from Corollaries 5.7 and 8.13. □

The above proof can be turned into an algorithm for computing $c_1,\ldots,c_n$. This algorithm uses Gröbner bases and is dealt with in Exercise 9.12.

Remark 8.20. In Exercise 8.10, the following stronger (but slightly less general) version of Noether normalization is shown: If the field K is infinite and $A = K[a_1,\ldots,a_m]$, then the elements $c_1,\ldots,c_n$ satisfying Theorem 8.19 can be chosen as linear combinations

$$c_i = a_i + \sum_{j=n+1}^{m} \gamma_{i,j} \cdot a_j \quad (\gamma_{i,j} \in K)$$

of the "original" generators a_i. ◁

It is not hard to give geometric interpretations of Noether normalization. In fact, Theorem 8.19 tells us that for an affine variety X of dimension n over a field K, there exists a morphism

$$f\colon X \to K^n$$

induced by the inclusion $C \subseteq K[X]$, and by Theorem 8.12, f is surjective and has finite fibers. So Noether normalization tells us that every affine variety may be interpreted as a "finite covering" of some K^n. A slightly different interpretation is that Noether normalization provides a new coordinate system such that the first n coordinates can be set to arbitrary values, which will be attained by finitely many points from the variety. So the first n coordinates act as "independent parameters." With both interpretations, it makes intuitive sense that X should have dimension n, which is a further indication that our definition of dimension is a good one. In Exercise 8.11, a further interpretation of Noether normalization as a "global system of parameters" is given.

Example 8.21. Consider the affine variety $X = \mathcal{V}_{K^2}(x_1x_2 - 1)$, which is a hyperbola as shown in Fig. 8.4. We write $\overline{x}_i$ for the image of x_i in the coordinate ring $K[X] = K[x_1, x_2]/(x_1x_2 - 1) = K[\overline{x}_1, \overline{x}_2]$. Notice that $K[X]$ is not integral over $K[\overline{x}_1]$ or over $K[\overline{x}_2]$. Motivated by Remark 8.20, we try $c = \overline{x}_1 - \overline{x}_2$ and find

$$0 = \overline{x}_1\overline{x}_2 - 1 = \overline{x}_1^2 - \overline{x}_1 c - 1,$$

so $K[X]$ is integral over $C := K[c]$. The morphism induced by the embedding $C \hookrightarrow K[X]$ is $f\colon X \to K^1$, $(\xi_1, \xi_2) \mapsto \xi_1 - \xi_2$. It is surjective, and all $\eta \in K$ with $\eta^2 \neq -4$ have two preimages, as indicated by the arrows in Fig. 8.4. ◁

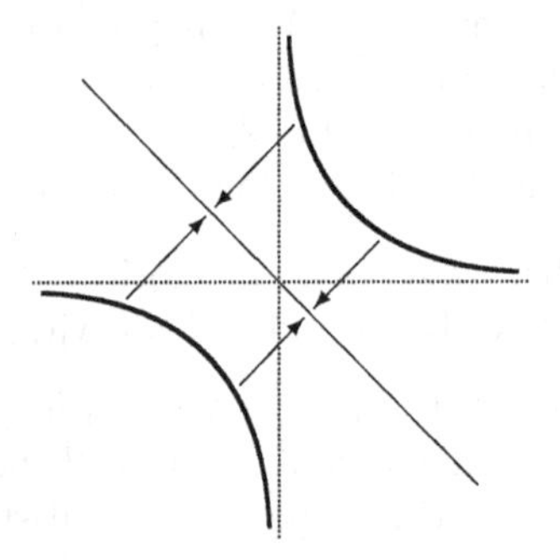

Fig. 8.4. A hyperbola and Noether normalization

We now turn our attention to chains of prime ideals in an affine algebra. Generally, in a set $\mathcal{M}$ whose elements are sets, a **maximal chain** is a subset $\mathcal{C} \subseteq \mathcal{M}$ that is totally ordered by inclusion "$\subseteq$" such that $\mathcal{C} \subseteq \mathcal{D} \subseteq \mathcal{M}$ with $\mathcal{D}$ totally ordered implies $\mathcal{C} = \mathcal{D}$. In particular, a chain

$$P_0 \subsetneqq P_1 \subsetneqq \cdots \subsetneqq P_n$$

of prime ideals $P_i \in \operatorname{Spec}(R)$ in some ring is maximal if no further prime ideal can be added into the chain by insertion or by appending at either end. In general rings, it is not true that all maximal chains of prime ideals have equal length. Examples for this are affine algebras that are not equidimensional, or, more subtly, the ring studied in Exercise 8.12. However, the following theorem says that this is the case for affine domains.

Theorem 8.22 (Chains of prime ideals in an affine algebra). *Let A be an affine algebra and let*

$$P_0 \subsetneq P_1 \subsetneq \cdots \subsetneq P_n \tag{8.6}$$

be a maximal chain of prime ideals $P_i \in \operatorname{Spec}(A)$. Then

$$n = \dim(A/P_0).$$

In particular, if A is equidimensional (which is always the case if A is an affine domain), then every maximal chain of prime ideals of A has length equal to $\dim(A)$.

Proof. We use induction on n. Substituting A by A/P_0, we may assume that A is an affine domain and $P_0 = \{0\}$. If $n = 0$, then P_0 is a maximal ideal, so A is a field and we are done. So we may assume $n > 0$. Applying Lemma 1.22 yields a maximal chain $P_1/P_1 \subsetneq P_2/P_1 \subsetneq \cdots \subsetneq P_n/P_1$ of prime ideals in A/P_1. Using induction, we obtain $n - 1 = \dim(A/P_1)$. So we need to show that $\dim(A/P_1) = \dim(A) - 1$.

Using Noether normalization (Theorem 8.19), we obtain $C \subseteq A$ with A integral over C and C isomorphic to a polynomial ring. By the maximality of (8.6), we have $\operatorname{ht}(P_1) = 1$. By Proposition 8.8, C is normal, so all hypotheses of Theorem 8.17 are satisfied. We obtain $\operatorname{ht}(C \cap P_1) = 1$. By the implication (b) $\Rightarrow$ (a) of Theorem 5.13, this implies $\dim(C/(C \cap P_1)) = \dim(C) - 1$. Since A/P_1 is integral over $C/(C \cap P_1)$, Corollary 8.13 yields

$$\dim(A/P_1) = \dim(C/(C \cap P_1)) = \dim(C) - 1 = \dim(A) - 1.$$

This finishes the proof. □

A ring R is called **catenary** if for two prime ideals $P \subseteq Q$ in $\operatorname{Spec}(R)$, all maximal chains of prime ideals between P and Q have the same length. So Theorem 8.22 implies that all affine algebras are catenary. It is not easy to find examples of noncatenary rings (see Nagata [41, Appendix, Example E2], Matsumura [37, Example 14E], or Hutchins [28, Example 27]). We get two immediate consequences of Theorem 8.22. The first one says that in affine domains, the height of an ideal and the dimension of the quotient ring behave complementarily.

Corollary 8.23 (Dimension and height). *Let A be an affine domain or, more generally, an equidimensional affine algebra. If $I \subseteq A$ is an ideal, then*

$$\operatorname{ht}(I) = \dim(A) - \dim(A/I).$$

Proof. If I is a prime ideal, there exists a maximal chain $\mathcal{C} \subseteq \operatorname{Spec}(A)$ with $I \in \mathcal{C}$, so the result follows from Theorem 8.22, Lemma 1.22, and Definition 6.10(a). For $I = A$, it follows from Definition 6.10(b). For all other I, Definition 6.10(b) and the fact that

$$\dim(A/I) = \max\left\{\dim(A/P) \mid P \in \mathcal{V}_{\operatorname{Spec}(A)}(I)\right\}$$

allow reduction to the case that I is a prime ideal. □

The following corollary is about the height of maximal ideals in affine algebras. In the case of a maximal ideal $\mathfrak{m} \in \operatorname{Spec}_{\max}(K[X])$ belonging to a point $x \in X$ of an affine variety, it says that the height of $\mathfrak{m}$ is equal to the largest dimension of an irreducible component of X containing x.

Corollary 8.24 (Height of maximal ideals). *Let A be an affine algebra with minimal prime ideals $P_1, \ldots, P_n$. (There are finitely many P_i by Corollaries 2.12 and 3.14*(a)*.) If $\mathfrak{m} \in \operatorname{Spec}_{\max}(A)$ is a maximal ideal, then*

$$\operatorname{ht}(\mathfrak{m}) = \max\left\{\dim(A/P_i) \mid P_i \subseteq \mathfrak{m}\right\}.$$

In particular, if A is an affine domain or, more generally, equidimensional, then all maximal ideals have $\operatorname{ht}(\mathfrak{m}) = \dim(A)$.

Proof. This is an immediate consequence of Theorem 8.22. □

To get a better appreciation of the last three results, it is important to see an example of a Noetherian domain (= a Noetherian integral domain) for which they fail. Such an example is given in Exercise 8.12.

The following result restates the principal ideal theorem (Theorem 7.5) for the special case of affine domains. Corollary 8.23 allows us to convert the statement from Theorem 7.5 on height into a statement on dimension. The theorem exemplifies the common paradigm that "imposing n further equations makes the dimension of the solution set go down by at most n."

Theorem 8.25 (Principal ideal theorem for affine domains). *Let A be an affine domain or, more generally, an equidimensional affine algebra, and let $I = (a_1, \ldots, a_n) \subseteq A$ be an ideal generated by n elements. Then every prime ideal $P \in \operatorname{Spec}(A)$ that is minimal over I satisfies*

$$\dim(A/P) \geq \dim(A) - n.$$

In particular, if $I \neq A$, then

$$\dim(A/I) \geq \dim(A) - n,$$

and if equality holds, then A/I is equidimensional.

Proof. By Theorem 7.5, every $P \in \operatorname{Spec}(A)$ that is minimal over I satisfies $\operatorname{ht}(P) \leq n$, so

$$\dim(A/P) \geq \dim(A) - n$$

by Corollary 8.23. The other claims follows directly from this. □

If $f_1, \ldots, f_n \in K[x_1, \ldots, x_m]$ are polynomials over an algebraically closed field, then by Theorem 8.25, the affine variety in $X = \mathcal{V}_{K^m}(f_1, \ldots, f_n)$ is empty or has dimension at least $m - n$. If the dimension is equal to $m - n$, then X is called a **complete intersection** ("intersection" referring to the intersection of the hypersurfaces given by the f_i). So the second assertion of Theorem 8.25 tells us that complete intersections are equidimensional. By a slight abuse of terminology, an affine K-algebra A is also called a complete intersection if $A \cong K[x_1, \ldots, x_m]/(f_1, \ldots, f_n)$ with $\dim(A) = m - n \geq 0$.

Geometrically, the first part of Theorem 8.25 gives a dimension bound for the intersection of affine varieties $X, Y \subseteq K^m$, where X is equidimensional and Y is given by n equations. A generalization is contained in Exercise 8.14.

We will close this chapter by proving that the normalization of an affine domain is again an affine domain, and applying this result to affine varieties. Although this material is interesting, it will be used in this book only in Chapter 14 to prove two results: the existence of a desingularization of an affine curve, and the fact that the integral closure of $\mathbb{Z}$ in a number field is Noetherian (which follows from Lemma 8.27). So readers may choose to skip the rest of this chapter.

Theorem 8.26. *Let A be an affine domain. Then the normalization $\widetilde{A}$ of A is an affine domain, too.*

Proof. By Noether normalization (Theorem 8.19), we have a subalgebra $R \subseteq A$ which is isomorphic to a polynomial algebra, such that A is integral over R. In particular, $N := \operatorname{Quot}(A)$ is a finite field extension of $\operatorname{Quot}(R)$, and $\widetilde{A}$ is the integral closure of R in N. So the result follows from the following lemma. □

Lemma 8.27 (Integral closure in a finite field extension)**.** *Let R be a Noetherian domain and N a finite field extension of $L := \operatorname{Quot}(R)$. Assume that*

(a) R is normal and N is separable over L, or
(b) R is isomorphic to a polynomial ring over a field.

Then the integral closure S of R in N is finitely generated as an R-module (and therefore also as an R-algebra).

Proof. Choose generators of N as an extension of L, and let N' be the splitting field of the product of the minimal polynomials of the generators. Then N' is a finite normal field extension of L with $N \subseteq N'$, and if N is separable over L, so is N'. Since S is a submodule of the integral closure S' of R in N', it suffices to show that S' is a finitely generated R-module (use Theorem 2.10). So we may assume that N is normal over L. Let $G := \operatorname{Aut}_L(N)$ and consider the trace map

$$\mathrm{Tr}\colon N \to N^G,\ x \mapsto \sum_{\sigma \in G} \sigma(x).$$

It follows from the linear independence of homomorphisms into a field (see Lang [33, Chapter VIII, Theorem 4.1]) that Tr is nonzero. It is clearly L-linear. Let $b_1, \ldots, b_m \in N$ be an L-basis of N. By Lemma 8.15, there exists a power q of the characteristic of L (with $q = 1$ if N is separable over L) such that $\mathrm{Tr}(b_i)^q \in L$ for all i.

We first treat the (harder) case that $R \cong K[x_1, \ldots, x_n]$ with K a field. In fact, we may assume $R = K[x_1, \ldots, x_n]$. Let K' be a finite field extension of K containing qth roots of all coefficients appearing in $\mathrm{Tr}(b_i)^q$ as rational functions in the x_j. Then $\mathrm{Tr}(b_i) \in K'(x_1^{1/q}, \ldots, x_n^{1/q}) =: L'$. (For this containment to make sense without any homomorphism linking L' and N, it is useful to embed both fields in an algebraic closure of L.) So $R' := K'[x_1^{1/q}, \ldots, x_n^{1/q}]$ satisfies the following properties: (i) R' is finitely generated as an R-module, (ii) R' is normal (by Example 8.9(1)), and (iii) $\mathrm{Tr}(b_i) \subseteq \mathrm{Quot}(R')$ for all i.

In the case that R is normal and N is separable over L, these three properties are satisfied for $R' := R$.

Since $L \subseteq \mathrm{Quot}(R')$, property (iii) implies $\mathrm{Tr}(N) \subseteq \mathrm{Quot}(R')$. For $s \in S$, $\mathrm{Tr}(s)$ is integral over R, and therefore $\mathrm{Tr}(s) \in R'$ by (ii). Every $x \in N$ is algebraic over L, so there exists $0 \neq a \in R$ with $ax \in S$. Indeed, choosing a common denominator $a \in R$ of the coefficients of an integral equation of degree n for x over $L = \mathrm{Quot}(R)$ and multiplying the equation by a^n yields an integral equation for ax over R. Therefore we may assume that the basis elements b_i lie in S. So S is contained in the R-module

$$M := \{x \in N \mid \mathrm{Tr}(xb_i) \in R' \text{ for all } i = 1, \ldots, m\} \subseteq N.$$

There is an R-linear map

$$\varphi\colon M \to (R')^m,\ x \mapsto (\mathrm{Tr}(xb_1), \ldots, \mathrm{Tr}(xb_m)).$$

To show that φ is injective, let $x \in M$ with $\varphi(x) = 0$. By the L-linearity of the trace map, this implies $\mathrm{Tr}(xy) = 0$ for all $y \in N$, so $x = 0$ since $\mathrm{Tr} \neq 0$. So S is isomorphic to a submodule of $(R')^m$. But $(R')^m$ is finitely generated over R by the property (i) of R', and the result follows by Theorem 2.10. □

It is tempting to hope that for every Noetherian domain R, the normalization $\widetilde{R}$ is finitely generated as an R-module. However, Nagata [41, Appendix, Example E5] has an example in which $\widetilde{R}$ is not even Noetherian.

Corollary 8.28 (Normalization of an affine variety). *Let X be an irreducible affine variety over an algebraically closed field K. Then there exists a normal affine variety $\widetilde{X}$ with a surjective morphism $f\colon \widetilde{X} \to X$ such that:*

(a) $\dim(\widetilde{X}) = \dim(X)$.
(b) All fibers of f *are finite, and if* $x \in X$ *is a point where the local ring* $K[X]_x$ *is normal, then the fiber of* x *consists of one point.*

Proof. By Theorem 8.26, the normalization $\widetilde{A}$ of the coordinate ring $A := K[X]$ is an affine domain, so by Theorem 1.25(b) there exists an affine variety $\widetilde{X}$ with $K[\widetilde{X}] \cong \widetilde{A}$. The inclusion $A \subseteq \widetilde{A}$ induces a morphism $f: \widetilde{X} \to X$. Clearly $\widetilde{X}$ is normal, and from Proposition 8.18 we obtain part (a), the surjectivity of f, and the statement on the fibers of points with A_x normal. The finiteness of the fibers follows from Theorem 8.12(c). □

The behavior of the morphism f from Corollary 8.28 can be observed very well in Example 8.9(4). Exercise 8.8 deals with a universal property of $\widetilde{X}$, as constructed in the above proof. Together with (a) and (b) of Corollary 8.28, this characterizes $\widetilde{X}$ up to isomorphism. The variety $\widetilde{X}$, or sometimes also $\widetilde{X}$ together with the morphism f, is called the **normalization** of X. In Section 14.1 we will see that if X is a curve, normalization is the same as desingularization.

Exercises for Chapter 8

8.1 (Rings of invariants of finite groups). In this exercise we prove that rings of invariants of finite groups are finitely generated under very general assumptions. The proof is due to Emmy Noether [43]. Let S be a ring with a subring $R \subseteq S$, and let $G \subseteq \mathrm{Aut}_R(S)$ be a finite group of automorphisms of S as an R-algebra (i.e., the elements of G fix R pointwise). Write

$$S^G := \{a \in S \mid \sigma(a) = a \text{ for all } \sigma \in G\} \subseteq S$$

for the **ring of invariants**. Observe that S^G is a sub-R-algebra of S.

(a) Show that S is integral over S^G. In particular, $\dim\left(S^G\right) = \dim(S)$.
(b) Assume that S is finitely generated as an R-algebra. Show that S^G has a finitely generated subalgebra $A \subseteq S^G$ such that S is integral over A.
(c) Assume in addition that R is Noetherian. Show that S^G is finitely generated as an R-algebra. In particular, S^G is Noetherian.

8.2 (Rings of invariants are normal). Let R be a normal ring, and let $G \subseteq \mathrm{Aut}(R)$ be a group of automorphisms of R. Show that R^G, the ring of invariants, is normal, too.

***8.3 (The intersection of localizations).** Let R be a normal Noetherian domain. Show that

$$R = \bigcap_{\substack{P \in \mathrm{Spec}(R),\\ \mathrm{ht}(P)=1}} R_P.$$

(Notice that all R_P are contained in $\mathrm{Quot}(R)$, so the intersection makes sense.)
Hint: For $a/b \in \mathrm{Quot}(R) \setminus R$, consider an ideal P that is maximal among all colon ideals $(b) : (a')$ with $a' \in (a) \setminus (b)$.

8.4 (Quadratic extensions of polynomial rings). Let $f \in K[x_1, \ldots, x_n]$ be a polynomial with coefficients in a field of characteristic not equal to 2. Assume that f is not a square of a polynomial. Show that the ring $R := K[x_1, \ldots, x_n, y]/(y^2 - f)$ (with y a further indeterminate) is normal if and only if f is square-free.

8.5 (A normality criterion). Let R be a ring with an element $a \in R$ such that

(1) a is not a zero divisor.
(2) the ideal (a) is a radical ideal.
(3) the localization R_a is a normal domain.

Show that R is a normal domain.

Use this to show that for every field K and every positive integer q, the ring

$$K[x_1, x_2, y_1, y_2, z]/(z^q - (x_1 y_1)^{q-1} z - x_1^q y_2 - y_1^q x_2)$$

(with x_1, x_2, y_1, y_2, and z indeterminates) is a normal domain.

8.6 (Normalization). Assume that K contains a primitive third root of unity. Compute the normalization $\widetilde{R}$ of $R = K[x_1^3, x_1^2 x_2, x_2^3]$.
Hint: You may use Exercise 8.2. Alternatively, you may do the exercise without using the hypothesis on K.

***8.7 (Normalization of polynomial rings).** Let R be a Noetherian domain. Show that

$$\widetilde{R[x]} = \widetilde{R}[x]$$

(i.e., the normalization of the polynomial ring over R is equal to the polynomial ring over the normalization). Conclude that $R[x]$ is normal if and only if R is normal.
Hint: The hard part is to show that a polynomial $f = \sum_{i=0}^{n} a_i x_i \in \mathrm{Quot}(R)[x]$ that is integral over $R[x]$ lies in $\widetilde{R}[x]$. This can be done as follows: Show that there exists $0 \neq u \in R$ such that $uf^k \in R[x]$ for all $k \geq 0$. Conclude that $R[a_n]$ is finitely generated as an R-module. Then use induction on n.
Remark: The result is also true if R is not Noetherian. In fact, one can reduce to the Noetherian case by substituting R with a finitely generated subring in the above proof idea.
(Solution on page 222)

8.8 (The universal property of normalization). Show that the variety $\widetilde{X}$ constructed in the proof of Corollary 8.28 satisfies the following universal property. If Y is a normal affine K-variety with a dominant morphism $g: Y \to X$ (this means that the image $g(Y)$ is dense in X), then there exists a unique morphism $h: Y \to \widetilde{X}$ with $f \circ h = g$.

8.9 (Where going down fails). In this exercise we study an example of an integral extension of rings in which going down fails. Let K be a field of characteristic $\neq 2$, $S = K[x,y]$ the polynomial ring in two indeterminates, and

$$R := K[a,b,y] \subset S \quad \text{with} \quad a = x^2 - 1 \quad \text{and} \quad b = xa.$$

(a) Show that S is the normalization of R.

(b) Show that

$$P := \left(a - (y^2-1), b - y(y^2-1)\right)_R \subset R$$

is a prime ideal, and P is contained in the prime ideal

$$Q' := (x-1, y+1)_S \in \operatorname{Spec}(S).$$

(c) Show that the unique ideal $Q \in \operatorname{Spec}(S)$ with $R \cap Q = P$ is

$$Q := (x-y)_S$$

and conclude that going down fails for the inclusion $R \hookrightarrow S$.

(d) Compare this example to Example 8.9(4). Try to give a geometric interpretation to the failure of going down for $R \hookrightarrow S$.
Hint: The generators of R satisfy the equation $b^2 - a^2 \cdot (a+1) = 0$.

8.10 (Noether normalization with linear combinations). Prove the statement in Remark 8.20.
Hint: Mimic the proof of Theorem 8.19, but set $y_i := x_i - \beta_i x_m$ with $\beta_i \in K$ $(i = 1, \dots, m-1)$.

***8.11 (Noether normalization and systems of parameters).** Let $X \neq \emptyset$ be an equidimensional affine variety over a field K and let $c_1, \dots, c_n \in A := K[X]$ be as in Theorem 8.19. Let $x \in X$ be a point with corresponding maximal ideal $\mathfrak{m} := \{f \in A \mid f(x) = 0\}$. Show that

$$a_i := \frac{c_i - c_i(x)}{1} \in A_{\mathfrak{m}} \quad (i = 1, \dots, n)$$

provides a system of parameters of the local ring $K[X]_x = A_{\mathfrak{m}}$ at x. An interpretation of this result is that Noether normalization provides a global system of parameters or, from a reverse angle, that systems of parameters are a local version of Noether normalization.

Hint: With $I := (c_1 - c_1(x), \ldots, c_n - c_n(x))_A$, first prove that A/I is Artinian. Then use Nakayama's lemma to show that $\mathfrak{m}_\mathfrak{m}^k \subseteq I_\mathfrak{m}$ for some k. *(Solution on page 223)*

8.12 (A Noetherian domain where Theorem 8.22 fails). Let $R = K[[x]]$ be a formal power series ring over a field, and $S = R[y]$ a polynomial ring. Exhibit two maximal ideals in $\operatorname{Spec}_{\max}(S)$ of different height. So S is a Noetherian domain for which Theorem 8.22 and Corollaries 8.23 and 8.24 fail.

8.13 (Hypotheses of Theorem 8.25). Use the following example to show that the hypothesis on equidimensionality cannot be dropped from Theorem 8.25:

$$A = K[x_1, x_2, x_3, x_4]/(x_1 - x_4, x_1^2 - x_2x_4, x_1^2 - x_3x_4)$$

and $a = \overline{x}_1 - 1$, the class of $x_1 - 1$ in A. Explain why this also shows that if $K[x_1, \ldots, x_m]/(f_1, \ldots, f_n)$ is a complete intersection, this need not imply that $K[x_1, \ldots, x_m]/(f_1, \ldots, f_{n-1})$ is a complete intersection, too.

8.14 (A dimension theorem). Let X and Y be two equidimensional affine varieties both of which lie in K^n. Show that every irreducible component Z of $X \cap Y$ satisfies

$$\dim(Z) \geq \dim(X) + \dim(Y) - n.$$

Hint: With $\Delta := \{(x, x) \mid x \in K^n\} \subset K^{2n}$ the diagonal, show that $X \cap Y \cong (X \times Y) \cap \Delta$ and conclude the result from that.

8.15 (Right or wrong?). Decide whether each of the following statements is true or false. Give reasons for your answers.

(a) Let K be a finite field and let X be a set. Then the ring $S = \{f\colon X \to K \mid f \text{ is a function}\}$ (with pointwise operations) is an integral extension of K (which is embedded into S as the ring of constant functions).
(b) If $R \subseteq S$ is an integral ring extension, then for every $P \in \operatorname{Spec}(R)$ the set $\{Q \in \operatorname{Spec}(S) \mid R \cap Q = P\}$ is finite.
(c) If A is an affine domain that can be generated by $\dim(A) + 1$ elements, then A is a complete intersection.
(d) If A is an affine algebra that can be generated by $\dim(A) + 1$ elements, then A is a complete intersection.
(e) If an affine domain is a complete intersection, it is normal.

Part III
Computational Methods

Chapter 9
Gröbner Bases

A large part of commutative algebra is formulated in nonconstructive ways. A typical example is Hilbert's basis theorem (Corollary 2.13), which guarantees the existence of finite ideal bases without giving a method to construct them. But commutative algebra also has a large computational part, which has developed into a field of research of its own, called *computational commutative algebra*. This field has its own conferences, its own research community, and it has produced a considerable number of books within a short period of time. The goal of this part of the book is to give readers a glimpse into this rich field. To learn more, readers should consult any of the following books, which I list roughly chronologically: Becker and Weispfenning [3], Cox et al. [12 and 13], Adams and Loustaunau [1], Vasconcelos [51], Kreuzer and Robbiano [31 and 32], Greuel and Pfister [22], and Decker and Lossen [15]. Eisenbud's book [17] also has a chapter on Gröbner bases.

However, readers who are not interested in computational matters can skip Chapters 9–11, which make up the third part of the book, almost entirely. In fact, only a small part of Chapter 11 needs to be incorporated in a modified way. How this can be done is the topic of Exercise 12.1.

In this chapter we introduce the notion of a Gröbner basis and present Buchberger's algorithm, which computes Gröbner bases. In commutative algebra, Gröbner bases and Buchberger's algorithm play a role similar to that of Gaussian elimination in linear algebra. In fact, virtually all computations in commutative algebra come down to the computation of one or several Gröbner bases, so Buchberger's algorithm is the common engine that they all have under the hood. For example, even for determining whether an ideal $I \subseteq K[x_1, \dots, x_n]$ in a polynomial ring over a field is proper, one normally uses a Gröbner basis. Several applications of Gröbner bases will be discussed in this and the following chapters. In this chapter, we will see how Gröbner bases can be used for testing membership in ideals, for computing the dimension of an affine algebra, for computing kernels of homomorphisms of affine algebras, for solving systems of polynomial equations, for computing intersections of ideals, and for making Noether normalization constructive. More applications will be discussed in Chapters 10 and 11.

G. Kemper, *A Course in Commutative Algebra*, Graduate Texts in Mathematics 256, DOI 10.1007/978-3-642-03545-6_10,

Of course this book presents only a small selection of the huge range of applications that Gröbner bases have. The most notable omissions are:

- The computation of radical ideals. This is surprisingly difficult, and methods can be found in Becker and Weispfenning [3, Chapter 9], Matsumoto [36], and Kemper [30].
- The computation of irreducible components of an affine variety (and, more generally, primary decomposition). This, too, is rather cumbersome, in part because it involves factorization of polynomials. See Becker and Weispfenning [3, Chapter 9].
- The computation of normalization and integral closure. There exist very nice algorithms for this, which can be found in de Jong [29], Vasconcelos [51, Chapter 6] and Derksen and Kemper [16, Section 1.3].
- The computation of syzygies, i.e., kernels of homomorphisms of free modules. This is the starting point of homological computations in commutative algebra. Algorithms for syzygy computation can be found in many sources, for example Eisenbud [17, Chapter 15.5].

9.1 Buchberger's Algorithm

In this chapter, K will always be a field and $K[x_1, \ldots, x_n]$ will be the polynomial ring in n indeterminates. A polynomial of the form

$$t = x_1^{e_1} \cdots x_n^{e_n} \quad (e_i \in \mathbb{N}_0)$$

will be called a **monomial**. A polynomial of the form $c \cdot t$, with $c \in K \setminus \{0\}$ and t a monomial, will be called a **term**. For $f \in K[x_1, \ldots, x_n]$ a polynomial, $T(f)$ denotes the set of all terms in f, so $f = \sum_{ct \in T(f)} ct$. Moreover, $\operatorname{Mon}(f)$ denotes the set of all monomials in f. In particular, $f = 0$ if and only if $\operatorname{Mon}(f) = \emptyset$. The reader should be advised that in part of the literature (e.g., Becker and Weispfenning [3]), the meanings of the words "monomial" and "term" are reversed; but recently the trend has gone towards using the same convention as we do in this book.

When dealing with univariate polynomials, we can compare monomials, which leads to such notions as degree, leading coefficient, and division with remainder. With multivariate polynomials, we have no canonical way of comparing monomials. As ever so often, mathematicians deal with this problem by making a definition.

Definition 9.1. *Let M be the set of all monomials in $K[x_1, \ldots, x_n]$.*

(a) A **monomial ordering** *on $K[x_1, \ldots, x_n]$ is an ordering "$\leq$" on M with the following properties:*

(1) "$\le$" is a total ordering, i.e., for if $s, t \in M$, then $s \le t$ or $t \le s$;
(2) if $t \in M$, then $1 \le t$;
(3) if $s, t_1, t_2 \in M$ with $t_1 \le t_2$, then $st_1 \le st_2$.

Observe that this implies that if $t_1 \in M$ divides $t_2 \in M$, it follows that $t_1 \le t_2$; so a monomial ordering refines the partial ordering given by divisibility. (Exercise 9.1 explores whether the converse is true.)

(b) *Assume that "$\le$" is a monomial ordering. If $f \in K[x_1, \dots, x_n]$is a nonzero polynomial, we write* $\mathrm{LM}(f)$ *for the greatest element of* $\mathrm{Mon}(f)$. *Moreover, we write* $\mathrm{LC}(f) \in K$ *for the coefficient of* $\mathrm{LM}(f)$ *in* f, *and* $\mathrm{LT}(f) := \mathrm{LC}(f) \cdot \mathrm{LM}(f)$.
$\mathrm{LM}(f)$ *is called the* **leading monomial**, $\mathrm{LT}(f)$ *the* **leading term**, *and* $\mathrm{LC}(f)$ *the* **leading coefficient** *of* f. *For* $f = 0$, *we set* $\mathrm{LM}(f) = \mathrm{LT}(f) = \mathrm{LC}(f) := 0$, *and we extend "$\le$" to $M \cup \{0\}$ by the convention* $0 < 1$.

It follows directly from the definition that if $f, g \in K[x_1, \dots, x_n]$ are two polynomials, then

$$\mathrm{LT}(f \cdot g) = \mathrm{LT}(f) \cdot \mathrm{LT}(g) \tag{9.1}$$

and

$$\mathrm{LM}(f + g) \le \max\{\mathrm{LM}(f), \mathrm{LM}(g)\}. \tag{9.2}$$

There are many different monomial orderings on $K[x_1, \dots, x_n]$ (provided that $n > 1$). As we will see soon, different monomial orderings often serve different purposes.

Example 9.2. We give some examples of monomial orderings. Let $t = x_1^{e_1} \cdots x_n^{e_n}$ and $t' = x_1^{e_1'} \cdots x_n^{e_n'}$ be monomials.

(1) The **lexicographic ordering** is given by saying $t \le t'$ if $t = t'$ or $e_i < e_i'$ for the smallest index i with $e_i \ne e_i'$. As we will see on page 131, the lexicographic ordering is useful for solving systems of polynomial equations. It is surely the most famous monomial ordering.
(2) A more complicated monomial ordering is the **graded reverse lexicographic ordering** (often nicknamed **grevlex**). It is given by saying that $t \le t'$ if $t = t'$ or $\deg(t) := \sum_{i=1}^n e_i < \deg(t') := \sum_{i=1}^n e_i'$, or $\deg(t) = \deg(t')$ and $e_i > e_i'$ for the *largest* index i with $e_i \ne e_i'$. For example, $x_1 x_3 < x_2^2$; with the lexicographic ordering, we would have the reverse inequality. As we will see in Chapter 11, the graded reverse lexicographic ordering is useful for computing Hilbert series. According to a vast amount of practical experience, it is also the ordering with which computations tend to be fastest, a phenomenon that is still not completely understood.
(3) Assume we are given two monomial orderings "$\le_1$" and "$\le_2$" on $K[x_1, \dots, x_k]$ and on $K[x_{k+1}, \dots, x_n]$, respectively. Then the **block ordering** (sometimes also called product ordering) is defined as an ordering on $K[x_1, \dots, x_n]$ by saying that $t \le t'$ if $x_1^{e_1} \cdots x_k^{e_k} <_1 x_1^{e_1'} \cdots x_k^{e_k'}$, or

$x_1^{e_1} \cdots x_k^{e_k} = x_1^{e'_1} \cdots x_k^{e'_k}$ and $x_{k+1}^{e_{k+1}} \cdots x_n^{e_n} \le_2 x_{k+1}^{e'_{k+1}} \cdots x_n^{e'_n}$. To be more precise, we speak of the block ordering with "$\le_1$" *dominating*, and it is clear how to define the block ordering with "$\le_2$" dominating. We will see in Section 9.2 that block orderings are useful for computing elimination ideals. ◁

For the rest of Section 9.1, we will fix a monomial ordering "$\le$" on $K[x_1, \ldots, x_n]$.

Recall that a set M with an ordering is called *well-ordered* if every nonempty subset $N \subseteq M$ has a *least element* $y \in N$, meaning that $y \le x$ for all $x \in N$. The most prominent example of a well-ordered set is the set $\mathbb{N}$ of natural numbers.

Lemma 9.3 (Monomial orderings are well-orderings). *The set M of all monomials in $K[x_1, \ldots, x_n]$ is well-ordered by the monomial ordering "$\le$". In particular, M satisfies the descending chain condition.*

Proof. Let $N \subseteq M$ be a nonempty subset. By Hilbert's basis theorem (Corollary 2.13), there exist $t_1, \ldots, t_m \in N$ generating the ideal $(N)_{K[x_1,\ldots,x_n]}$. Since "$\le$" is a total ordering, there exists i with $t_i \le t_j$ for $1 \le j \le m$. Let $t \in N$. Then $t = f_1 t_1 + \cdots + f_m t_m$ with $f_i \in K[x_1, \ldots, x_n]$, so t occurs as a monomial in at least one of the $f_j t_j$. It follows that t is a multiple of t_j, so $t \ge t_j \ge t_i$. Therefore t_i is the desired least element of N. □

Definition 9.4.

(a) Let $S \subseteq K[x_1, \ldots, x_n]$ be a set of polynomials. The ideal

$$L(S) := (\mathrm{LM}(f) \mid f \in S)_{K[x_1,\ldots,x_n]}$$

is called the **leading ideal** *of S.*

(b) Let $I \subseteq K[x_1, \ldots, x_n]$ be an ideal. A finite subset $G \subseteq I$ is called a **Gröbner basis** *(with respect to the chosen monomial ordering "$\le$") of I if*

$$L(I) = L(G).$$

This condition can be expressed more explicitly by saying that for each nonzero $f \in I$ there exists $g \in G$ such that $\mathrm{LM}(g)$ *divides* $\mathrm{LM}(f)$.

We will see in Corollary 9.10 that every Gröbner basis of I generates I as an ideal. (This also follows by an easy argument using Lemma 9.3; we postpone the proof to save space.) Observe that it follows from the Noether property of $K[x_1, \ldots, x_n]$ that every ideal has a Gröbner basis. It is Buchberger's algorithm that makes this existence statement constructive.

Example 9.5. $K[x_1, \ldots, x_n]$ (as an ideal in itself) has the Gröbner basis $G = \{1\}$. However, the generating set $S := \{x_1, x_1 + 1\}$ is not a Gröbner basis, since $L(S) = (x_1)$. ◁

The above example seems to suggest that Gröbner bases tend to be nice and small. Unfortunately, this is not the case: They tend to be large and ugly! In general, it depends on the choice of the monomial ordering whether a given generating subset $G \subseteq I$ is a Gröbner basis or not.

Definition 9.6. *Let $S = \{g_1, \dots, g_r\} \subset K[x_1, \dots, x_n]$ be a finite set of polynomials, and $f \in K[x_1, \dots, x_n]$.*

(a) We say that f is in **normal form** *with respect to S if no $t \in \mathrm{Mon}(f)$ is divisible by the leading monomial $\mathrm{LM}(g_i)$ of any $g_i \in S$.*
(b) A polynomial $f^* \in K[x_1, \dots, x_n]$ is said to be a **normal form** *of f with respect to S if the following conditions hold:*

(1) f^* is in normal form with respect to S.
(2) There exist $h_1, \dots, h_r \in K[x_1, \dots, x_n]$ with

$$f - f^* = \sum_{i=1}^{r} h_i g_i \quad \text{and} \quad \mathrm{LM}(h_i g_i) \leq \mathrm{LM}(f) \quad \text{for all } i \tag{9.3}$$

(in particular, f and f^* are congruent modulo the ideal generated by S);

Example 9.7. Let $S = \{x_1, x_1 + 1\}$, as in Example 9.5. Then 1 is congruent to 0 modulo (S), but 0 is *not* a normal form of 1. Moreover, $f = x_1$ has two normal forms: 0 and -1. So in general, normal forms are not uniquely determined. ◁

Observe that the set S from the above example is not a Gröbner basis. We will see that normal forms with respect to a Gröbner basis are unique (Theorem 9.9). But first we present an algorithm for computing a normal form, thereby also proving its existence. To actually run the algorithm on a computer, we need to assume that there exists a subfield $K' \subseteq K$ containing the coefficients of all polynomials from the input of the algorithm, so that we can perform the field operations of K' on a computer. This remark applies to all algorithms from this chapter.

Algorithm 9.8 (Normal form).

Input: A finite set $S = \{g_1, \dots, g_r\} \subseteq K[x_1, \dots, x_n]$, and a polynomial $f \in K[x_1, \dots, x_n]$.
Output: A normal form f^* of f with respect to S and, if desired, polynomials $h_1, \dots, h_r \in K[x_1, \dots, x_n]$ satisfying (9.3).

(1) Set $f^* := f$ and $h_i := 0$ for all $i \in \{1, \dots, r\}$.
(2) Repeat steps 3–6.
(3) Set

$$\mathcal{M} := \{(t, i) \mid t \in \mathrm{Mon}(f^*),\ i \in \{1, \dots, r\} \text{ such that } \mathrm{LM}(g_i) \text{ divides } t\}.$$

(4) If $\mathcal{M} = \emptyset$, terminate and return f^* and, if desired, the h_i.

(5) Choose $(t, i) \in \mathcal{M}$ with t maximal, and let $c \in K$ be the coefficient of t in f^*.

(6) Set
$$f^* := f^* - \frac{ct}{\mathrm{LT}(g_i)} g_i \quad \text{and} \quad h_i := h_i + \frac{ct}{\mathrm{LT}(g_i)}.$$

In Step 6, the term ct is deleted from f^*, and only monomials that are smaller than t may be added to f^*. It follows that the monomials t from each pass through the loop form a strictly descending sequence, so Lemma 9.3 guarantees that Algorithm 9.8 terminates after finitely many steps. It is clear that (9.3) is satisfied during the entire run, and that f^* is in normal form when the algorithm terminates.

Theorem 9.9 (The normal form map). *Let G be a Gröbner basis of an ideal $I \subseteq K[x_1, \ldots, x_n]$.*

(a) Every $f \in K[x_1, \ldots, x_n]$ has precisely one normal form with respect to G. So there is a map $\mathrm{NF}_G : K[x_1, \ldots, x_n] \to K[x_1, \ldots, x_n]$ assigning to each polynomial its normal form with respect to G.

(b) The map NF_G is K-linear, and $\ker(\mathrm{NF}_G) = I$.

(c) If $\widetilde{G}$ is another Gröbner basis of I (but with respect to the same monomial ordering), then $\mathrm{NF}_{\widetilde{G}} = \mathrm{NF}_G$. So the normal form map NF_G depends only on I and the chosen monomial ordering.

Proof. We prove (a) and (c) together. To this end, let f^* and $\widetilde{f}$ be normal forms of f with respect to G and $\widetilde{G}$, respectively. It follows from (9.3) that $f^* - \widetilde{f} \in I$, so
$$\mathrm{LM}\left(f^* - \widetilde{f}\right) \in L(I) = L(G) = L(\widetilde{G}).$$

Assume $f^* \neq \widetilde{f}$. Then there exist $g \in G$ and $\widetilde{g} \in \widetilde{G}$ such that $\mathrm{LM}(g)$ and $\mathrm{LM}(\widetilde{g})$ divide $\mathrm{LM}(f^* - \widetilde{f})$. But $\mathrm{LM}(f^* - \widetilde{f})$ lies in $\mathrm{Mon}(f^*)$ or in $\mathrm{Mon}(\widetilde{f})$, contradicting the first part of Definition 9.6(b). So $f^* = \widetilde{f}$, and (a) and (c) follow.

The following argument for linearity of the normal form map was shown to me by Martin Kohls. Let $f, g \in K[x_1, \ldots, x_n]$ and $c \in K$. Then $h := \mathrm{NF}_G(f + cg) - \mathrm{NF}_G(f) - c\,\mathrm{NF}_G(g)$ is congruent to $f + cg - f - cg = 0$ modulo (G), so $h \in I$. If $h \neq 0$, $\mathrm{LM}(h)$ would be divisible by $\mathrm{LM}(g)$ for some $g \in G$, contradicting the fact that h is in normal form with respect to G. So $h = 0$, and the linearity follows.

Finally, for $f \in \ker(\mathrm{NF}_G)$ we have $f = f - \mathrm{NF}_G(f) \in I$. Conversely, if $f \in I$, then also $f^* := \mathrm{NF}_G(f) \in I$. If $f^* \neq 0$, there would exist $g \in G$ such that $\mathrm{LM}(g)$ divides $\mathrm{LM}(f^*)$, contradicting the first part of Definition 9.6(b). □

Theorem 9.9 tells us that if we have a Gröbner basis G of an ideal $I \subseteq K[x_1, \ldots, x_n]$, then we also have a membership test: $f \in I$ if and only if $\mathrm{NF}_G(f) = 0$. In the special case $f = 1$, we obtain the equivalence

$$1 \in I \iff G \text{ contains a nonzero constant polynomial.} \tag{9.4}$$

So we have a test for the properness of an ideal. By Hilbert's Nullstellensatz (Corollary 1.8), this yields a method to test an affine variety over an algebraically closed field for emptiness.

Apart from providing a membership test, the normal form map NF_G induces an embedding $A := K[x_1, \ldots, x_n]/I \hookrightarrow K[x_1, \ldots, x_n]$, so the Gröbner basis provides a way to make explicit computations in the affine algebra A. This is one of the most important applications of Gröbner bases.

Corollary 9.10 (Gröbner bases are ideal bases). *Let G be a Gröbner basis of an ideal $I \subseteq K[x_1, \ldots, x_n]$. Then $I = (G)_{K[x_1,\ldots,x_n]}$.*

Proof. By definition, $G \subseteq I$, so $(G) \subseteq I$. Conversely, for $f \in I$ we have $\mathrm{NF}_G(f) = 0$ by Theorem 9.9(b), so $f \in (G)$ by (9.3). □

Remark 9.11 (Gröbner bases over rings). Part of what we have done so far in this section carries over to the case that K is an arbitrary ring, not a field. First of all, Definition 9.1 really has nothing to do with the properties of K. So Lemma 9.3 carries over to polynomial rings over arbitrary rings as well. We can and will also use Definitions 9.4 and 9.6 in the more general situation. However, Algorithm 9.8 needs to be modified by replacing Step 6 with

(6') Set

$$f^* := \mathrm{LC}(g_i) \cdot f^* - \frac{ct}{\mathrm{LM}(g_i)} \cdot g_i \quad \text{and} \quad h_i := \mathrm{LC}(g_i) \cdot h_i + \frac{ct}{\mathrm{LM}(g_i)}.$$

With this modification, Algorithm 9.8 computes a normal form not of f, but of some $u \cdot f$ with $u \in K$ a product formed from leading coefficients of polynomials from S. Moreover, Theorem 9.9 and Corollary 9.10 do *not* carry over to the case that K is a ring. So the term Gröbner *basis* is a bit misleading in this case. Gröbner bases over rings will be used in this book only in Proposition 9.18 and Lemma 10.1. In what follows, we go back to assuming that K is a field. ◁

We will shortly present Buchberger's algorithm for computing Gröbner bases. This is based on Buchberger's criterion, which we prove first. To formulate it, we need the following construction: For $f, g \in K[x_1, \ldots, x_n]$ two nonzero polynomials, let t be the gcd of $\mathrm{LM}(f)$ and $\mathrm{LM}(g)$ (so t is a monomial). Then

$$\mathrm{spol}(f, g) := \frac{\mathrm{LT}(g)}{t} \cdot f - \frac{\mathrm{LT}(f)}{t} \cdot g$$

is called the **s-polynomial** of f and g. Observe that the s-polynomial is formed in such a way that the leading terms of the two summands cancel. For example, if $f = x_1^2 + x_2^2$ and $g = x_1 x_2$, then (assuming $x_1 > x_2$)

$$\mathrm{spol}(f, g) = x_2 \cdot f - x_1 \cdot g = x_2^3.$$

The following theorem gives a test for Gröbner bases that can be performed in finitely many steps. This is the centerpiece of Buchberger's algorithm (Algorithm 9.13).

Theorem 9.12 (Buchberger's criterion). *Let $G \subseteq K[x_1, \ldots, x_n]$ be a finite set of nonzero polynomials. Then the following statements are equivalent:*

(a) G is a Gröbner basis of the ideal $I \subseteq K[x_1, \ldots, x_n]$ generated by G.
(b) For all $g, h \in G$, 0 is a normal form of $\operatorname{spol}(g, h)$ *with respect to G.*

Proof. Clearly all s-polynomials of elements from G lie in I, so if G is a Gröbner basis, the s-polynomials have normal form 0 by Theorem 9.9(b). So (a) implies (b).

To prove the converse, assume that (b) holds but G is not a Gröbner basis. Then there exists $f \in I$ with $\operatorname{LM}(f) \notin L(G)$. Writing $G = \{g_1, \ldots, g_r\}$, we have $h_i \in K[x_1, \ldots, x_n]$ with

$$f = \sum_{i=1}^{r} h_i g_i. \tag{9.5}$$

By Lemma 9.3 we may choose the h_i in such a way that

$$t := \max\{\operatorname{LM}(h_i g_i) \mid i = 1, \ldots, r\}$$

becomes minimal. Because of (9.5) there exists i with $\operatorname{LM}(f) \in \operatorname{Mon}(h_i g_i)$, and since $\operatorname{LM}(f) \notin L(G)$, this implies $\operatorname{LM}(h_i g_i) > \operatorname{LM}(f)$. So $t > \operatorname{LM}(f)$. Therefore the coefficient of t in the right-hand side of (9.5) is zero, so with

$$c_i := \begin{cases} \operatorname{LC}(h_i) & \text{if } \operatorname{LM}(h_i g_i) = t, \\ 0 & \text{otherwise,} \end{cases}$$

we have

$$\sum_{i=1}^{r} c_i \operatorname{LC}(g_i) = 0. \tag{9.6}$$

By reordering the g_i, we may assume $c_1 \neq 0$.

Let $i \in \{2, \ldots, r\}$ with $c_i \neq 0$. Then $\operatorname{LM}(g_i)$ divides t. If t_i is the least common multiple of $\operatorname{LM}(g_i)$ and $\operatorname{LM}(g_1)$, then also t_i divides t. The definition of the s-polynomial gives

$$\operatorname{spol}(g_i, g_1) = \frac{\operatorname{LC}(g_1) t_i}{\operatorname{LM}(g_i)} \cdot g_i - \frac{\operatorname{LC}(g_i) t_i}{\operatorname{LM}(g_1)} \cdot g_1,$$

and we have $\operatorname{LM}(\operatorname{spol}(g_i, g_1)) < t_i$. By the hypothesis (b), there exist $h_{i,j} \in K[x_1, \ldots, x_n]$ with

$$\operatorname{spol}(g_i, g_1) = \sum_{j=1}^{r} h_{i,j} g_j \quad \text{and} \quad \operatorname{LM}(h_{i,j} g_j) \le \operatorname{LM}\left(\operatorname{spol}(g_i, g_1)\right) < t_i$$

for all j. Set $s_i := t/t_i \cdot \operatorname{spol}(g_i, g_1)$. Since

$$\operatorname{LM}(h_i)\operatorname{LM}(g_i) = t = \operatorname{LM}(h_1)\operatorname{LM}(g_1),$$

we get

$$s_i = \operatorname{LC}(g_1)\operatorname{LM}(h_i) g_i - \operatorname{LC}(g_i)\operatorname{LM}(h_1) g_1, \tag{9.7}$$

and on the other hand,

$$s_i = \sum_{j=1}^{r} \frac{t}{t_i} h_{i,j} g_j \quad \text{with} \quad \operatorname{LM}\left(\frac{t}{t_i} h_{i,j} g_j\right) < t \quad \text{for all } j. \tag{9.8}$$

Now we set $g := \sum_{i=1}^{r} c_i \operatorname{LM}(h_i) g_i$ and write $\operatorname{LC}(g_1) \cdot g$ as

$$\begin{aligned} \operatorname{LC}(g_1) \cdot g = & \sum_{i=2}^{r} c_i \left(\operatorname{LC}(g_1)\operatorname{LM}(h_i) g_i - \operatorname{LC}(g_i)\operatorname{LM}(h_1) g_1\right) \\ & + \left(\sum_{i=2}^{r} c_i \operatorname{LC}(g_i) + c_1 \operatorname{LC}(g_1)\right) \operatorname{LM}(h_1) g_1. \end{aligned}$$

With (9.7), (9.6), and (9.8), this yields

$$g = \sum_{i=2}^{r} \frac{c_i}{\operatorname{LC}(g_1)} \cdot s_i = \sum_{j=1}^{r} \widetilde{h}_j g_j \quad \text{with} \quad \operatorname{LM}\left(\widetilde{h}_j g_j\right) < t \quad \text{for all } j.$$

By (9.5), we have

$$f = (f - g) + g = \sum_{j=1}^{r} \left(h_j - c_j \operatorname{LM}(h_j) + \widetilde{h}_j\right) g_j,$$

and it follows from the definition of t and c_j that $\operatorname{LM}\left(\left(h_j - c_j \operatorname{LM}(h_j)\right) g_j\right) < t$ for all j, so

$$\operatorname{LM}\left(\left(h_j - c_j \operatorname{LM}(h_j) + \widetilde{h}_j\right) g_j\right) < t,$$

contradicting the minimality of t. This contradiction shows that G is a Gröbner basis. □

Having proved Buchberger's criterion, we are ready to present an algorithm for computing Gröbner bases.

Algorithm 9.13 (Buchberger's algorithm).

Input: A finite set $S \subseteq K[x_1, \ldots, x_n]$ of polynomials.
Output: A Gröbner basis G (with respect to the chosen monomial ordering "$\leq$") of the ideal $I \subseteq K[x_1, \ldots, x_n]$ generated by S.

(1) Set $G := S \setminus \{0\}$.
(2) For all $g, h \in G$, perform steps 3–4.
(3) Compute the s-polynomial $s := \text{spol}(g, h)$ and a normal form s^* of s with respect to G.
(4) If $s^* \neq 0$, set $G := G \cup \{s^*\}$ and go to step 2.
(5) *(This step is reached only if no nonzero s^* occurred in the previous loop.)* Terminate the computation and return G.

Each time that a new polynomial s^* is added into G in Algorithm 9.13, the ideal $L(G)$ increases strictly. Therefore the termination of the algorithm is guaranteed by Hilbert's basis theorem (Corollary 2.13). Since clearly all s^* lie in I, the correctness of the algorithm follows from Buchberger's criterion (Theorem 9.12).

Algorithm 9.13 has numerous optimizations. Some of them are rather trivial (such as not considering pairs $(g, h) \in G \times G$ where (g, h) or (h, g) have been considered before) and would occur to any reasonable programmer, but many others are much more subtle and artful. A good implementation has criteria for discarding superfluous pairs $(g, h) \in G \times G$ (see Exercise 9.3 for such a criterion), a good strategy for first choosing pairs with a high potential for pushing up $L(G)$ fast, and good heuristics for choosing polynomials g_i for the reduction step (step 6 in Algorithm 9.8). Apart from this, there is almost no limit to the creativity of a good programmer for finding variants of the algorithm and implementation tricks for speeding up the performance.

Algorithm 9.13 has a number of variants. One of them, sometimes called the *extended Buchberger algorithm*, keeps track of how the new elements s^* of the Gröbner basis arise as $K[x_1, \ldots, x_n]$-linear combinations of the original generators. This information is useful for some purposes, e.g., the computation of syzygies, which we will not treat in this book. A further variant computes a *reduced Gröbner basis* G, which by definition has the additional property that every $g \in G$ has leading coefficient 1 and is in normal form with respect to $G \setminus \{g\}$. Reduced Gröbner bases are uniquely determined by the ideal I they generate and, of course, by the choice of the monomial ordering (see Exercise 9.4). A third variant of Buchberger's algorithm computes Gröbner bases of submodules of a free module $K[x_1, \ldots, x_n]^m$ over the polynomial ring. Of course, this requires extending the definitions of a monomial ordering and a Gröbner basis. Gröbner bases of submodules in $K[x_1, \ldots, x_n]^m$ are useful for computations in homological algebra, particularly free resolutions.

Buchberger's algorithm is the workhorse of computational commutative algebra. It is therefore part of every computer algebra system that specializes in commutative algebra, such as CoCoA [9], MACAULAY (2) [21],

MAGMA [5], or SINGULAR [23]. The competition between these systems is strong, and it is rare for a conference on computational commutative algebra to pass without at least one talk that reports that one system outperforms all others.

The cost, in terms of running time and memory requirement, of Buchberger's algorithm tends to be extremely high. Its complexity is usually described as "doubly exponential," although the true story is a bit more complicated. More details on the complexity can be found in the book by von zur Gathen and Gerhard [19, Section 21.7]. However, practical experience shows that the algorithm often does terminate after a reasonable time. Whether a particular Gröbner basis computation is feasible is usually hard to predict in advance, so Gröbner basis computations are still an adventurous business.

9.2 First Application: Elimination Ideals

As in the previous section, $K[x_1, \ldots, x_n]$ will denote a polynomial ring over a field throughout this section. Let $I \subseteq K[x_1, \ldots, x_n]$ be an ideal. In the previous section we have already seen that having a Gröbner basis of I yields a membership test for I and a way to make explicit computations in the affine algebra $K[x_1, \ldots, x_n]/I$ (see before Corollary 9.10). In particular, if we have another ideal $J \subseteq K[x_1, \ldots, x_n]$ given by a finite set of generators, we can also test whether J is contained in I. These applications work independently of the chosen monomial ordering.

In this section we will study some further "immediate" applications. These are all linked to the computation of elimination ideals, which we define now.

Definition 9.14. *Let $S = \{x_{i_1}, \ldots, x_{i_k}\}$ be a set of indeterminates.*

*(a) For an ideal $I \subseteq K[x_1, \ldots, x_n]$, the S-**elimination ideal** of I is defined to be the intersection*

$$I_S := K[x_{i_1}, \ldots, x_{i_k}] \cap I$$

(where $I_\emptyset$ is understood to be the set of constants lying in I).

*(b) A monomial ordering "$\leq$" on $K[x_1, \ldots, x_n]$ is called an S-**elimination ordering** if*

$$t < x_j \quad \text{for all monomials } t \in K[x_{i_1}, \ldots, x_{i_k}] \text{ and all } x_j \in \overline{S},$$

where $\overline{S} := \{x_1, \ldots, x_n\} \setminus S$.

Elimination orderings exist, as the following example shows.

Example 9.15. (1) Let "$\leq$" be an arbitrary monomial ordering on $K[x_1, \ldots, x_n]$ and let S be a set of indeterminates with complement $\{x_1, \ldots, x_n\} \setminus S = \{x_{j_1}, \ldots, x_{j_r}\}$. We define a new monomial ordering "$\preceq$" by saying $t = x_1^{e_1} \cdots x_n^{e_n} \preceq t' = x_1^{e'_1} \cdots x_n^{e'_n}$ if $e_{j_1} + \cdots + e_{j_r} < e'_{j_1} + \cdots + e'_{j_r}$

or if $e_{j_1} + \cdots + e_{j_r} = e'_{j_1} + \cdots + e'_{j_r}$ and $t \leq t'$. It is easy to check that "$\preceq$" is an S-elimination ordering.

(2) The block ordering from Example 9.2(3) is an $\{x_{k+1}, \ldots, x_n\}$-elimination ordering.

(3) In particular, the lexicographic ordering is an $\{x_{k+1}, \ldots, x_n\}$-elimination ordering for every k.

(4) On the other hand, the grevlex ordering is an S-elimination ordering only for $S = \emptyset$ and $S = \{1, \ldots, n\}$, and in fact every monomial ordering is an elimination ordering for these two extremes. ◁

The following theorem tells us how elimination ideals can be computed with Gröbner bases. Notice that restricting a monomial ordering "$\leq$" on $K[x_1, \ldots, x_n]$ to the set of all monomials in $K[x_{i_1}, \ldots, x_{i_k}]$ gives a monomial ordering on $K[x_{i_1}, \ldots, x_{i_k}]$, which we call the **restricted monomial ordering**.

Theorem 9.16 (Computing elimination ideals). *Let $I \subseteq K[x_1, \ldots, x_n]$ be an ideal and $S = \{x_{i_1}, \ldots, x_{i_k}\}$ a set of indeterminates. Let G be a Gröbner basis of I with respect to an S-elimination ordering "$\leq$". Then*

$$G_S := K[x_{i_1}, \ldots, x_{i_k}] \cap G$$

is a Gröbner basis of the S-elimination ideal I_S with respect to the restricted monomial ordering.

Proof. Clearly $G_S \subseteq I_S$. To prove that $L(I_S) = L(G_S)$, let $f \in I_S$ be nonzero. Since $\mathrm{LM}(f) \in L(I)$, there exists $g \in G$ such that $\mathrm{LM}(g)$ divides $\mathrm{LM}(f)$. This implies that $\mathrm{LM}(g)$ lies in $K[x_{i_1}, \ldots, x_{i_k}]$. But then every monomial $t \in \mathrm{Mon}(g)$ lies in $K[x_{i_1}, \ldots, x_{i_k}]$, too, since otherwise t would be greater than $\mathrm{LM}(g)$ by the hypothesis on "$\leq$". So $g \in G_S$, and the proof is finished. □

So all we need to do for getting the elimination ideal is to compute a Gröbner basis G with respect to an elimination ordering, and pick out those polynomials from G that involve only the indeterminates from S.

Elimination ideals can be used to compute the dimension of an affine variety given as $A = K[x_1, \ldots, x_n]/I$, with I an ideal. Indeed, Theorem 5.9 and Proposition 5.10 tell us that $\dim(A)$ is the maximal size of a set $S = \{x_{i_1}, \ldots, x_{i_k}\}$ of indeterminates such that $\{x_{i_1} + I, \ldots, x_{i_k} + I\} \subseteq A$ is algebraically independent. But this condition is equivalent to $I_S = \{0\}$, so it can be checked by means of Gröbner bases. This gives the desired method for computing $\dim(A)$. Unfortunately, even after some optimizations this method requires the computation of a considerable number of Gröbner bases of I with respect to different monomial orderings. A test for dimension zero, which is much cheaper than this method, is discussed in Exercise 9.7. A better way to calculate the dimension of an affine algebra is discussed in Section 11.2 (see Corollary 11.14 on page 159).

A further application of elimination ideals is the computation of kernels of homomorphisms between affine K-algebras. Let $A = K[y_1, \ldots, y_m]/I$ and $B = K[x_1, \ldots, x_n]/J$ be two affine K-algebras (with I and J ideals in polynomial rings), and let $\varphi\colon B \to A$ be a homomorphism of K-algebras. Composing φ with the canonical map $K[x_1, \ldots, x_n] \to B$ yields a homomorphism $\psi\colon K[x_1, \ldots, x_n] \to A$, and $\ker(\varphi) = \ker(\psi)/J$. So it is enough to compute $\ker(\psi)$. In other words, for computing the kernel of a homomorphism between affine K-algebras, we may assume that the first algebra is a polynomial ring. The following proposition tells us that in this situation the kernel can be calculated as an elimination ideal.

Proposition 9.17 (Kernel of a homomorphism of affine algebras). *Let*

$$\varphi\colon K[x_1, \ldots, x_n] \to A := K[y_1, \ldots, y_m]/I$$

be a homomorphism of K-algebras, given by $\varphi(x_i) = g_i + I$ with $g_i \in K[y_1, \ldots, y_m]$. Consider the ideal

$$J := \Big(I \cup \{g_1 - x_1, \ldots, g_n - x_n\}\Big)_{K[x_1, \ldots, x_n, y_1, \ldots, y_m]}.$$

Then

$$\ker(\varphi) = K[x_1, \ldots, x_n] \cap J.$$

Proof. It follows from the definition of J that for every $f \in K[x_1, \ldots, x_n]$ we have

$$f(g_1, \ldots, g_n) - f \in J. \tag{9.9}$$

Assume $f \in \ker(\varphi)$. Then $f(g_1, \ldots, g_n) \in I$, which with (9.9) yields $f \in J$, so $f \in K[x_1, \ldots, x_n] \cap J$.

Conversely, if $f \in K[x_1, \ldots, x_n] \cap J$, then (9.9) yields $f(g_1, \ldots, g_n) \in J$, so

$$f(g_1, \ldots, g_n) = \sum_{i=1}^{r} h_i f_i + \sum_{j=1}^{n} p_j (g_j - x_j)$$

with $h_i, p_j \in K[x_1, \ldots, x_n, y_1, \ldots, y_m]$ and $f_i \in I$. Setting $x_i = g_i$ on both sides of the above equation yields $f(g_1, \ldots, g_n) \in I$, so $f \in \ker(\varphi)$. □

If we put together Proposition 9.17 and Theorem 9.16, we obtain the following algorithm: With the notation from Proposition 9.17, form the ideal $J \subseteq K[x_1, \ldots, x_n, y_1, \ldots, y_m]$ and choose an $\{x_1, \ldots, x_n\}$-elimination ordering "$\le$" on $K[x_1, \ldots, x_n, y_1, \ldots, y_m]$. Compute a Gröbner basis G of J with respect to "$\le$", and set $G_x := K[x_1, \ldots, x_n] \cap G$. Then G_x is a Gröbner basis of the kernel of φ. It may seem odd that only one part of the Gröbner basis G is used, and the rest is "thrown away," and one might wonder whether the other part of G has any significance in the context of the map φ. This is indeed the case, as we will see in Proposition 9.18.

Observe that the kernel in Proposition 9.17 is nothing else but the ideal of relations between the elements $g_i + I \in A$. So we have an algorithm for computing relation ideals of elements of affine algebras. In particular, we can compute relation ideals between polynomials. A nice application is a constructive version of Noether normalization, which is explored in Exercise 9.12.

A homomorphism $\varphi\colon R \to S$ of rings induces a map $\varphi^*\colon \operatorname{Spec}(S) \to \operatorname{Spec}(R)$, $P \mapsto \varphi^{-1}(P)$. We claim that the Zariski closure of the image can be expressed as

$$\overline{\operatorname{im}(\varphi^*)} = \mathcal{V}_{\operatorname{Spec}(R)}\left(\ker(\varphi)\right). \tag{9.10}$$

This gives a geometric interpretation of the kernel. To prove (9.10), observe that $\overline{\operatorname{im}(\varphi^*)} = \mathcal{V}_{\operatorname{Spec}(R)}\left(\bigcap_{P \in \operatorname{Spec}(S)} \varphi^*(P)\right)$, and

$$\bigcap_{P \in \operatorname{Spec}(S)} \varphi^*(P) = \varphi^{-1}\Big(\bigcap_{P \in \operatorname{Spec}(S)} P\Big) = \varphi^{-1}\left(\sqrt{\{0\}}\right) = \sqrt{\ker(\varphi)},$$

where the second equality follows from Corollary 1.12.

If $f\colon X \to Y$ is a morphism of affine varieties over an algebraically closed field, given by a homomorphism $\varphi\colon K[Y] \to K[X]$, then (9.10) becomes

$$\overline{f(X)} = \mathcal{V}_Y\left(\ker(\varphi)\right).$$

So using Proposition 9.17, we can compute image closures of morphisms of affine varieties. At first glance it may seem disappointing that the kernel of φ describes only the image closure and not the image itself. But the image closure is in fact the best we can reasonably expect, since the variety given by the kernel is always closed, but the image of a morphism is, in general, not closed. A typical example of this phenomenon is a hyperbola X with a morphism f given by projecting to the x-axis. This is shown in Fig. 9.1.

In Chapter 10 we will develop an algorithm that computes the image of a morphism, and we will learn more about the nature of images of morphisms.

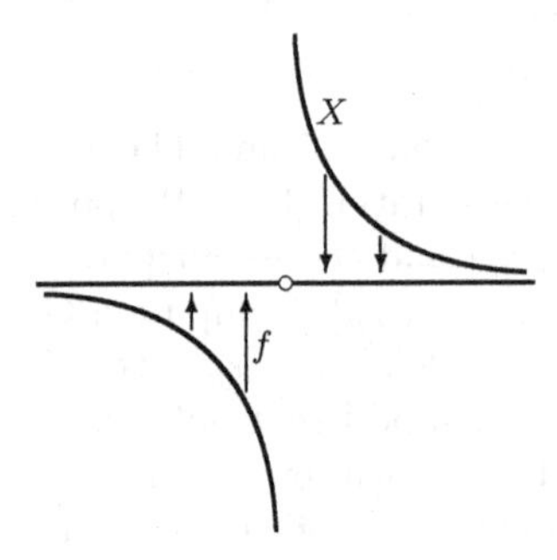

Fig. 9.1. A morphism whose image is not closed

By a very similar (but simpler) argument, we obtain a geometric interpretation of elimination ideals themselves: If $X \subseteq K^n$ is an affine variety over an algebraically closed field, given by an ideal $I \subseteq K[x_1, \ldots, x_n]$, and if $S = \{x_{i_1}, \ldots, x_{i_k}\}$, then the S-elimination ideal describes the closure of the image of X under the projection $\pi_S\colon K^n \to K^{|S|}$, $(\xi_1, \ldots, \xi_n) \mapsto (\xi_{i_1}, \ldots, \xi_{i_k})$:

$$\overline{\pi_S(X)} = \mathcal{V}_{K^{|S|}}(I_S). \tag{9.11}$$

This leads to a further application of elimination ideals: solving systems of polynomial equations. Suppose that $I \subseteq K[x_1, \ldots, x_n]$ is an ideal in a polynomial ring over an algebraically closed field, and suppose we know that the variety $X := \mathcal{V}(I)$ is finite. (This is equivalent to $\dim(K[x_1, \ldots, x_n]/I) \leq 0$, so it can be checked with elimination ideals.) Then in (9.11) the Zariski closure can be omitted, and in particular all $I_{\{x_k, \ldots, x_n\}}$ are nonzero. By Theorem 9.16, all $I_{\{x_k, \ldots, x_n\}}$ can be computed from a single lexicographic Gröbner basis of I. Since $K[x_n]$ is a principal ideal ring, we have $I_{\{x_n\}} = (g)$ with $g \in K[x_n]$ nonzero. Equation (9.11) tells us that the zeros of g are precisely those $\xi_n \in K$ for which there exists at least one point of X having ξ_n as last component. For each such ξ_n, substituting $x_n = \xi_n$ in the generators of $I_{\{x_{n-1}, x_n\}}$ yields some polynomials in $K[x_{n-1}]$, and every common zero ξ_{n-1} of these polynomials yields a pair (ξ_{n-1}, ξ_n) that can be extended to at least one point from X. Continuing in this way, we can work our way down until we reach $I_{\{x_1, \ldots, x_n\}} = I$. Then we have found all points from X. In other words, we have solved the system of polynomial equations given by I. What is required for this method to work in practice is that we be able to compute zeros of polynomials in $K[x]$. But it also works if K is not algebraically closed, provided that we know that $\dim(K[x_1, \ldots, x_n]/I) \leq 0$. For K not algebraically closed, it may happen that a zero ξ_n of g does not extend to a point $(\xi_1, \ldots, \xi_n)$ of X.

This last application probably points to the origin of the term "elimination ideal": It can be used for eliminating unknowns from a system of equations.

The following proposition may be seen as a sequel to Proposition 9.17 and Theorem 9.16. It answers the question about the significance of the part of a Gröbner basis that is "thrown away" in computing an elimination ideal. The proposition is rather technical, but it is crucial for getting a constructive version of the generic freeness lemma in Chapter 10. The proposition will be used only in Chapter 10, so readers who plan to skip that chapter can also skip the rest of this section and go directly to Chapter 11.

Proposition 9.18 (The forgotten part of the Gröbner basis). *Let*

$$\varphi\colon K[x_1, \ldots, x_n] \to A := K[y_1, \ldots, y_m]/I$$

be a homomorphism of K-algebras, given by $\varphi(x_i) = g_i + I$ with $g_i \in K[y_1, \ldots, y_m]$. With $R := \operatorname{im}(\varphi) \subseteq A$, consider the homomorphism

$$\psi\colon R[y_1,\ldots,y_m] \to A,\ y_i \mapsto y_i + I,$$

of R-algebras. Also consider the homomorphism

$$\Phi\colon K[x_1,\ldots,x_n,y_1,\ldots,y_m] \to R[y_1,\ldots,y_m]$$

given by applying φ *coefficientwise. Let "*$\le$*" be an* $\{x_1,\ldots,x_n\}$*-elimination ordering on* $K[x_1,\ldots,x_n,y_1,\ldots,y_m]$*, and let* G *be a Gröbner basis with respect to "*$\le$*" of the ideal*

$$J := \Big(I \cup \{g_1 - x_1,\ldots,g_n - x_n\}\Big)_{K[x_1,\ldots,x_n,y_1,\ldots,y_m]}.$$

If $G_x := K[x_1,\ldots,x_n] \cap G$ *and* $G_y := G \setminus G_x$ *(the "rest" of* G*), then:*

(a) G_x *is a Gröbner basis of* $\ker(\varphi)$ *with respect to the restriction of "*$\le$*" to* $K[x_1,\ldots,x_n]$*.*
(b) $\ker(\psi) = (\Phi(G_y))_{R[y_1,\ldots,y_m]}$*.*
(c) If "$\le$*" is the block ordering of monomial orderings "*$\le_x$*" on* $K[x_1,\ldots,x_n]$ *and "*$\le_y$*" on* $K[y_1,\ldots,y_m]$ *with "*$\le_y$*" dominating, then* $\Phi(G_y)$ *is a Gröbner basis of* $\ker(\psi)$ *with respect to "*$\le_y$*". (See Remark 9.11 for Gröbner bases over a ring.)*

Proof. Part (a) follows from Proposition 9.17 and Theorem 9.16.

To prove (b), take $g \in G_y$. Then $g \in J$, and our definitions imply $\psi(\Phi(g)) = 0$. So $(\Phi(G_y))_{R[y_1,\ldots,y_m]} \subseteq \ker(\psi)$. Conversely, take $f \in \ker(\psi)$. We can write $f = \Phi(F)$ with $F \in K[x_1,\ldots,x_n,y_1,\ldots,y_m]$. Then $F \in J = (G_x \cup G_y)_{K[x_1,\ldots,x_n,y_1,\ldots,y_m]}$. So

$$f \in (\Phi(G_x) \cup \Phi(G_y))_{R[y_1,\ldots,y_m]}.$$

But by (a), every $g \in G_x$ lies in $\ker(\varphi)$, so $\Phi(g) = \varphi(g) = 0$. This completes the proof of (b).

For proving part (c) we need the following lemma. □

Lemma 9.19. *In the situation of Proposition 9.18*(c)*, let* $f \in K[x_1,\ldots,x_n,y_1,\ldots,y_m]$ *be such that there exists no* $g \in G_x$ *with* $\mathrm{LM}(g)$ *dividing* $\mathrm{LM}(f)$*. Write* $\mathrm{LM}_y(f)$ *and* $\mathrm{LC}_y(f)$ *for the leading monomial and leading coefficient of* f *considered as a polynomial in the indeterminates* y_i *and with coefficients in* $K[x_1,\ldots,x_n]$*. Then*

$$\mathrm{LC}\left(\Phi(f)\right) = \varphi\left(\mathrm{LC}_y(f)\right) \tag{9.12}$$

and

$$\mathrm{LM}(f) = \mathrm{LM}_y\left(\Phi(f)\right) \cdot \mathrm{LM}\left(\mathrm{LC}_y(f)\right). \tag{9.13}$$

Proof. We may assume $f \neq 0$. Since "$\leq$" is a block ordering, we obtain

$$\mathrm{LM}(f) = \mathrm{LM}_y(f) \cdot \mathrm{LM}\left(\mathrm{LC}_y(f)\right). \tag{9.14}$$

By way of contradiction, assume that $\mathrm{LM}_y\left(\Phi(f)\right) \neq \mathrm{LM}_y(f)$. Then $\varphi\left(\mathrm{LC}_y(f)\right) = 0$, and by Proposition 9.18(a) this implies the existence of $g \in G_x$ such that $\mathrm{LM}(g)$ divides $\mathrm{LM}\left(\mathrm{LC}_y(f)\right)$. So by (9.14), $\mathrm{LM}(g)$ divides $\mathrm{LM}(f)$, contradicting our hypothesis. We conclude that $\mathrm{LM}_y\left(\Phi(f)\right) = \mathrm{LM}_y(f)$. This implies (9.12) directly, and together with (9.14) it implies (9.13). □

Proof of Proposition 9.18(c). We need to show that for every nonzero $f \in \ker(\psi)$ there exists $g \in G_y$ such that $\mathrm{LM}_y\left(\Phi(g)\right)$ divides $\mathrm{LM}_y(f)$. As shown in the proof of part (b), there exists $F \in J$ with $f = \Phi(F)$. By part (a), we may substitute F by a normal form of F with respect to G_x. Since $F \neq 0$, there exists $g \in G_y$ such that $\mathrm{LM}(g)$ divides $\mathrm{LM}(F)$. Applying (9.13) to F and to g shows that $\mathrm{LM}_y\left(\Phi(g)\right)$ divides $\mathrm{LM}_y\left(\Phi(F)\right) = \mathrm{LM}_y(f)$. This completes the proof. □

The last part of Proposition 9.18 works under the stronger hypothesis that the elimination ordering is a block ordering. This raises the question whether in fact every elimination ordering is a block ordering (formed from orderings on the set of indeterminates that are eliminated and on its complement). After trying in vain to prove this for several days, I posed the question to my students Kathi Binder and Tobias Kamke, who immediately answered it in the negative. Exercise 9.8 deals with this.

In Exercise 9.9 it is shown how the Gröbner basis G from Proposition 9.18 can be used to obtain a membership test for the subalgebra R.

Exercises for Chapter 9

In the following exercises, $K[x_1, \ldots, x_n]$ stands for a polynomial ring over a field. If not stated otherwise, $K[x_1, \ldots, x_n]$ is equipped with a monomial ordering.

9.1 (Refining the ordering by divisibility). We have seen that monomial orderings refine the partial ordering given by divisibility. Conversely, let "$\leq$" be a total ordering on the set M of monomials such that if $t_1 \in M$ divides $t_2 \in M$, it follows that $t_1 \leq t_2$. Does this imply that "$\leq$" is a monomial ordering?

***9.2 (The convex cone of a monomial ordering, weight vectors).** In this exercise we will study a fundamental geometric object, called the convex cone, that belongs to a monomial ordering. If $\mathbf{e} = (e_1, \ldots, e_n)$ and

$\mathbf{f} = (f_1, \ldots, f_n) \in \mathbb{N}_0^n$ are two tuples of nonnegative integers, we write $\mathbf{e} \leq \mathbf{f}$ if $\prod_{i=1}^n x_i^{e_i} \leq \prod_{i=1}^n x_i^{f_i}$ (using a given monomial ordering "$\leq$" on $K[x_1, \ldots, x_n]$). Define the set

$$\mathcal{C} := \{\mathbf{e} - \mathbf{f} \mid \mathbf{e}, \mathbf{f} \in \mathbb{N}_0^n \text{ with } \mathbf{f} < \mathbf{e}\}.$$

$\mathcal{C}$ is called the **convex cone** associated to the monomial ordering "$\leq$". Prove the following.

(a) If $\mathbf{c}_1, \ldots, \mathbf{c}_m \in \mathcal{C}$, and if $\alpha_1, \ldots, \alpha_m \in \mathbb{R}_{>0}$ are positive real numbers such that $\mathbf{c} := \sum_{i=1}^m \alpha_i \mathbf{c}_i \in \mathbb{Z}^n$, then $\mathbf{c} \in \mathcal{C}$.
Hint: Show that $\mathcal{C}$ is closed under addition, and that if $k \cdot \mathbf{c} \in \mathcal{C}$ for $k \in \mathbb{N}_{>0}$ and $\mathbf{c} \in \mathbb{Z}^n$, then $\mathbf{c} \in \mathcal{C}$. Conclude the result for $\alpha_i \in \mathbb{Q}_{>0}$, then for $\alpha_i \in \mathbb{R}_{>0}$.

(b) If $\mathbf{c}_1, \ldots, \mathbf{c}_m \in \mathcal{C}$, there exist positive integers $w_1, \ldots, w_n \in \mathbb{N}_{>0}$ such that $\sum_{j=1}^n w_j c_{i,j} > 0$ for all $i = 1, \ldots, m$.
Hint: Conclude from part (a) that $\mathbf{0} \in \mathbb{R}^n$ does not lie in the convex hull $\mathcal{H} := \left\{\sum_{i=1}^m \alpha_i \mathbf{c}_i \mid \alpha_i \in \mathbb{R}_{\geq 0} \text{ with } \sum_{i=1}^m \alpha_i = 1\right\}$ of the $\mathbf{c}_i$. Then consider a vector $\mathbf{w}' \in \mathcal{H}$ that is closest to 0.

Before formulating the next statement, we need to introduce a new monomial ordering "$\leq_\mathbf{w}$" that depends on a "weight vector" $\mathbf{w} = (w_1, \ldots, w_n) \in \mathbb{N}_{>0}^n$. This is defined by saying that $\mathbf{e} \leq_\mathbf{w} \mathbf{f}$ if $\sum_{j=1}^n w_j e_j < \sum_{j=1}^n w_j f_j$, or $\sum_{j=1}^n w_j e_j = \sum_{j=1}^n w_j f_j$ and $\mathbf{e} \leq \mathbf{f}$.

(c) Let $I \subseteq K[x_1, \ldots, x_n]$ be an ideal, and let G be a Gröbner basis of I with respect to "$\leq$". Then there exists a weight vector $\mathbf{w} \in \mathbb{N}_{>0}^n$ such that G is also a Gröbner basis with respect to $\leq_\mathbf{w}$, and the leading ideals with respect to the two monomial orderings coincide: $L_\leq(I) = L_{\leq_\mathbf{w}}(I)$.
Hint: You can use (b) to make sure that the leading monomials of some polynomials (of your choice) do not change in going from "$\leq$" to "$\leq_\mathbf{w}$".

Remark: It is part (a) that earns $\mathcal{C}$ the name "convex cone." To get a better appreciation of part (c), notice that if "$\leq$" is the lexicographic ordering, there is no weight vector $\mathbf{w}$ with $\leq = \leq_\mathbf{w}$. So $\mathbf{w}$ has to depend on G. The ordering "$\leq_\mathbf{w}$" has the special property that every monomial has only finitely many monomials below it. Part (c) will be used in Exercise 11.7 to generalize Corollary 11.14. *(Solution on page 223)*

9.3 (S-polynomials). Let g and $h \in K[x_1, \ldots, x_n]$ be nonzero polynomials such that $\mathrm{LM}(g)$ and $\mathrm{LM}(h)$ are coprime. Show that 0 is a normal form of $\mathrm{spol}(g, h)$ with respect to $G := \{g, h\}$. So pairs of polynomials with coprime leading monomials need not be considered in Buchberger's algorithm (Algorithm 9.13).
Hint: Show that

$$\mathrm{spol}(g, h) = (\mathrm{LT}(h) - h) \cdot g - (\mathrm{LT}(g) - g) \cdot h, \tag{9.15}$$

and conclude the result from that.

9.4 (Reduced Gröbner bases). Let $I \subseteq K[x_1, \ldots, x_n]$ be an ideal.

(a) Find an algorithm that converts a Gröbner basis of I into a reduced Gröbner basis of I.
(b) Let G and G' be two reduced Gröbner bases of I. Show that $G = G'$.

9.5 (Another proof of $\dim(A) \le \operatorname{trdeg}(A)$). Use Exercise 6.8(c) to give another proof of Theorem 5.5 and the first part of Corollary 5.7.

9.6 (Radical membership test). It is rather complicated to compute the radical ideal $\sqrt{I}$ of an ideal $I \subseteq K[x_1, \ldots, x_n]$. However, it is much easier to decide the membership of a given polynomial $f \in K[x_1, \ldots, x_n]$ in the radical ideal. In fact, take an additional indeterminate y and form the ideal

$$J := (I \cup \{y \cdot f - 1\})_{K[x_1,\ldots,x_n,y]} \subseteq K[x_1, \ldots, x_n, y].$$

Prove the equivalence

$$f \in \sqrt{I} \iff 1 \in J.$$

This yields the desired test for radical membership, since the condition $1 \in J$ can be tested by computing a Gröbner basis of J.

9.7 (Testing affine algebras for dimension zero). Let $I \subsetneqq K[x_1, \ldots, x_n]$ be a proper ideal, and let G be a Gröbner basis of I. Show that the following statements are equivalent:

(a) $\dim (K[x_1, \ldots, x_n]/I) = 0$.
(b) For every $i \in \{1, \ldots, n\}$, there exists a positive integer d_i and $g_i \in G$ with $\operatorname{LM}(g_i) = x_i^{d_i}$.

This yields a test for determining whether the affine algebra $A := K[x_1, \ldots, x_n]/I$ has dimension 0 by computing just one Gröbner basis with respect to an arbitrary monomial ordering.
Remark: A generalization of this test is contained in Algorithm 11.15, which computes the dimension of an affine algebra.

9.8 (Elimination orderings vs. block orderings). Give an example of an $\{x_1, \ldots, x_k\}$-elimination ordering on $K[x_1, \ldots, x_n]$ that is not a block ordering formed from monomial orderings on $K[x_1, \ldots, x_k]$ and on $K[x_{k+1}, \ldots, x_n]$.

9.9 (Subalgebra membership test). Let $A = K[y_1, \ldots, y_m]/I$ be an affine K-algebra and $R = K[g_1 + I, \ldots, g_n + I] \subseteq A$ a subalgebra given by polynomials $g_i \in K[y_1, \ldots, y_m]$. With $x_1, \ldots, x_n$ additional indeterminates, form the ideal $J := (I \cup \{g_1 - x_1, \ldots, g_n - x_n\})_{K[x_1,\ldots,x_n,y_1,\ldots,y_m]}$ (as in Proposition 9.18). Let G be a Gröbner basis of J with respect to an $\{x_1, \ldots, x_n\}$-elimination ordering "$\le$" on $K[x_1, \ldots, x_n, y_1, \ldots, y_m]$. Show that for $f \in K[y_1, \ldots, y_m]$ the equivalence

$$f + I \in R \iff \widetilde{f} := \mathrm{NF}_G(f) \in K[x_1, \dots, x_n]$$

holds. Furthermore, show that if the above conditions are satisfied, then $f + I = \widetilde{f}(g_1 + I, \dots, g_n + I)$. So we have an algorithm for testing membership in R.

9.10 (Computing intersections of ideals). Let $I_1, I_2 \subseteq K[x_1, \dots, x_n]$ be two ideals. With y an additional indeterminate, form the ideal

$$J := (y \cdot I_1 \cup (1 - y) \cdot I_2)_{K[x_1,\dots,x_n,y]} \subseteq K[x_1, \dots, x_n, y].$$

Show that

$$I_1 \cap I_2 = K[x_1, \dots, x_n] \cap J.$$

Since the right-hand side is an elimination ideal, this yields an algorithm for computing intersections of polynomial ideals. Can you think of an extension of this algorithm that computes intersections of m ideals by computing just one elimination ideal and using just one additional indeterminate? (You may assume that K contains at least m elements.)

9.11 (Colon ideals). Find an algorithm for computing the colon ideal $I : J$ of two ideals $I, J \subseteq K[x_1, \dots, x_n]$. Assume that the ideals are given by finitely many generators.

9.12 (A constructive version of Noether normalization). Find an algorithm that finds the elements $c_1, \dots, c_n$ from Theorem 8.19. Can the version of Noether normalization discussed in Remark 8.20 be made constructive in similar ways?
Hint: Take another look at the proof of Theorem 8.19, and turn this proof into an algorithm.

9.13 (Elementary symmetric polynomials). In this exercise it is shown that the ring of invariants of the symmetric group is generated by the so-called **elementary symmetric polynomials**. These are defined as

$$s_k := \sum_{1 \le i_1 < i_2 < \cdots < i_k \le n} x_{i_1} \cdots x_{i_k} \in K[x_1, \dots, x_n] \quad (k = 1, \dots, n).$$

The symmetric group S_n acts on $K[x_1, \dots, x_n]$ by algebra automorphisms given by $\pi(x_i) = x_{\pi(i)}$ for $\pi \in S_n$. Show that

$$K[x_1, \dots, x_n]^{S_n} = K[s_1, \dots, s_n].$$

Does this also hold if K is a ring, not a field?
Hint: Determine the leading monomials of the s_i with respect to the lexicographic ordering, and use this information in the proof.

Chapter 10
Fibers and Images of Morphisms Revisited

In this chapter we will continue the investigation that was started in Section 7.2. First we use Gröbner basis theory to prove the generic freeness lemma. This leads to an algorithm for computing the image of a morphism of affine varieties. Then we will draw more theoretical consequences on the images of morphisms and the dimension of fibers. Finally, we will apply our results to the topic of invariant theory. As mentioned before, the results of this chapter will not be used anywhere else in the book, so there is an option to skip it.

10.1 The Generic Freeness Lemma

Roughly speaking, the generic freeness lemma asserts that (under suitable hypotheses) a ring extension becomes a free module after localizing almost everywhere. The following lemma is a constructive version of the generic freeness lemma. When reading it, one should bear Proposition 9.18 in mind, since this proposition tells us how the Gröbner basis G appearing in the lemma can be constructed in the case that R and S are affine algebras.

Lemma 10.1 (Generic freeness, constructive version). *Let $R \subseteq S$ be a finitely generated ring extension, so that there is an epimorphism*

$$\psi \colon R[x_1, \ldots, x_n] \to S$$

(with x_i indeterminates). Let $G \subseteq R[x_1, \ldots, x_n]$ be a Gröbner basis of $\ker(\psi)$ with respect to some monomial ordering (see Remark 9.11 for Gröbner bases over rings). If $U \subseteq R$ is a multiplicative subset containing the product $\prod_{g \in G} \mathrm{LC}(g)$, then $U^{-1}S$ is free as a $U^{-1}R$-module, and there exists a basis containing $1 \in U^{-1}S$.

Remark. In Lemma 10.1, if R is not an integral domain, it is possible that the product $\prod_{g \in G} \mathrm{LC}(g)$ becomes zero. If that happens, the lemma is meaningless. ◁

G. Kemper, *A Course in Commutative Algebra*, Graduate Texts in Mathematics 256, DOI 10.1007/978-3-642-03545-6_11,

Proof of Lemma 10.1. Let $B \subset R[x_1, \dots, x_n]$ be the set of all monomials that are not divisible by any leading monomial $\mathrm{LM}(g)$ with $g \in G$. Since ψ is injective on R, we have $1 \in B$. Moreover, $\psi(B) \subseteq S$ is linearly independent over R, since if $\sum_{i=1}^m a_i \psi(t_i) = 0$ with $t_1, \dots, t_m \in B$ pairwise distinct and $a_i \in R$, then $h := \sum_{i=1}^m a_i t_i \in \ker(\psi)$, so $h = 0$ since no monomial of h is divisible by an $\mathrm{LM}(g)$ for $g \in G$.

Let $M := (\psi(B))_R \subseteq S$ be the (free) R-module generated by $\psi(B)$. We claim that for every $s \in S$ there exists a $u \in U$ such that $us \in M$. To prove this, take $f \in R[x_1, \dots, x_n]$ with $s = \psi(f)$. By the modification of the normal form algorithm (Algorithm 9.8) discussed in Remark 9.11, there exists $f^* \in R[x_1, \dots, x_n]$ that is a normal form with respect to G of some $u \cdot f$, where u is a product formed from leading coefficients of elements of G. By multiplying u by further leading coefficients of elements of G, we can achieve that u is a power of $\prod_{g \in G} \mathrm{LC}(g)$, so $u \in U$. The definition of a normal form implies that f^* lies in $(B)_R$ and $u \cdot f - f^* \in (G)_{R[x_1, \dots, x_n]} \subseteq \ker(\psi)$. So

$$u \cdot s = u \cdot \psi(f) = \psi(u \cdot f) = \psi(f^*) \in M,$$

which proves the claim.

Now it is straightforward to check that $\widetilde{B} := \left\{ \frac{\psi(t)}{1} \mid t \in B \right\} \subseteq U^{-1}S$ is a basis of $U^{-1}S$ as a $U^{-1}R$-module. This completes the proof. □

By combining Lemma 10.1 with Proposition 9.18 and Lemma 7.16, we get an algorithm for computing the image of a morphism between the spectra of affine algebras (which, in the case of an algebraically closed ground field, comes down to computing the image of a morphism of affine varieties). Before we state the algorithm, we derive the "existence version" of the generic freeness lemma.

Corollary 10.2 (Generic freeness lemma). *Let R be an integral domain and let S be a ring extension of R that is finitely generated as an R-algebra. Then there exists a nonzero element $a \in R$ such that for every multiplicative subset $U \subseteq R$ with $a \in U$, the localization $U^{-1}S$ is free as a $U^{-1}R$-module, and there exists a basis containing $1 \in U^{-1}S$.*

Proof. We have an epimorphism $\psi\colon R[x_1, \dots, x_n] \to S$. Let $I := \ker(\psi)$, $K := \mathrm{Quot}(R)$, and $J := (I)_{K[x_1, \dots, x_n]} = K \cdot I$. Choose a monomial ordering on $K[x_1, \dots, x_n]$, and let $G \subseteq J \setminus \{0\}$ be a Gröbner basis of J. Multiplying the polynomials in G by suitable nonzero elements of R, we can achieve that G is contained in I, so G is a Gröbner basis of I. With $a := \prod_{g \in G} \mathrm{LC}(g) \in R \setminus \{0\}$, the result follows from Lemma 10.1. □

Loosely speaking, Corollary 10.2 says that freeness holds "almost everywhere" (assuming the hypotheses of the corollary). More precisely, it holds after localization at all $P \in \mathrm{Spec}(R)$ with $a \notin P$. These P form an open, dense subset of $\operatorname{Spec}(R)$. (Proof of density: We have $a \notin \{0\} \in \operatorname{Spec}(R)$, and the closure of $\{0\}$ is $\operatorname{Spec}(R)$.) The generic freeness lemma is due to Grothendieck, and it has been traditionally referred to as the *generic flatness lemma*. In fact, in its original version, the assertion is flatness of the map $U^{-1}R \to U^{-1}S$, a property that is weaker than freeness and is not treated in this book. In Exercise 10.1 we explore the necessity of the hypotheses of Corollary 10.2. Exercise 10.2 contains a version of the generic freeness lemma for modules over S. Exercise 10.3 contains a surprising application of the generic freeness lemma: A special case of this application says that a subalgebra of an affine K-domain has a localization that is again an affine K-domain.

As announced above, we now get to the algorithm for computing the image of a morphism $\varphi^*\colon \operatorname{Spec}(A) \to \operatorname{Spec}(B)$ of spectra of affine algebras. Defining B as a quotient ring of a polynomial ring is the same as giving an embedding $\operatorname{Spec}(B) \hookrightarrow \operatorname{Spec}(K[x_1,\ldots,x_n])$. Therefore we may assume that B is a polynomial ring. In the following, $K[x_1,\ldots,x_n]$ and $K[y_1,\ldots,y_m]$ are polynomial rings over a field.

Algorithm 10.3 (Image of a morphism of spectra).

Input: An ideal $I \subseteq K[y_1,\ldots,y_m]$ defining an affine algebra $A := K[y_1,\ldots,y_m]/I$, and polynomials $g_1,\ldots,g_n \in K[y_1,\ldots,y_m]$ defining a K-algebra homomorphism $\varphi\colon K[x_1,\ldots,x_n] \to A$, $x_i \mapsto g_i + I$.

Output: Ideals $J_1,\ldots,J_l \subseteq K[x_1,\ldots,x_n]$ and polynomials $f_1,\ldots,f_l \in K[x_1,\ldots,x_n]$ such that the image of the induced morphism $\varphi^*\colon \operatorname{Spec}(A) \to \operatorname{Spec}(K[x_1,\ldots,x_n]) =: Y$ is

$$\operatorname{im}(\varphi^*) = \bigcup_{i=1}^{l} \Big(\mathcal{V}_Y(J_i) \setminus \mathcal{V}_Y(f_i)\Big) \tag{10.1}$$

and the image closure is

$$\overline{\operatorname{im}(\varphi^*)} = \mathcal{V}_Y(J_1). \tag{10.2}$$

(1) Choose monomial orderings "$\le_x$" on $K[x_1,\ldots,x_n]$ and "$\le_y$" on $K[y_1,\ldots,y_m]$, and let "$\le$" be the block ordering on $K[x_1,\ldots,x_n,y_1,\ldots,y_m]$ with "$\le_y$" dominating.

(2) Form the ideal

$$J := \Big(I \cup \{g_1 - x_1,\ldots,g_n - x_n\}\Big)_{K[x_1,\ldots,x_n,y_1,\ldots,y_m]}$$

and compute a Gröbner basis G of J with respect to "$\le$".

(3) Set

$$G_x := K[x_1,\ldots,x_n] \cap G, \quad G_y := \{\operatorname{NF}_{G_x}(g) \mid g \in G\} \setminus \{0\},$$

and

$$M := \{\mathrm{LC}_y(g) \mid g \in G_y\} \setminus K \subseteq K[x_1, \ldots, x_n].$$

Here $\mathrm{LC}_y(g)$ denotes the leading coefficient with respect to "$\leq_y$" of g considered as a polynomial in the y_i-variables.

(4) Initialize the lists $J_1, \ldots, J_l$ and $f_1, \ldots, f_l$ by setting $l = 1$,

$$J_1 := (G_x)_{K[x_1,\ldots,x_n]}, \quad \text{and} \quad f_1 := \prod_{f \in M \cup \{1\}} f.$$

(5) For all $f \in M$, perform step (6).

(6) Apply the algorithm recursively with $(I \cup \{f(g_1, \ldots, g_n)\})_{K[y_1,\ldots,y_m]}$ as first argument and $g_1, \ldots, g_n$ as second argument. Append the resulting lists of ideals and polynomials to the current lists $J_1, \ldots, J_l$ and $f_1, \ldots, f_l$.

Theorem 10.4. *Algorithm 10.3 terminates after finitely many steps and calculates the image of φ^* and its closure correctly.*

Proof. We use the notation from the algorithm. By way of contradiction, assume that there exists an ideal $I \subseteq K[y_1, \ldots, y_m]$ such that the algorithm applied to I does not terminate after finitely many steps. By Hilbert's basis theorem (Corollary 2.13), we may assume I to be maximal with this property. Since all steps except (6) clearly terminate after finitely many steps, there exists $f = \mathrm{LC}_y(g) \in M$ (with $g \in G_y$) such that step (6) does not terminate for f. By the maximality of I, this implies $f(g_1, \ldots, g_n) \in I$, so $f \in J$ by the definition of J. Since $f \in K[x_1, \ldots, x_n] \setminus \{0\}$, Theorem 9.16 yields a $g' \in G_x$ whose leading monomial divides $\mathrm{LM}(f)$. Since "$\leq$" is a block ordering, $\mathrm{LM}(g) = \mathrm{LM}_y(g) \cdot \mathrm{LM}(f)$, so $\mathrm{LM}(g')$ divides $\mathrm{LM}(g)$, too. This contradicts the fact that all $g \in G_y$ are in normal form with respect to G_x, establishing the termination of the algorithm.

We proceed with proving the correctness. The correctness of (10.2) follows from (9.10), Proposition 9.17, and Theorem 9.16. To prove (10.1), consider the decomposition

$$\mathrm{im}(\varphi^*) = \Big(\mathrm{im}(\varphi^*) \setminus \mathcal{V}_Y(f_1)\Big) \cup \bigcup_{f \in M} \Big(\mathrm{im}(\varphi^*) \cap \mathcal{V}_Y(f)\Big), \tag{10.3}$$

which follows from the definition of f_1 in step (4). We claim that

$$\mathrm{im}(\varphi^*) \setminus \mathcal{V}_Y(f_1) = \mathcal{V}_Y(J_1) \setminus \mathcal{V}_Y(f_1). \tag{10.4}$$

Indeed $\mathrm{im}(\varphi^*) \setminus \mathcal{V}_Y(f_1) \subseteq \overline{\mathrm{im}(\varphi^*)} \setminus \mathcal{V}_Y(f_1) = \mathcal{V}_Y(J_1) \setminus \mathcal{V}_Y(f_1)$, where we used (10.2). Conversely, take $P \in \mathcal{V}_Y(J_1) \setminus \mathcal{V}_Y(f_1)$. In the following we use the notation from Proposition 9.18. By (9.12), we obtain

$$\prod_{g \in G_y} \mathrm{LC}\left(\Phi(g)\right) = \prod_{g \in G_y} \varphi\left(\mathrm{LC}_y(g)\right) = \varphi(cf_1) \in \varphi\left(K[x_1,\ldots,x_n] \setminus P\right) =: U,$$

where $c \in K \setminus \{0\}$. Moreover, the definition of G_y in step (3) implies $\Phi(G_y) = \Phi(G \setminus G_x) \setminus \{0\}$, so Proposition 9.18(c) tells us that $\Phi(G_y)$ is a Gröbner basis of the kernel of ψ. So by Lemma 10.1, $U^{-1}A$ is free as a $U^{-1}R$-module, and there exists a basis containing 1. By Lemma 7.16, this implies that the map $\mathrm{Spec}(U^{-1}A) \to \mathrm{Spec}(U^{-1}R)$ is surjective. By Theorem 6.5, this means that for every $\mathfrak{p} \in \mathrm{Spec}(R)$ with $U \cap \mathfrak{p} = \emptyset$, there exists $Q \in \mathrm{Spec}(A)$ with $R \cap Q = \mathfrak{p}$. Particularly, $\mathfrak{p} := \varphi(P) \in \mathrm{Spec}(R)$ satisfies the condition $U \cap \mathfrak{p} = \emptyset$, since otherwise there would exist $h \in P$ and $u \in K[x_1,\ldots,x_n] \setminus P$ with $\varphi(h) = \varphi(u)$, leading to the contradiction $u = (u-h) + h \in \ker(\varphi) + P = P$, since $\ker(\varphi) = J_1 \subseteq P$. So we have $Q \in \mathrm{Spec}(A)$ with $R \cap Q = \mathfrak{p} = \varphi(P)$, which is equivalent to $P = \varphi^{-1}(Q)$. So $P \in \mathrm{im}(\varphi^*) \setminus \mathcal{V}_Y(f_1)$, and (10.4) is proved.

By induction on the recursion depth, we can assume that for every $f \in M$ the recursive call of the algorithm computes the image of φ_f^*, where $\varphi_f\colon K[x_1,\ldots,x_n] \to A/(\varphi(f))_A$ is given by $x_i \mapsto \varphi(x_i) + (\varphi(f))_A$. So in view of (10.3) and (10.4), it suffices to show that

$$\mathrm{im}(\varphi^*) \cap \mathcal{V}_Y(f) = \mathrm{im}(\varphi_f^*). \tag{10.5}$$

Indeed, $\mathrm{im}(\varphi_f^*)$ consists of all $P \in \mathrm{Spec}\left(K[x_1,\ldots,x_n]\right)$ such that $P = \varphi_f^{-1}(\mathfrak{q})$ with $\mathfrak{q} \in \mathrm{Spec}\left(A/(\varphi(f))_A\right)$. This condition is equivalent to $P = \varphi^{-1}(Q)$ with $Q \in \mathrm{Spec}(A)$, $\varphi(f) \in Q$. This in turn is equivalent to $P = \varphi^{-1}(Q)$ and $f \in P$, i.e., $P \in \mathrm{im}(\varphi^*) \cap \mathcal{V}_Y(f)$. This establishes (10.5), and the proof is complete. □

Exercise 10.4 contains an explicit example to which the algorithm is applied.

Algorithm 10.3 also computes the image of a morphism $f\colon X \to Y$ of affine varieties over an algebraically closed field K. In fact, Y (just like X) is embedded in some K^n, so for computing the image one may assume $Y = K^n$. The morphism f induces a homomorphism $\varphi\colon K[x_1,\ldots,x_n] \to K[X] =: A$. Applying Algorithm 10.3 to φ yields $J_1,\ldots,J_l \subseteq K[x_1,\ldots,x_n]$ and $f_1,\ldots,f_l \in K[x_1,\ldots,x_n]$ such that

$$\mathrm{im}(\varphi^*) = \bigcup_{i=1}^{l} \left(\mathcal{V}_{\mathrm{Spec}(K[x_1,\ldots,x_n])}(J_i) \setminus \mathcal{V}_{\mathrm{Spec}(K[x_1,\ldots,x_n])}(f_i)\right).$$

Using the algebra–geometry lexicon, it is easy to see that this implies

$$\mathrm{im}(f) = \bigcup_{i=1}^{l} \left(\mathcal{V}_{K^n}(J_i) \setminus \mathcal{V}_{K^n}(f_i)\right).$$

10.2 Fiber Dimension and Constructible Sets

Corollary 10.2 is exactly what we need to draw consequences on the fibers of morphisms. By putting it together with Lemmas 7.15 and 7.16, we obtain the following result.

Theorem 10.5. *Let $\varphi: R \to S$ be a ring homomorphism such that*

(1) R is a Noetherian integral domain,
(2) S is finitely generated as an R-algebra, and
(3) φ is injective.

Then there exists a nonzero $a \in R$ such that for all $P \in \operatorname{Spec}(R)$ with $a \notin P$, the fiber $\mathcal{M}_P := \left\{Q \in \operatorname{Spec}(S) \mid \varphi^{-1}(Q) = P\right\}$ is nonempty. If $Q \in \mathcal{M}_P$, then

$$\operatorname{ht}(\mathfrak{Q}) = \operatorname{ht}(Q) - \operatorname{ht}(P), \tag{10.6}$$

where $\mathfrak{Q} \in \operatorname{Spec}\left(S_{[P]}\right)$ is the image of Q in the fiber ring $S_{[P]}$ (i.e., $\mathfrak{Q} = U^{-1}(Q/I)$ with the notation used before Proposition 7.11). In particular, the fiber dimension is

$$\dim\left(S_{[P]}\right) = \max\left\{\operatorname{ht}(Q) \mid Q \in \mathcal{M}_P\right\} - \operatorname{ht}(P). \tag{10.7}$$

Proof. Corollary 10.2 yields $a \in R \setminus \{0\}$ such that for $P \in \operatorname{Spec}(R)$ with $a \notin P$ the localization $U^{-1}S$ (with $U := R \setminus P$) is a free R_P-module with 1 contained in a basis. Moreover, S and $U^{-1}S$ are Noetherian by Corollary 2.12 and Corollary 6.4, so Lemma 7.16 applies. By Lemma 7.16(b), there exists $Q' \in \operatorname{Spec}(U^{-1}S)$ with $R_P \cap Q' = P_P$. It is routine to check that the preimage $Q \in \operatorname{Spec}(S)$ of Q' satisfies $\varphi^{-1}(Q) = P$. Now Lemma 7.16(a) and Lemma 7.15 yield (10.6), and (10.7) follows directly. □

We now specialize Theorems 7.12 and 10.5 to the case of coordinate rings of affine varieties.

Corollary 10.6. *Let $f: X \to Y$ be a morphism of equidimensional affine varieties over an algebraically closed field. For a point $y \in Y$, every irreducible component $Z \subseteq f^{-1}(\{y\})$ of the fiber has dimension*

$$\dim(Z) \geq \dim(X) - \dim(Y). \tag{10.8}$$

If f is dominant and Y is irreducible, there exists an open, dense subset $U \subseteq Y$ such that for every $y \in U$, the fiber $f^{-1}(\{y\})$ is nonempty and equidimensional of dimension $\dim(X) - \dim(Y)$.

Proof. Let $P \in \operatorname{Spec}_{\max}(K[Y])$ be the maximal ideal corresponding to a point $y \in Y$. The fiber over y is an affine variety, so by Corollary 8.24, the inequality (10.8) follows if we can show that every maximal ideal in the coordinate ring of the fiber has height at least $d := \dim(X) - \dim(Y)$. By

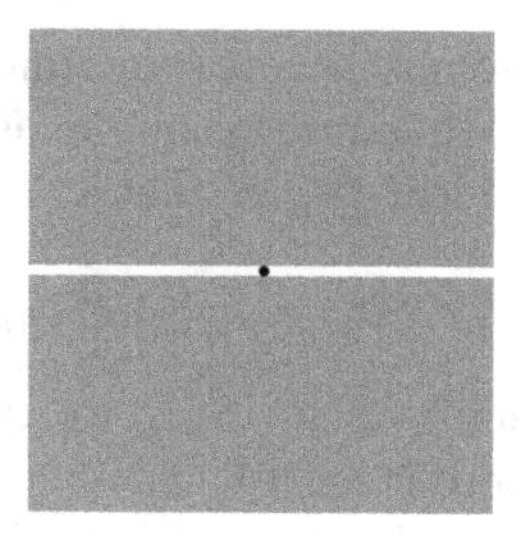

Fig. 10.1. An image of a morphism

Proposition 7.11, this is the same as showing that every maximal ideal $\mathfrak{Q}$ in the fiber ring over P has height at least d. Such a $\mathfrak{Q}$ corresponds to a maximal ideal $Q \in \operatorname{Spec}_{\max}(K[X])$, so $\operatorname{ht}(Q) = \dim(X)$ by Corollary 8.24. Since $\operatorname{ht}(P) = \dim(Y)$ (again by Corollary 8.24), the inequality (7.7) in Theorem 7.12 guarantees $\operatorname{ht}(\mathfrak{Q}) \geq d$, so (10.8) is proved.

If f is dominant and Y is irreducible, the induced homomorphism φ: $K[Y] \to K[X]$ is injective and $K[Y]$ is an integral domain, so Theorem 10.5 is applicable. This yields $a \in K[Y] \setminus \{0\}$, so $U := \{y \in Y \mid a(y) \neq 0\}$ is open and nonempty, and therefore dense because of the irreducibility of Y. If y lies in U, then $a \notin P$, so Theorem 10.5 tells us that $f^{-1}(\{y\})$ is nonempty and has dimension d. By (10.8), it must be equidimensional. □

Corollary 10.6 has a consequence that is sometimes referred to as the "upper semicontinuity of fiber dimension." For more, see Exercise 10.5.

An important consequence of Theorem 10.5 is Chevalley's theorem on images of morphisms. A typical example for the nature of an image of a morphism is Example 7.14(2): The image consists of all $(\alpha, \beta) \in K^2$ with $\beta \neq 0$ or $\alpha = \beta = 0$, as shown in Fig. 10.1.

So the image is neither closed nor open. However, according to the following definition, it is a constructible subset of K^2.

Definition 10.7. *Let X be a topological space. A subset $L \subseteq X$ is called* **locally closed** *if L is the intersection of an open and a closed subset. A subset $C \subseteq X$ is called* **constructible** *if C is the union of finitely many locally closed subsets.*

In K^n with the Zariski topology, a constructible set can be described by giving finitely many polynomial equations and using the logical operators "and" and "not." It can be shown that the constructible sets are precisely the sets that have such a description.

Corollary 10.8 (Chevalley's theorem on images of morphisms).
Let φ: $R \to S$ be a homomorphism of Noetherian rings making S into a finitely generated R-algebra. Then the image $\operatorname{im}(\varphi^*)$ of the induced map φ^*: $\operatorname{Spec}(S) \to \operatorname{Spec}(R)$ is a constructible subset of $\operatorname{Spec}(R)$.

Proof. The proof technique we use here is sometimes called Noetherian induction. This works as follows. We assume that the assertion is false. Since S is Noetherian, there exists an ideal $I \subseteq S$ that is maximal with the property that

$$Y(I) := \varphi^* \left(\mathcal{V}_{\operatorname{Spec}(S)}(I)\right) \subseteq \operatorname{Spec}(R)$$

is not constructible. Replacing S by S/I, we may assume that $Y(J)$ is constructible for every nonzero ideal $J \subseteq S$.

Let $Q_1, \dots, Q_n$ be the minimal prime ideals of S. Since $\operatorname{im}(\varphi^*) = \bigcup_{i=1}^n Y(Q_i)$, there exists i such that $Y(Q_i)$ is not constructible. It follows that $Q_i = \{0\}$, so S is an integral domain. With $P := \ker(\varphi)$, the map φ^* factors through $\operatorname{Spec}(R/P)$. Since the natural map $\operatorname{Spec}(R/P) \to \operatorname{Spec}(R)$ takes constructible subsets of $\operatorname{Spec}(R/P)$ to constructible subsets of $\operatorname{Spec}(R)$, the map $\operatorname{Spec}(S) \to \operatorname{Spec}(R/P)$ has a nonconstructible image. So we may assume that φ is injective and therefore R is an integral domain. Now Theorem 10.5 yields a nonzero $a \in R$ with $\operatorname{Spec}(R) \setminus \mathcal{V}_{\operatorname{Spec}(R)}(a) \subseteq \operatorname{im}(\varphi^*)$, so

$$\varphi^* \left(\operatorname{Spec}(S) \setminus \mathcal{V}_{\operatorname{Spec}(S)}(\varphi(a))\right) = \operatorname{Spec}(R) \setminus \mathcal{V}_{\operatorname{Spec}(R)}(a)$$

is open in $\operatorname{Spec}(R)$. But $Y\left(\varphi(a)\right)$ is constructible, and it follows that $\operatorname{im}(\varphi^*)$ is constructible, too. □

Specializing Corollary 10.8 to the case of coordinate rings of affine varieties, we obtain that the image of a morphism $f\colon X \to Y$ of affine varieties over an algebraically closed field is constructible. The example of the natural embedding $\varphi\colon \mathbb{Z} \to \mathbb{Q}$ shows that the finite generation hypothesis cannot be dropped from Corollary 10.8: The image of φ^* consists only of the zero ideal, and it is easily checked that this is not a constructible subset of $\operatorname{Spec}(\mathbb{Z})$. An extended version of Chevalley's theorem says that images of constructible sets under morphisms are always constructible (see Exercise 10.9). Corollary 10.8 and Exercise 10.7 imply that the image of a morphism has a subset that is open and dense in the image closure. This "thickness" result is much stronger and more useful than it appears at first glance.

10.3 Application: Invariant Theory

In this section we consider the theory of algebraic group actions and invariant theory as an example to which our result on fiber dimension (Corollary 10.6) is applied several times. This yields some fundamental information on the dimensions of orbits, fixed groups, and the invariant ring.

In the following, let K be an algebraically closed field. A **linear algebraic group** over K is an affine K-variety G, together with morphisms $G \times G \to G$, $(g_1, g_2) \mapsto g_1 \cdot g_2$, and $G \to G$, $g \mapsto g^{-1}$, making G into a group. Typical examples are the classical groups. A G**-variety** is an affine K-variety X

together with a morphism $G \times X \to X$, $(g, x) \mapsto g(x)$ defining an action of G on X. Typical examples are the natural module K^n of a classical group, symmetric powers of the natural module, and direct sums of such modules. For the sake of simplicity, we assume that G and X are both irreducible varieties. (The "opposite" case, in which G is a finite group, was considered in Exercise 8.1.) Readers who are interested in learning more about invariant theory will find a vast amount of literature. Good sources with which to begin are Springer [48], Sturmfels [50], and Popov and Vinberg [44]. Here we set ourselves the goal to find out as much as possible about the dimensions of G-orbits, point stabilizers, and of the ring of invariants. We proceed in several steps.

First, fix a point $x \in X$ and consider the morphism

$$f_x \colon G \to X, \; g \mapsto g(x),$$

whose image is the orbit $G(x)$. Since G is irreducible, the Zariski closure $\overline{G(x)}$ of the orbit is also irreducible. For $x' = g_0(x) \in G(x)$, the fiber is

$$f_x^{-1}(\{x'\}) = \{g \in G \mid g(x) = g_0(x)\} = g_0 \cdot G_x,$$

where G_x stands for the point stabilizer. Since multiplication by g_0 is a topological automorphism of G, we have $\dim\left(f_x^{-1}(\{x'\})\right) = \dim(G_x)$ for all $x' \in G(x)$. By Corollary 10.6, there exist points $x' \in f_x(G)$ for which (10.8) is an equality, so we obtain

$$\dim(G_x) = \dim(G) - \dim\left(\overline{G(x)}\right). \tag{10.9}$$

This is a fundamental connection between the orbit dimension and the stabilizer dimension. Notice that the orbit $G(x)$ often fails to be closed. An example of this phenomenon is the natural action of the multiplicative group $(K \setminus \{0\}, \cdot)$ on K, which has the nonclosed orbit $K \setminus \{0\}$.

From what we have seen up to now, the function $X \to \mathbb{N}_0$, $x \mapsto \dim(G_x)$ could behave in a totally erratic way. To study this function, we consider the morphism

$$h \colon G \times X \to X \times X, \; (g, x) \mapsto (x, g(x)).$$

The image $\Gamma := h(G \times X)$ is sometimes called the **graph of the action.** $G \times X$ is irreducible by Exercise 3.9, so the same holds for the image closure $\overline{\Gamma}$. For $(x, g(x)) \in \Gamma$, the fiber is

$$h^{-1}\left(\{(x, g(x))\}\right) = (g \cdot G_x) \times \{x\} \cong G_x.$$

Applying Corollary 10.6 and Theorem 5.15 yields

$$\dim(G_x) \geq \dim(G) + \dim(X) - \dim\left(\overline{\Gamma}\right) =: d_0$$

for all $x \in X$, and there exists an open, dense subset $U' \subseteq \overline{\Gamma}$ such that for $(x, x') \in U'$ we have equality. If $\pi: X \times X \to X$ is the first projection, then

$$\overline{U'} \subseteq \pi^{-1}\left(\overline{\pi(U')}\right),$$

so

$$X = \pi(\Gamma) \subseteq \pi\left(\overline{U'}\right) \subseteq \overline{\pi\left(U'\right)} \subseteq X.$$

By Exercises 10.7 and 10.9, there exists a subset $U \subseteq \pi(U')$ that is open and dense in $\overline{\pi\left(U'\right)} = X$. By the above, all $x \in U$ satisfy

$$\dim(G_x) = d_0 = \min\left\{\dim\left(G_{x'}\right) \mid x' \in X\right\}. \tag{10.10}$$

So the minimal value d_0 for $\dim(G_x)$ is attained on an open, dense subset; in other words, points with a stabilizer of greater dimension are exceptional. It also follows that if we have found a dense subset of X where $\dim(G_x)$ takes a constant value, then this value is d_0.

We now consider the ring of invariants, which is the main object of study in invariant theory. We have a G-action on the coordinate ring $K[X]$ given by $g(f) = f \circ g^{-1}$ for $g \in G$ and $f \in K[X]$. The **ring of invariants** is

$$K[X]^G := \{f \in K[X] \mid g(f) = f \text{ for all } g \in G\}.$$

So a regular function is an invariant if and only if it is constant on every G-orbit. Let $f_1, \ldots, f_n \in K[X]^G$ be invariants. They generate a subalgebra $A := K[f_1, \ldots, f_n] \subseteq K[X]$. By Theorem 1.25, A is the coordinate ring of an irreducible affine variety Y, and the inclusion $A \subseteq K[X]$ induces a dominant morphism $F: X \to Y$. (Explicitly, $Y \subseteq K^n$ is given by the ideal of relations of the f_i, and F is given by $x \mapsto (f_1(x), \ldots, f_n(x))$.) For $y = F(x)$ with $x \in X$, the invariance and continuity of the f_i implies $\overline{G(x)} \subseteq F^{-1}(\{y\})$, so

$$\dim\left(F^{-1}(\{y\})\right) \geq \dim\left(\overline{G(x)}\right) = \dim(G) - \dim(G_x),$$

where we used (10.9). It is easy to see that the dominance of F implies that $\overline{F(U)} = Y$. By Exercises 10.7 and 10.9, $F(U)$ has a subset O that is open and dense in $\overline{F(U)} = Y$. Applying Corollary 10.6 to F, we can shrink O further such that for all $y \in O$, the dimension of $F^{-1}(\{y\})$ equals $\dim(X) - \dim(Y)$. So choose $y \in O$ and $x \in U$ with $F(x) = y$. Then

$$\dim(X) - \dim(Y) = \dim\left(F^{-1}(\{y\})\right) \geq \dim(G) - d_0$$

(with d_0 the minimal, and typical, dimension of a point stabilizer G_x), so

$$\dim(Y) \leq \dim(X) - \dim(G) + d_0 := d.$$

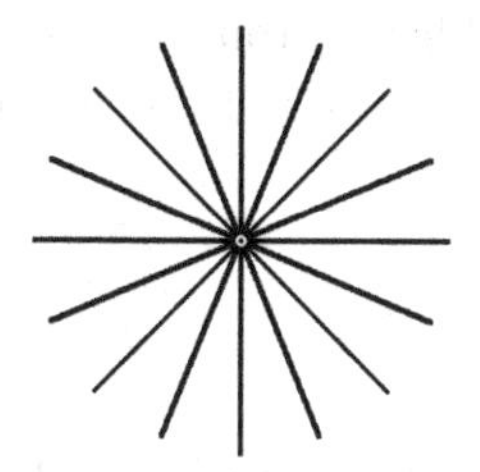

Fig. 10.2. Orbits of an action of the multiplicative group

Therefore $\operatorname{trdeg}(A) \leq d$ by Theorem 5.9. Since this holds for any choice of $f_1, \ldots, f_n \in K[X]^G$, it follows that $\operatorname{trdeg}\left(K[X]^G\right) \leq d$. So applying Theorem 5.9 again (or Exercise 5.3 in the case that $K[X]^G$ is not finitely generated) yields the nice inequality

$$\dim\left(K[X]^G\right) \leq \dim(X) - \dim(G) + \dim(G_x), \tag{10.11}$$

with $x \in X$ a point where $\dim(G_x)$ becomes minimal.

It is an interesting question when (10.11) is an equality. This is the case in many examples. However, counterexamples are also easy to find. For instance, if the multiplicative group $(K \setminus \{0\}, \cdot)$ acts on $X = K^2$ by normal multiplication, the zero vector lies in all orbit closures, as shown in Fig. 10.2. It follows that all invariants are constant, and we get a strict inequality

$$\dim\left(K[X]^G\right) = 0 < 2 - 1 + 0 = \dim(X) - \dim(G) + \dim(G_x).$$

This example can be interpreted by saying that the invariant $f := x_1/x_2$ is missing, since it is not a regular function. But if we exclude $K \times \{0\}$ from X, f becomes a regular function and (10.11) becomes an equality. By a famous theorem of Rosenlicht (see Popov and Vinberg [44, Theorem 2.3] or Springer [49, Satz 2.2]), this behavior is universal: One can always restrict X in such a way that (10.11) becomes an equality, and even such that every fiber of F is precisely one G-orbit (provided one chooses enough invariants for forming F). In particular, for the field of invariants $K(X)^G$, we always have the equality

$$\operatorname{trdeg}\left(K(X)^G\right) = \dim(X) - \dim(G) + \dim(G_x)$$

with $x \in X$ a point with $\dim(G_x)$ minimal. Notice that $K(X)^G$ is not always the field of fractions of $K[X]^G$; but if it is, the above equation implies equality in (10.11). For example, if $X = K^n$ is affine n-space and there exists no surjective homomorphism from G to the multiplicative group $\operatorname{GL}_1(K)$, then it is not hard to show that $K(X)^G = \operatorname{Quot}\left(K[X]^G\right)$, so (10.11) is an equality. Typical examples of groups G that have no surjective homomorphism to $\operatorname{GL}_1(K)$

are $G = \mathrm{SL}_n(K)$ and the additive group $G = G_a := (K, +)$. Exercise 10.10 studies two examples in which (10.11) is an equality.

Exercises for Chapter 10

10.1 (Hypotheses of the generic freeness lemma). Give an example showing that the hypothesis on finite generation of S as an R-algebra cannot be dropped from the generic freeness lemma (Corollary 10.2).

***10.2 (Generic freeness for modules).** Prove the following version of the generic freeness lemma: If R is an integral domain, S a finitely generated R-algebra, and M a Noetherian S-module, then there exists $a \in R \setminus \{0\}$ such that for every multiplicative subset $U \subseteq R$ with $a \in U$, the localization $U^{-1}M$ is free as a $U^{-1}R$-module.

10.3 (Subalgebras of finitely generated algebras). We know that subalgebras of finitely generated algebras need not be finitely generated (see Exercise 2.1). However, in this exercise it is shown that under very general hypotheses there exists a localization that is finitely generated.

Let R be a ring, A an R-domain (= an integral domain that is a finitely generated R-algebra), and $B \subseteq A$ a subalgebra. Show that there exists $a \in B \setminus \{0\}$ such that B_a is an R-domain.
Hint: You may use the fact that subextensions of finitely generated field extensions are also finitely generated (see Bourbaki [7, Chapter IV, § 15, Corollary 3]). *(Solution on page 225)*

10.4 (Computing the image of a morphism). Let K be an algebraically closed field of characteristic not 2. Let $X := \{(\xi_1, \xi_2, \xi_3) \in K^3 \mid \xi_1^2 + \xi_2^2 - \xi_3^2 = 0\}$ and consider the morphism

$$f\colon X \to K^2,\ (\xi_1, \xi_2, \xi_3) \mapsto (\xi_1, \xi_2 + \xi_3)$$

(see Example 7.14(2)). Use Algorithm 10.3 to determine the image of f.

10.5 (Upper semicontinuity of fiber dimension). Let $f\colon X \to Y$ be a morphism of affine varieties over an algebraically closed field. For a nonnegative integer d, show that the set

$$X_d := \{x \in X \mid f^{-1}(\{f(x)\}) \text{ has an irreducible component } Z \text{ with } x \in Z \text{ and } \dim(Z) \geq d\}$$

is closed in X.

Remark: In the jargon of the trade, this result is often referred to as the "upper semicontinuity of fiber dimension." Intuitively speaking (and oversimplifying), this means that the fiber dimension goes up only at exceptional points. Unfortunately, the definition of X_d is a bit complicated. In Exercise 10.6, two tempting, but false simplifications are explored.

10.6 (Two false statements on fiber dimension). Find examples that show that the following two statements (which are nicer than the result of Exercise 10.5) are false:

(a) If $f\colon X \to Y$ is a morphism of affine varieties over an algebraically closed field, then for every nonnegative integer d the set

$$Y_d := \left\{y \in Y \mid \dim\left(f^{-1}(\{y\})\right) \geq d\right\}$$

is closed in Y.

(b) If $f\colon X \to Y$ is a morphism of affine varieties over an algebraically closed field, then for every nonnegative integer d the set

$$X_d := \left\{x \in X \mid \dim\left(f^{-1}(\{f(x)\})\right) \geq d\right\}$$

is closed in X.

Is (b) true if X is assumed to be irreducible?

***10.7 (Constructible subsets).** Let X be a Noetherian topological space and let $Y \subseteq X$ be a constructible subset. Show that there exists a subset $U \subseteq Y$ that is open and dense in the closure $\overline{Y}$.
Hint: You can proceed as follows. Let Y be the union of locally closed sets L_i. Show that each irreducible component Z_j of $\overline{Y}$ is contained in at least one $\overline{L_i}$. Exclude all components other than Z_j from the open set belonging to L_i. If the resulting open set is U_j', show that $U_j' \cap Z_j \subseteq Y$ and $\overline{U_j' \cap Z_j} = Z_j$. Then form $U := \overline{Y} \cap \left(\bigcup_j U_j'\right)$, and show that this is a subset of Y that is open and dense in $\overline{Y}$. *(Solution on page 225)*

10.8 (Open and dense subsets of image closures). Let $f\colon X \to Y$ be a map of topological spaces. From which of the following hypotheses does it follow that the image $\operatorname{im}(f)$ has a subset U that is open and dense in the image closure $\overline{\operatorname{im}(f)}$?

(a) X and Y are affine varieties (over a field that need not be algebraically closed), and f is a morphism.
(b) X and Y are affine varieties over an algebraically closed field, and f is continuous (with respect to the Zariski topology).
(c) $X = Y = \mathbb{R}$ with the usual Euclidean topology, and f is continuous.

10.9 (Images of constructible sets). Let $\varphi\colon R \to S$ be a homomorphism of Noetherian rings making S into a finitely generated R-algebra, and let $X \subseteq \operatorname{Spec}(S)$ be a constructible subset. Show that $\varphi^*(X) \subseteq \operatorname{Spec}(R)$ is constructible, too.
Hint: Reduce to the case that $X := \mathcal{V}_{\operatorname{Spec}(S)}(I) \setminus \mathcal{V}_{\operatorname{Spec}(S)}(a)$ with $I \subseteq S$ an ideal and $a \in S$. *(Solution on page 226)*

10.10 (Some invariant theory). Consider the following examples of a linear algebraic group G over an algebraically closed field K, together with a G-variety X.

(a) $G = \operatorname{GL}_n(K)$ and $X = K^{n \times n}$ (the space of $n \times n$ matrices, which as a variety is just K^{n^2}) with G acting by $g(A) := g \cdot A \cdot g^{-1}$.
(b) $G = \operatorname{SL}_n(K)$ and $X = K^{n \times m}$ (the space of $n \times m$ -matrices) with G acting by $g(A) := g \cdot A$ (matrix product).

For each example, determine the minimal dimension of a point stabilizer G_x and an open, dense subset of X where this dimension is attained. Try to find enough invariants $f_1, \dots, f_m \in K[X]^G$ to show that (10.11) is an equality.

Chapter 11
Hilbert Series and Dimension

An affine algebra A of positive Krull dimension is always infinite-dimensional as a vector space (see Theorem 5.11). The goal of introducing the Hilbert series is nevertheless to measure the size in some way. The trick is to break up A into finite-dimensional pieces, given by the degrees. The Hilbert series then is the power series whose coefficients are the dimensions of the pieces. So instead of measuring the dimension by a number, we measure its growth as the degrees rise, and encode that information into a power series. In the first section of this chapter, we prove the surprising fact that the Hilbert series can always be written as a rational function, and almost all its coefficients are given by a polynomial, the Hilbert polynomial. We also learn how the Hilbert series can be computed algorithmically. In fact, as in the last chapter, algorithms and theory go hand in hand here. In the second section we show that the degree of the Hilbert polynomial is equal to the Krull dimension of A. Apart from being an interesting result in itself, this leads to a new and better algorithm for computing the dimension. The result also plays an important role in Chapter 12.

Throughout this chapter, $K[x_1, \dots, x_n]$ will denote a polynomial ring over a field.

11.1 The Hilbert–Serre Theorem

The following definition sets the theme of this chapter.

Definition 11.1. *For a monomial $t = x_1^{e_1} \cdots x_n^{e_n}$, the* **degree** *of t is defined as* $\deg(t) := e_1 + \cdots + e_n$. *For a nonzero polynomial $f \in K[x_1, \dots, x_n]$ we set*

$$\deg(f) := \max \{\deg(t) \mid t \in \operatorname{Mon}(f)\},$$

and $\deg(0) := -1$.

G. Kemper, *A Course in Commutative Algebra*, Graduate Texts in Mathematics 256, DOI 10.1007/978-3-642-03545-6_12,

For $I \subseteq K[x_1,\dots,x_n]$ an ideal, let $A := K[x_1,\dots,x_n]/I$, and for d a nonnegative integer, set

$$A_{\leq d} := \{f + I \mid f \in K[x_1,\dots,x_n],\ \deg(f) \leq d\}.$$

Observe that $A_{\leq d}$ is a finite-dimensional K-vector space. The function h_I: $\mathbb{N}_0 \to \mathbb{N}_0$ defined by

$$h_I(d) := \dim_K(A_{\leq d})$$

is called the **Hilbert function** *of I. The formal power series*

$$H_I(t) := \sum_{d=0}^{\infty} h_I(d) t^d \in \mathbb{Z}[[t]]$$

is called the **Hilbert series** *of I.*

Example 11.2. (1) Let $I = (x_1,\dots,x_n)$. Then $h_I(d) = 1$ for all d, so

$$H_I(t) = \sum_{d=0}^{\infty} t^d = \frac{1}{1-t}.$$

(2) Let $I = (x_1 - x_2^2) \subset K[x_1,x_2]$ and $A = K[x_1,x_2]/I$. For $d \in \mathbb{N}_0$, the residue classes of $1, x_1, \dots, x_1^d, x_2, x_1x_2, \dots, x_1^{d-1}x_2$ form a basis of $A_{\leq d}$, so $h_I(d) = 2d + 1$. We obtain

$$H_I(t) = \frac{1+t}{(1-t)^2}.$$

◁

Remark 11.3. (a) In the definition of the Hilbert series, we need not worry about convergence issues, since $H_I(t)$ is defined as an element of the formal power series ring over $\mathbb{Z}$. This also applies to the representations of the Hilbert series as "rational functions" in Example 11.2: The polynomial $1 - t$ is an invertible element of $\mathbb{Z}[[t]]$, and its inverse is $\sum_{d=0}^{\infty} t^d$. Of course, these representations could also be interpreted as identities of real or complex functions that are defined for $|t| < 1$.

(b) It is tempting to define the Hilbert function and Hilbert series of an affine algebra A as follows. By choosing generators of A, we obtain a presentation of A as $A \cong K[x_1,\dots,x_n]/I$. Then set $h_A(d) := h_I(d)$ and $H_A(t) := H_I(t)$. However, these objects will depend on the choice of the generators. For instance, choosing the (rather unusual) generators x^2 and x for the polynomial ring $A = K[x]$ yields the Hilbert function $2d + 1$ by Example 11.2(2). But choosing just x yields $d+1$ by Remark 11.5 below. So the Hilbert function and Hilbert series are *not* invariants of an affine algebra.

(c) Our definition of the $A_{\leq d}$ provides an ascending *filtration* of A, in the sense that $A_{\leq d} \subseteq A_{\leq d+1}$ for all d and $A = \bigcup_{d \in \mathbb{N}_0} A_{\leq d}$. In the literature,

Hilbert series are often defined for *graded vector spaces*, i.e., vector spaces V that have a direct sum decomposition

$$V = \bigoplus_{d \in \mathbb{N}_0} V_d$$

with V_d finite-dimensional K-vector spaces. A special case of a graded vector space is a *graded algebra*, where the grading provides a structure of a graded ring. In this setting, the graded Hilbert series is defined as

$$H_V^{\text{grad}}(t) := \sum_{d=0}^{\infty} \dim_K(V_d) t^d \in \mathbb{Z}[[t]].$$

But this is strongly related to our definition of the Hilbert series. In fact, a grading can be turned into an ascending filtration by setting $V_{\leq d} := \bigoplus_{i=0}^{d} V_i$. Then $H_V^{\text{grad}}(t)$ and $H_V(t) := \sum_{d=0}^{\infty} \dim_K(V_{\leq d}) t^d$ are obviously connected by

$$H_V^{\text{grad}}(t) = (1-t) H_V(t).$$

Exercise 12.3 studies the Hilbert series of a graded module over a graded ring. ◁

We now calculate the Hilbert series of a principal ideal. As we will see later, this has much more importance than just providing a further example.

Proposition 11.4 (The Hilbert series of a principal ideal). *If $I = (f) \subseteq K[x_1, \ldots, x_n]$ is a principal ideal, then*

$$H_I(t) = \frac{1 - t^{\deg(f)}}{(1-t)^{n+1}} \quad \textit{if} \quad f \neq 0$$

and

$$H_I(t) = \frac{1}{(1-t)^{n+1}} \quad \textit{if} \quad f = 0.$$

Proof. We start with the case $f = 0$. Since the Hilbert function and Hilbert series of the zero ideal depend on the number n of indeterminates, we will write them in this proof as $h_n(d)$ and $H_n(t)$, respectively. We use induction on n, starting with $n = 0$. We have $h_0(d) = 1$ for all d, so $H_0(t) = \frac{1}{1-t}$. For $n > 0$, we use the direct sum decomposition

$$K[x_1, \ldots, x_n]_{\leq d} = \bigoplus_{\substack{i,j \in \mathbb{N}_0, \\ i+j=d}} K[x_1, \ldots, x_{n-1}]_{\leq i} \cdot x_n^j. \tag{11.1}$$

With the induction hypothesis, this implies

$$H_n(t) = H_{n-1}(t) \cdot \left(\sum_{j=0}^{\infty} t^j \right) = H_{n-1}(t) \cdot \frac{1}{1-t} = \frac{1}{(1-t)^{n+1}}.$$

Now assume that $f \neq 0$. For every $d \in \mathbb{N}_0$ we have

$$(K[x_1,\dots,x_n]/(f))_{\leq d} \cong K[x_1,\dots,x_n]_{\leq d}/\left(f \cdot K[x_1,\dots,x_n]_{\leq d-\deg(f)}\right).$$

Since multiplication by f is injective on $K[x_1,\dots,x_n]$, we obtain

$$H_I(t) = (1 - t^{\deg(f)}) \cdot H_{\{0\}}(t) = \frac{1 - t^{\deg(f)}}{(1-t)^{n+1}}.$$

□

Remark 11.5. We can also determine the Hilbert function $h_{\{0\}}(d)$ of the zero ideal. Since $h_{\{0\}}(d)$ equals the number of monomials of degree at most d, it can be determined combinatorially. Taking a different route, we expand the Hilbert series as a binomial series. This yields

$$H_{\{0\}}(t) = (1-t)^{-n-1} = \sum_{d=0}^{\infty} \binom{-n-1}{d} (-t)^d,$$

so

$$h_{\{0\}}(d) = (-1)^d \binom{-n-1}{d} = \binom{d+n}{d} = \binom{d+n}{n}.$$

In particular, we see that $h_{\{0\}}(d)$ is given by a polynomial of degree n in d. This will be generalized in Corollary 11.10. ◁

Our next goal is to link the topic of Hilbert series to the theory of Gröbner bases. For this, we need the concept of a **total degree ordering**. By definition, this is a monomial ordering on $K[x_1,\dots,x_n]$ such that two monomials t, t' with $t \leq t'$ satisfy $\deg(t) \leq \deg(t')$. The most important example of a total degree ordering is the grevlex ordering. A counterexample is the lexicographic ordering (if $n > 1$). Recall that if $I \subseteq K[x_1,\dots,x_n]$ is an ideal, we write $L(I)$ for the leading ideal, which depends on the choice of the monomial ordering.

Theorem 11.6 (Hilbert series and leading ideal)**.** *Suppose that the polynomial ring $K[x_1,\dots,x_n]$ is equipped with a total degree ordering, and let $I \subseteq K[x_1,\dots,x_n]$ be an ideal. Then*

$$H_I(t) = H_{L(I)}(t).$$

Proof. Set $A := K[x_1,\dots,x_n]/I$. By Theorem 9.9, the normal form map NF_G, given by a Gröbner basis G of I, induces an injective linear map $\varphi\colon A \to K[x_1,\dots,x_n]$. For every $d \in \mathbb{N}_0$ we have a restriction $\varphi_d\colon A_{\leq d} \to K[x_1,\dots,x_n]$. Let $V_d \subseteq K[x_1,\dots,x_n]$ be the subspace spanned by all monomials t with $\deg(t) \leq d$ and $t \notin L(I)$. Since all $f \in V_d$ are in normal form with respect to G, we get $f = \mathrm{NF}_G(f) = \varphi_d(f + I)$, so $V_d \subseteq \mathrm{im}(\varphi_d)$. On the other hand, Definition 9.6(b) and the hypothesis on the monomial ordering imply $\mathrm{im}(\varphi_d) \subseteq V_d$. We conclude that

$$h_I(d) = \dim(V_d).$$

Observe that the definition of V_d depends only on $L(I)$. So two ideals with the same leading ideal have the same Hilbert function and Hilbert series. Since $L\left(L(I)\right) = L(I)$, the result follows. □

Exercise 11.2 shows that the hypothesis on the monomial ordering cannot be dropped from Theorem 11.6.

A polynomial $f \in K[x_1, \dots, x_n]$ is called **homogeneous** if all monomials of f have the same degree. So every polynomial can be written uniquely as a sum of homogeneous polynomials of pairwise distinct degrees, its **homogeneous parts**. An ideal $I \subseteq K[x_1, \dots, x_n]$ is called **homogeneous** if it is generated by homogeneous polynomials. For example, the leading ideal $L(I)$ of any ideal I is homogeneous. For more on homogeneous ideals, see Exercise 11.3.

Lemma 11.7 (Hilbert series of the sum and intersection of ideals). *Let $I, J \subseteq K[x_1, \dots, x_n]$ be homogeneous ideals. Then*

$$H_{I+J}(t) + H_{I \cap J}(t) = H_I(t) + H_J(t).$$

Proof. Let d be a nonnegative integer. For an ideal $L \subseteq K[x_1, \dots, x_n]$ we write $L_{\leq d} := \{f \in L \mid \deg(f) \leq d\}$. It follows from the hypothesis that $I + J$ is generated by homogeneous polynomials $g_1, \dots, g_m \in I \cup J$, so every $f \in (I + J)_{\leq d}$ can be written as $f = \sum_{i=1}^m h_i g_i$, with $h_i \in K[x_1, \dots, x_n]$. This equation still holds if from each h_i we delete all homogeneous parts of degree $> d - \deg(g_i)$. This shows that the map $I_{\leq d} \to (I + J)_{\leq d}/J_{\leq d}$, $f \mapsto f + J_{\leq d}$ is surjective. Its kernel is $(I \cap J)_{\leq d}$, so

$$\dim_K(I_{\leq d}) - \dim_K\left((I \cap J)_{\leq d}\right) = \dim_K\left((I + J)_{\leq d}\right) - \dim_K(J_{\leq d}).$$

Passing to the dimensions of the quotient spaces in $K[x_1, \dots, x_n]_{\leq d}$ and forming power series yields the result. □

The reduction step given by Theorem 11.6 is crucial in the following algorithm for computing the Hilbert series of an ideal.

Algorithm 11.8 (Hilbert series of a polynomial ideal).

Input: An ideal $I \subseteq K[x_1, \dots, x_n]$, given by generators.
Output: The Hilbert series $H_I(t)$.

(1) Choose a total degree ordering "$\leq$" on $K[x_1, \dots, x_n]$ and compute a Gröbner basis G of I with respect to "$\leq$". Let $m_1, \dots, m_r$ be the leading monomials of the nonzero elements of G.
(2) If $r = 0$, return $H_I(t) := \frac{1}{(1-t)^{n+1}}$.

(3) Set

$$J := (m_2, \ldots, m_r) \quad \text{and} \quad \widetilde{J} := (\operatorname{lcm}(m_1, m_2), \ldots, \operatorname{lcm}(m_1, m_r)).$$

(4) Compute the Hilbert series $H_J(t)$ and $H_{\widetilde{J}}(t)$ by a recursive call of the algorithm. *(Notice that J and $\widetilde{J}$ are generated by monomials, so there is nothing to do when performing step* (1) *on J and $\widetilde{J}$.)*

(5) Return

$$H_I(t) := \frac{1 - t^{\deg(m_1)}}{(1-t)^{n+1}} + H_J(t) - H_{\widetilde{J}}(t).$$

Notice that Algorithm 11.8 requires just one Gröbner basis computation, and we can use the grevlex ordering, which tends to make computations fastest.

Theorem 11.9. *Algorithm 11.8 terminates after finitely many steps and calculates $H_I(t)$ correctly.*

Proof. With each recursive call of the algorithm, the number r decreases strictly. This guarantees termination.

Let $\widetilde{I} := (m_1, \ldots, m_r) = L(I)$. By Theorem 11.6, we need to show that steps (2) through (5) calculate $H_{\widetilde{I}}(t)$ correctly. We use induction on r.

For $r = 0$, the Hilbert series in step (2) is correct by Proposition 11.4. Assume $r > 0$. By induction, the Hilbert series of J and $\widetilde{J}$ are calculated correctly. We claim that

$$\widetilde{J} = J \cap (m_1). \tag{11.2}$$

Clearly every least common multiple of m_1 and an m_i ($i \geq 2$) lies in J and in (m_1), so $\widetilde{J} \subseteq J \cap (m_1)$. Conversely, take $f \in J \cap (m_1)$. Then $f = g_1 m_1$ and $f = \sum_{i=2}^r g_i m_i$ with $g_1, \ldots, g_r \in K[x_1, \ldots, x_n]$. For every monomial $t \in \operatorname{Mon}(g_1)$ there exists $i \geq 2$ such that $tm_1 \in \operatorname{Mon}(g_i m_i)$, so m_i divides tm_1. This implies that $\operatorname{lcm}(m_1, m_i)$ divides tm_1, so $tm_1 \in \widetilde{J}$. We conclude that $f \in \widetilde{J}$, so (11.2) is established.

Since $\widetilde{I} = J + (m_1)$, Lemma 11.7 and Proposition 11.4 yield

$$H_{\widetilde{I}}(t) = H_{(m_1)}(t) + H_J(t) - H_{\widetilde{J}}(t) = \frac{1 - t^{\deg(m_1)}}{(1-t)^{n+1}} + H_J(t) - H_{\widetilde{J}}(t),$$

completing the proof. □

A consequence of the correctness of Algorithm 11.8 is that the Hilbert series $H_I(t)$ can be written as a rational function with $(1-t)^{m+1}$ as denominator. Going one step further, we can extract information about the Hilbert function from this. The results are stated in the following corollary.

Corollary 11.10 (Hilbert–Serre theorem). *Let $I \subseteq K[x_1, \ldots, x_n]$ be an ideal. Then the Hilbert series has the form*

$$H_I(t) = \frac{a_0 + a_1 t + \cdots + a_k t^k}{(1-t)^{n+1}} \tag{11.3}$$

with $k \in \mathbb{N}_0$ and $a_i \in \mathbb{Z}$. Moreover, the Hilbert function $h_I(d)$ is a polynomial for large d. More precisely, the polynomial

$$p_I := \sum_{i=0}^{k} a_i \binom{x+n-i}{n} \in \mathbb{Q}[x] \tag{11.4}$$

satisfies

$$h_I(d) = p_I(d) \tag{11.5}$$

for all sufficiently large integers d.

Proof. Induction on the recursion depth in Algorithm 11.8 immediately yields (11.3). By Remark 11.5, we can write $\frac{1}{(1-t)^{n+1}} = \sum_{d=0}^{\infty} \binom{d+n}{n} t^d$, so

$$H_I(t) = \sum_{i=0}^{k} \sum_{d=i}^{\infty} a_i \binom{d+n-i}{n} t^d = \sum_{d=0}^{\infty} \sum_{i=0}^{\min\{d,k\}} a_i \binom{d+n-i}{n} t^d, \tag{11.6}$$

and we see that the definition of p_I according to (11.4) yields (11.5) for $d \geq k$. □

It is not hard to determine the largest integer d for which (11.5) fails. This is done in Exercise 11.5. Corollary 11.10 prompts the following definition.

Definition 11.11. *The polynomial $p_I \in \mathbb{Q}[x]$ from Corollary 11.10 is called the* **Hilbert polynomial** *of I.*

The Hilbert polynomial can be calculated using Algorithm 11.8 and then applying (11.4). Having assigned a polynomial p_I to an ideal I, it is natural to ask whether such numbers as the degree and the leading coefficient of p_I mean anything interesting for I. These questions will be addressed in the following section.

11.2 Hilbert Polynomials and Dimension

We have seen in Remark 11.3(b) that the Hilbert series and the Hilbert polynomial are *not* invariants of an affine algebra. However, the following lemma tells us that the *degree* of the Hilbert polynomial is an invariant.

Lemma 11.12 (The degree of the Hilbert polynomial is an invariant). *Let $I \subseteq K[x_1, \ldots, x_n]$ and $J \subseteq K[y_1, \ldots, y_m]$ be ideals in polynomial rings such that the K-algebras $A := K[x_1, \ldots, x_n]/I$ and $B := K[y_1, \ldots, y_m]/J$ are isomorphic. Then*

$$\deg(p_I) = \deg(p_J).$$

Proof. We have an isomorphism $\varphi: A \to B$ of K-algebras, so there exist polynomials $g_1, \ldots, g_m \in K[x_1, \ldots, x_n]$ such that $\varphi(g_i + I) = y_i + J$. Set $k := \max\{\deg(g_1), \ldots, \deg(g_m)\}$. Then for every $d \in \mathbb{N}_0$, $B_{\leq d}$ is contained in $\varphi(A_{\leq kd})$, so

$$h_J(d) \leq \dim_K(\varphi(A_{\leq kd})) = h_I(kd).$$

This implies that p_J cannot have a degree greater than p_I. By symmetry, the degrees are equal. □

So if A is an affine algebra, we may choose generators $a_1, \ldots, a_n$ and consider the kernel I of the map $K[x_1, \ldots, x_n] \to A$, $x_i \mapsto a_i$. Then $A \cong K[x_1, \ldots, x_n]/I$, so by Lemma 11.12, $\deg(p_I)$ is independent of the choice of the generators. We will use this by choosing a very convenient generating set, coming from Noether normalization, to prove that this degree is actually equal to the Krull dimension of A.

Theorem 11.13 (Degree of the Hilbert polynomial and Krull dimension). *Let $A \cong K[x_1, \ldots, x_n]/I$ be an affine algebra. Then*

$$\deg(p_I) = \dim(A).$$

Proof. The result is correct (by our various conventions) if A is the zero ring, so we may assume that $A \neq \{0\}$. By Noether normalization (Theorem 8.19) and by Theorem 8.4, there exist algebraically independent elements $c_1, \ldots, c_m \in A$ with $m = \dim(A)$, and further elements $b_1, \ldots, b_r \in A$ such that $A = \sum_{j=1}^{r} C \cdot b_j$, where $C := K[c_1, \ldots, c_m] \subseteq A$. We may assume that $b_1 = 1$. Let $y_1, \ldots, y_m$ and $z_1, \ldots, z_r$ be indeterminates and let $J \subset K[y_1, \ldots, y_m, z_1, \ldots, z_r]$ be the kernel of the map $K[y_1, \ldots, y_m, z_1, \ldots, z_r] \to A$, $y_i \mapsto c_i$, $z_j \mapsto b_j$. By Lemma 11.12, $\deg(p_I) = \deg(p_J)$, so we need to show that $\deg(p_J) = m$. Write

$$B_{\leq d} := \{f + J \mid f \in K[y_1, \ldots, y_m, z_1, \ldots, z_r], \ \deg(f) \leq d\}$$

and

$$C_{\leq d} := \{f + J \mid f \in K[y_1, \ldots, y_m], \ \deg(f) \leq d\}.$$

Since $C_{\leq d} \subseteq B_{\leq d}$ for every $d \in \mathbb{N}_0$, we obtain

$$h_J(d) \geq \dim_K(C_{\leq d}) = \binom{d+m}{m},$$

where we used the algebraic independence of the c_i and Remark 11.5. This implies $\deg(p_J) \geq m$.

To prove the reverse inequality, observe that for $0 \leq i \leq j \leq r$, the product $b_i b_j$ can be written as $b_i b_j = \sum_{k=1}^{r} a_{i,j,k} b_k$ with $a_{i,j,k} \in C$. There exists a positive integer e such that $a_{i,j,k} \in C_{\leq e}$ for all i, j, k. So $b_i b_j \in \sum_{k=1}^{r} C_{\leq e} \cdot b_k$, and by induction we see that the product of s of the b_i lies in $\sum_{k=1}^{r} C_{\leq (s-1)e} \cdot b_k$ (for $s > 0$). It follows that

$$B_{\leq d} \subseteq C_{\leq d} \cdot b_1 + \sum_{s=1}^{d} \sum_{k=1}^{r} C_{\leq d-s} \cdot C_{\leq (s-1)e} \cdot b_k \subseteq \sum_{k=1}^{r} C_{\leq ed} \cdot b_k =: V_d$$

for all $d \geq 0$. We obtain that

$$h_J(d) \leq \dim_K(V_d) \leq r \cdot \dim_K(C_{\leq ed}) = r \cdot \binom{ed+m}{m},$$

where we used Remark 11.5 again. As a polynomial in d, this upper bound has degree m, so we conclude $\deg(p_J) \leq m$. This completes the proof. □

Theorem 11.13 provides a new interpretation of the concept of the dimension of an affine variety X. Indeed, if X is given by an ideal I, then $h_I(d)$ is a measure for the quantity of regular functions on X of degree at most d. So the dimension of X may be seen as the rate at which the quantity of regular functions grows with the degree.

Since the Hilbert polynomial can be calculated with just one Gröbner basis computation, we also have an improved algorithm for computing the dimension of an affine algebra. Recall that the first method for this, which we discussed in Section 9.2 on page 128, requires several Gröbner basis computations. The following corollary will enable us to make a further optimization.

Corollary 11.14 (Computing dimension via the leading ideal). *Let $I \subseteq K[x_1, \ldots, x_n]$ be an ideal, and let $L(I)$ be its leading ideal with respect to a total degree ordering. Then*

$$\dim\left(K[x_1, \ldots, x_n]/I\right) = \dim\left(K[x_1, \ldots, x_n]/L(I)\right).$$

Proof. This follows from Theorems 11.6 and 11.13. □

Corollary 11.14 is actually true for arbitrary monomial orderings. This is shown in Exercise 11.7 by generalizing the results from this chapter to the case of weighted degrees. Another way of proving Corollary 11.14 for arbitrary monomial orderings is by working with a so-called *flat deformation*. This method is more difficult but conceptually very interesting, so let us say a few words about it. Geometrically speaking, one constructs an affine variety Z together with a morphism $f \colon Z \to K^1$ such that the fiber over $1 \in K^1$ is $X := \mathcal{V}(I)$, the variety of the given ideal, and the fiber over $0 \in K^1$ is $Y := \mathcal{V}(L(I))$. The variety Z together with f can be considered as a *family* of varieties, given by the fibers of f, and since X and Y occur as fibers, we can view the passage from X to Y as a *deformation*. The important point is that Z is constructed in such a way that the homomorphism $K[t] \to K[Z]$ induced by f makes $K[Z]$ into a free $K[t]$-module. So we could speak of a free family and a free deformation. Since freeness implies flatness, we may follow the traditional way of speaking of flatness, enabling us to utter very nice sentences such as: "The passage from an ideal to its leading ideal constitutes a flat deformation." This flatness, or freeness, property means that an ideal and its leading ideal are strongly related, and may be seen as one of the sources of the usefulness of Gröbner bases. But why does it imply equality of the dimensions of X and Y? This follows from applying the tools that we developed in Section 7.2, in particular Lemmas 7.15 and 7.16, but more considerations are needed. Let us give a hint how Z is constructed. One uses a weighted degree as given by Exercise 9.2(c), and then forms an ideal $\widetilde{I} \subseteq K[t, x_1, \ldots, x_n]$ from a Gröbner basis of I by homogenization. The freeness then follows by applying the constructive version of the generic freeness lemma (Lemma 10.1). See Eisenbud [17, Section 15.8] or Greuel and Pfister [22, Section 7.5] for more on Gröbner bases and flatness.

The leading ideal $J := L(I)$ is a monomial ideal, i.e., it is generated by monomials $m_1, \ldots, m_r$. It is especially simple to compute the dimension of $A := K[x_1, \ldots, x_n]/J$ for a monomial ideal J using Theorem 5.9 and Proposition 5.10. In fact, a set $M \subseteq \{x_1, \ldots, x_n\}$ of indeterminates is algebraically dependent modulo J if and only if there exists a monomial m_j among the generators of J that involves only indeterminates from M. So M is algebraically independent modulo J if and only if every m_j involves an indeterminate x_i that is not in M. And the complement $\{x_1, \ldots, x_n\} \setminus M$ is algebraically independent modulo J if and only if every m_j involves an $x_i \in M$. This leads to the following algorithm.

Algorithm 11.15 (Dimension of an affine algebra).

Input: An ideal $I \subseteq K[x_1, \ldots, x_n]$ defining an affine algebra $A := K[x_1, \ldots, x_n]/I$.
Output: The Krull dimension $\dim(A)$.

(1) Choose a total degree ordering "$\leq$" on $K[x_1, \ldots, x_n]$ and compute a Gröbner basis G of I with respect to "$\leq$". Let $m_1, \ldots, m_r$ be the leading monomials of the nonzero elements of G. *(In fact, by Exercise 11.7 the algorithm works for arbitrary monomial orderings.)*
(2) If $m_j = 1$ for some j, return $\dim(A) = -1$.
(3) By an exhaustive search, find a set $M \subseteq \{x_1, \ldots, x_n\}$ of minimal size such that every m_j involves at least one indeterminate from M.
(4) Return $\dim(A) = n - |M|$.

Let us emphasize again that Algorithm 11.15 requires just one Gröbner basis computation, and this can be performed with respect to the grevlex ordering, which tends to be the fastest. After the Gröbner basis computation, the algorithm is purely combinatorial. Even when implemented in a crude way, the cost of step (3) (where the combinatorics happens) will in most cases be dwarfed by the cost of the preceding Gröbner basis computation. Nevertheless, it is interesting to think about optimizing this step. An optimized version can be found in Becker and Weispfenning [3, Algorithm 9.6].

We close this chapter with a definition.

Definition 11.16. *For $I \subseteq K[x_1, \ldots, x_n]$ a proper ideal, let $m = \deg(p_I)$ be the degree and $\mathrm{LC}(p_I)$ the leading coefficient of the Hilbert polynomial. Then*

$$\deg(I) := m! \cdot \mathrm{LC}(p_I)$$

is called the **degree** *of I.*

This is certainly a valid definition, but does the degree of an ideal have any meaning, and is it useful? Exercises 11.8–11.10 deal with the degree, and there we learn that the degree is a positive integer, and how the classical theorem of Bézout can be proved by using it. Suffice it to make a few additional comments here. The degree of a principal ideal $I = (f)$ with $f \in K[x_1, \ldots, x_n]$ nonconstant is $\deg(I) = \deg(f)$. The easiest way to see this is to use Proposition 11.4 and Exercise 11.8. This may be a first indication that the definition of the degree is a good one. Moreover, if $X \subseteq K^n$ is a finite set of points and $I := \mathcal{I}(X)$ its ideal, then $\deg(I) = |X|$. This follows from Exercise 5.4. Finally, notice that the degree is not an invariant of an affine algebra (see the example in Remark 11.3(b)).

Exercises for Chapter 11

11.1 (Comparing ideals by their Hilbert functions). Let I and J be ideals in $K[x_1, \ldots, x_n]$ such that $I \subseteq J$ and $h_I = h_J$. Show that $I = J$.

11.2 (Hypotheses of Theorem 11.6). Find an example that shows that the hypothesis on the monomial ordering cannot be dropped from Theorem 11.6.

11.3 (Homogeneous ideals). In this exercise we assume K to be infinite.

(a) Let $I \subseteq K[x_1, \ldots, x_n]$ be an ideal. For $a \in K$, let $\varphi_a\colon K[x_1, \ldots, x_n] \to K[x_1, \ldots, x_n]$ be the algebra endomorphism given by $\varphi_a(x_i) = a x_i$. Show that the following statements are equivalent:

   (1) I is homogeneous.
   (2) The inclusion

$$\varphi_a(I) \subseteq I \tag{11.7}$$

   holds for all $a \in K$.

(3) The inclusion (11.7) holds for infinitely many $a \in K$.
(4) For every $f \in I$, all homogeneous parts of f also lie in I.

(b) Let $I \subseteq K[x_1, \dots, x_n]$ be a homogeneous ideal. Show that all prime ideals $P \subset K[x_1, \dots, x_n]$ that are minimal over I are homogeneous.

Remark: Part (b) and the equivalence of (1) and (4) also hold if K is a finite field (see Eisenbud [17, Theorem 3.1(a) and Proposition 3.12] for part (b)).

11.4 (Hypotheses of Lemma 11.7). Give an example that shows that the homogeneity hypothesis in Lemma 11.7 cannot be dropped.

11.5 (For which d is the Hilbert polynomial correct?). Let $I \subsetneqq K[x_1, \dots, x_n]$ be a proper ideal. Use the proof of Corollary 11.10 to show that the largest integer d_{fail} for which (11.5) fails is given by

$$d_{\text{fail}} = \deg\left(H_I(t)\right),$$

where the degree of the rational function is defined as the difference between the degrees of the numerator and denominator. (Since we are talking about integer values for d here, we need to extend h_I to $\mathbb{Z}$ by setting $h_I(d) := 0$ for $d < 0$.) Notice that d_{fail} may be negative.
Remark: The number d_{fail} is closely related to the so-called a-invariant of a graded algebra A. This is defined as the degree of the graded Hilbert series $H_A^{\text{grad}}(t)$. So if in the above setting I is homogeneous, then the a-invariant is $d_{\text{fail}} + 1$.

11.6 (Computing the Hilbert series). Let $I \subseteq K[x_1, x_2, x_3]$ be the ideal given by

$$I = (x_1^2 x_2^2, x_1^2 x_3^2, x_2^2 x_3^2).$$

Determine the Hilbert series, Hilbert function, and Hilbert polynomial of I. For which d do the Hilbert function and the Hilbert polynomial coincide? Determine $\dim\left(K[x_1, x_2, x_3]/I\right)$ and $\deg(I)$.

11.7 (Weighted degrees and Hilbert series). Let $\mathbf{w} = (w_1, \dots, w_n)$ be a "weight vector" with $w_i \in \mathbb{N}_{>0}$ positive integers. Define the **weighted degree** of a monomial $t = x_1^{e_1} \cdots x_n^{e_n}$ as $\deg_{\mathbf{w}}(t) := \sum_{i=1}^n w_i e_i$. Starting with this definition, go through the definitions and results of this chapter, and adjust everything to the "weighted situation." Which modifications are necessary? Develop the theory to include a version of Corollary 11.14 for "weighted degree orderings." Then use Exercise 9.2 to conclude that Corollary 11.14 holds for an arbitrary monomial ordering. *(Solution on page 227)*

11.8 (Extracting dimension and degree from the Hilbert series). For $I \subsetneqq K[x_1, \dots, x_n]$ a proper ideal, prove the following statements.

(a) The dimension of $K[x_1, \dots, x_n]/I$ is the smallest integer m such that the Hilbert series can be written as

$$H_I(t) = \frac{g(t)}{(1-t)^{m+1}} \tag{11.8}$$

with $g(t) \in \mathbb{Z}[t]$.

(b) If $m = \dim(K[x_1, \dots, x_n]/I)$ and $g(t)$ as in (11.8), the degree of I is

$$\deg(I) = g(1).$$

In particular, the degree is a positive integer.

Remark: The results of this exercise can be restated as follows: One plus the dimension m is the pole order at $t = 1$ of the Hilbert series, and the degree is $(-1)^m$ times the first nonzero coefficient in a Laurent series expansion of $H_I(t)$ about $t = 1$.

11.9 (The degree of intersections of ideals). In this exercise we deal with the degree of intersections of homogeneous ideals. If $I \subseteq K[x_1, \dots, x_n]$ is an ideal, we will write $\dim(I)$ instead of $\dim(K[x_1, \dots, x_n]/I)$ for brevity.

(a) Let $I, J \subseteq K[x_1, \dots, x_n]$ be homogeneous ideals such that $\dim(I) = \dim(J) < \dim(I + J)$. Show that

$$\deg(I \cap J) = \deg(I) + \deg(J).$$

(b) Let $I, J \subseteq K[x_1, \dots, x_n]$ be homogeneous ideals such that $\dim(I) > \dim(J)$. Show that

$$\deg(I \cap J) = \deg(I).$$

(c) Let $I \subseteq K[x_1, \dots, x_n]$ be a homogeneous ideal, and let $P_1, \dots, P_r \in \operatorname{Spec}(K[x_1, \dots, x_n])$ be those minimal prime ideals over I that satisfy $\dim(P_i) = \dim(I)$. Show that

$$\deg\left(\sqrt{I}\right) = \sum_{i=1}^{r} \deg(P_i).$$

In this part, you may assume K to be infinite, so that Exercise 11.3(b) can be used, or you may use the remark at the end of Exercise 11.3.

(d) Let $I \subsetneqq K[x_1, \dots, x_n]$ be a proper ideal. Show that

$$\deg\left(\sqrt{I}\right) \le \deg(I).$$

***11.10 (The degree and Bézout's theorem).** In this exercise we use the degree of an ideal for proving the theorem of Bézout, which is a well-known, classical result from algebraic geometry. If necessary in order to use the results from Exercise 11.3 and Exercise 11.9, assume K to be infinite. Prove the following statements.

(a) If $I \subseteq K[x_1, \dots, x_n]$ is a homogeneous prime ideal and $f \in K[x_1, \dots, x_n] \setminus I$ a homogeneous polynomial, then

$$\deg\left(I + (f)\right) = \deg(I) \cdot \deg(f).$$

(b) Let $I = (f_1, \dots, f_m)$ with $f_i \in K[x_1, \dots, x_n]$ homogeneous, and let $P_1, \dots, P_r \in \operatorname{Spec}\left(K[x_1, \dots, x_n]\right)$ be the prime ideals that are minimal over I. If $\dim\left(K[x_1, \dots, x_n]/I\right) = n - m$, then

$$\sum_{i=1}^{r} \deg(P_i) \le \prod_{j=1}^{m} \deg(f_j).$$

(c) Let $f_1, \dots, f_{n-1} \in K[x_1, \dots, x_n]$ be homogeneous polynomials such that $\dim\left(K[x_1, \dots, x_n]/(f_1, \dots, f_{n-1})\right) = 1$. Then $f_1, \dots, f_{n-1}$ have at most $\prod_{j=1}^{n-1} \deg(f_j)$ common zeros in the projective space $\mathbb{P}^{n-1}(K)$.

(d) Let $f_1, f_2 \in K[x_1, x_2]$ be two coprime polynomials (which need not be homogeneous). Then the number of common zeros (in K^2) of f_1 and f_2 is at most $\deg(f_1) \cdot \deg(f_2)$.

Remark: Part (c) is Bézout's theorem. There are a number of refinements. Most importantly, if we assume K to be algebraically closed and count the common zeros of the f_i by (adequately defined) multiplicities, then we obtain an equality instead of an upper bound. It should also be noted that in the situation of part (b), the equality $\deg(I) = \prod_{j=1}^{m} \deg(f_j)$ holds. The proof requires some results from the theory of Cohen–Macaulay rings, which is not treated in this book.

Part IV
Local Rings

Chapter 12
Dimension Theory

Recall that a ring is called *local* if it has precisely one maximal ideal. One of the reasons for the interest in local rings is the idea that they describe the local behavior of a global object (such as an affine variety). A large part of the research in commutative algebra is devoted to local rings. A typical talk at a typical commutative algebra conference has the speaker start with, "Let R be a Noetherian local ring with maximal ideal $\mathfrak{m}$ ..." Therefore it seems more than appropriate to devote the last part of this book to local rings.

In this chapter we develop a central element in the theory of local rings, often referred to as *dimension theory*. To R (as introduced by our typical speaker) we associate a graded algebra $\mathrm{gr}(R)$ over the field $R/\mathfrak{m}$. The main result is that R and $\mathrm{gr}(R)$ have the same Krull dimension. This result is obtained by comparing "sizes" of R and $\mathrm{gr}(R)$. For $\mathrm{gr}(R)$, the notion of "size" is given by the Hilbert function, which in Chapter 11 was shown to be eventually equal to a polynomial function. For R, "size" is measured by the so-called *lengths* of the modules $R/\mathfrak{m}^d$. So we need to discuss the concept of length first. This will be done in the following section.

Readers who have skipped Chapters 9–11 will find instructions in Exercise 12.1 on what is needed to continue without tearing any holes in the proofs.

12.1 The Length of a Module

The goal of the following definition is to measure the size of a module M over a ring R. In the theory of vector spaces, size is measured by the number of vectors in a basis. But in general, modules have no basis (if they do, they are called free). In view of this, we resort to the idea of considering chains of submodules.

Definition 12.1. *Let M be a module over a ring R. The* **length** *of M, written as* $\mathrm{length}(M)$, *is the supremum of the lengths n of chains*

G. Kemper, *A Course in Commutative Algebra*, Graduate Texts in Mathematics 256, DOI 10.1007/978-3-642-03545-6_13,

$$M_0 \subsetneqq M_1 \subsetneqq \cdots \subsetneqq M_n$$

of submodules $M_i \subseteq M$. *So* $\text{length}(M) \in \mathbb{N}_0 \cup \{\infty\}$.

Example 12.2. (1) For m a positive integer, the $\mathbb{Z}$-module $M = \mathbb{Z}/(m)$ has length equal to the number of prime factors (with multiplicities) of m.
(2) $\mathbb{Z}$ has infinite length as a module over itself.
(3) If $R = K$ is a field and V a vector space, then $\text{length}(V) = \dim_K(V)$.
(4) An R-module M is simple (i.e., $M \neq \{0\}$ and there exists no nonzero, proper submodule) if and only if $\text{length}(M) = 1$. Moreover, $M = \{0\}$ if and only if $\text{length}(M) = 0$. ◁

Recall the definition of a maximal chain on page 106. In particular, a finite chain $M_0 \subsetneqq M_1 \subsetneqq \cdots \subsetneqq M_n$ of submodules of M is maximal if no further submodule can be added into the chain by insertion or by appending at either end. This means that $M_0 = \{0\}$, $M_n = M$, and all M_i/M_{i-1} ($i = 1, \ldots, n$) are simple modules. A finite maximal chain of submodules of M is also called a *composition series* of M.

Theorem 12.3 (Basic facts about length). *Let M be a module over a ring R.*

(a) If M has a finite maximal chain $M_0 \subsetneqq M_1 \subsetneqq \cdots \subsetneqq M_n$ of submodules, then $\text{length}(M) = n$. *So in particular, all maximal chains have the same length.*
(b) M has finite length if and only if it is Artinian and Noetherian. In particular, R has finite length as a module over itself if and only if it is Artinian.
(c) Let $N \subseteq M$ be a submodule. Then

$$\text{length}(M) = \text{length}(N) + \text{length}(M/N).$$

Proof. (a) We use induction on n. If $n = 0$, then $M = \{0\}$ and so $\text{length}(M) = 0$. Therefore we may assume $n > 0$. Let $N \subsetneqq M$ be a proper submodule, and set $N_i := N \cap M_i$. The N_i need not be distinct, but the set $\mathcal{C} := \{N_0, \ldots, N_n\}$ is a chain of submodules of N. We have $N_0 = \{0\}$ and $N_n = N$. Moreover, for $1 \leq i \leq n$, the natural map $N_i \to M_i/M_{i-1}$ induces an isomorphism from N_i/N_{i-1} to a submodule of M_i/M_{i-1}, so N_i/N_{i-1} is either simple or zero. Therefore $\mathcal{C}$ is a maximal chain of submodules of N. Clearly $\text{length}(\mathcal{C}) \leq n$. By way of contradiction, assume that $\text{length}(\mathcal{C}) = n$. We will show by induction on i that this implies $N_i = M_i$ for all i. This is true for $i = 0$. For $i > 0$, N_i/N_{i-1} is nonzero by assumption, so the map $N_i \to M_i/M_{i-1}$ is surjective. Using induction, we conclude that $M_i \subseteq N_i + M_{i-1} = N_i + N_{i-1} = N_i \subseteq M_i$, proving our claim. In particular, we obtain $N = N_n = M_n = M$, a contradiction. So $\text{length}(\mathcal{C}) < n$. By induction on n, this implies $\text{length}(N) < n$. Since this holds for all proper submodules N, we conclude that $\text{length}(M) \leq n$. On the other hand, we are given a chain of submodules of M of length n, so part (a) follows.

(b) If is clear from the definition that a module of finite length has to be Artinian and Noetherian. Conversely, if M is Noetherian and $M \neq \{0\}$, there exists a maximal proper submodule $M_1 \subsetneqq M$. If $M_1 \neq \{0\}$, we can continue and find $M_2 \subsetneqq M_1$ with no submodules in between, and so on. If M is Artinian, this process stops, so we end up with a finite chain $\{0\} = M_k \subsetneqq M_{k-1} \subsetneqq \cdots \subsetneqq M_1 \subsetneqq M_0 := M$. This chain is maximal by construction, so $\operatorname{length}(M) = k$ by part (a). The last statement follows since every Artinian ring is Noetherian by Theorem 2.8.

(c) If $\operatorname{length}(N) = \infty$ or $\operatorname{length}(M/N) = \infty$, then by part (b) at least one of these modules fails to be Noetherian or Artinian, so by Proposition 2.4 the same is true for M, and $\operatorname{length}(M) = \infty$, too. So we may assume that N and M/N have finite length. Taking maximal chains of submodules of N and of M/N, lifting the latter into M, and putting these chains together yields a maximal chain of submodules of M. So (c) follows by part (a). □

By applying Theorem 12.3(c) several times, we see that if $\{0\} = M_0 \subseteq M_1 \subseteq \cdots \subseteq M_k = M$ is a chain of submodules, then

$$\operatorname{length}(M) = \sum_{i=1}^{k} \operatorname{length}\left(M_i/M_{i-1}\right). \tag{12.1}$$

A special case of Theorem 12.3(c) says that a direct sum $M \oplus N$ of R-modules has $\operatorname{length}(M \oplus N) = \operatorname{length}(M) + \operatorname{length}(N)$. With Theorem 12.3(b), this shows that the free module $M = R^n$ over an Artinian ring R has finite length. Since any finitely generated module is a factor module of some R^n, another application of Theorem 12.3(c) shows that a finitely generated module over an Artinian ring has finite length. Exercise 12.2 deals with a further consequence of Theorem 12.3(c) on exact sequences of modules.

An important example to which Theorem 12.3(b) applies is the following. If R is a Noetherian local ring with maximal ideal $\mathfrak{m}$ and $\mathfrak{q} \subseteq R$ is an ideal with $\sqrt{\mathfrak{q}} = \mathfrak{m}$, then $R/\mathfrak{q}$ has dimension 0. Therefore it is Artinian and Noetherian (as a ring and therefore also as an R-module) by Theorem 2.8, so $\operatorname{length}(R/\mathfrak{q}) < \infty$. In particular, this applies to $\mathfrak{q} = \mathfrak{m}^d$, a power of the maximal ideal. The following result is about the lengths of the modules $R/\mathfrak{m}^d$.

Lemma 12.4. *Let R be a Noetherian local ring with maximal ideal $\mathfrak{m}$. Then there exists a polynomial $p \in \mathbb{Q}[x]$ of degree $n = \dim(R)$ such that*

$$\operatorname{length}\left(R/\mathfrak{m}^d\right) \leq p(d) \quad \text{for all} \quad d \in \mathbb{N}_0.$$

Proof. The main idea is to use a system of parameters $a_1, \ldots, a_n \in \mathfrak{m}$, whose existence is guaranteed by Corollary 7.9. Setting $\mathfrak{q} := (a_1, \ldots, a_n)$, we have

$\mathfrak{q}^d \subseteq \mathfrak{m}^d$ for all $d \in \mathbb{N}_0$, so there is an epimorphism $R/\mathfrak{q}^d \to R/\mathfrak{m}^d$. By Theorem 12.3(c), this implies

$$\text{length}\left(R/\mathfrak{m}^d\right) \leq \text{length}\left(R/\mathfrak{q}^d\right). \tag{12.2}$$

Consider the chain

$$\{0\} = \mathfrak{q}^d/\mathfrak{q}^d \subseteq \mathfrak{q}^{d-1}/\mathfrak{q}^d \subseteq \cdots \subseteq \mathfrak{q}^2/\mathfrak{q}^d \subseteq \mathfrak{q}/\mathfrak{q}^d \subseteq R/\mathfrak{q}^d,$$

which has factors $\mathfrak{q}^i/\mathfrak{q}^{i+1}$. For every i, $\mathfrak{q}^i$ is generated (as an R-module) by the monomials of degree i in $a_1, \ldots, a_n$. By Remark 11.5, there are $k_i := \binom{i+n}{n} - \binom{i-1+n}{n}$ such monomials. So $\mathfrak{q}^i/\mathfrak{q}^{i+1}$ can be generated by k_i elements as an $R/\mathfrak{q}$-module, giving an epimorphism $(R/\mathfrak{q})^{k_i} \to \mathfrak{q}^i/\mathfrak{q}^{i+1}$. Using (12.2), (12.1), and Theorem 12.3(c), we conclude that

$$\begin{aligned} \text{length}\left(R/\mathfrak{m}^d\right) \leq \text{length}\left(R/\mathfrak{q}^d\right) &= \sum_{i=0}^{d-1} \text{length}\left(\mathfrak{q}^i/\mathfrak{q}^{i+1}\right) \\ &\leq \sum_{i=0}^{d-1} \text{length}\left((R/\mathfrak{q})^{k_i}\right) \\ &= \sum_{i=0}^{d-1} k_i \cdot \text{length}\,(R/\mathfrak{q}) = \binom{d-1+n}{n} \cdot \text{length}\,(R/\mathfrak{q}). \end{aligned}$$

By Corollary 7.9 we have $\sqrt{\mathfrak{q}} = \mathfrak{m}$, so $\text{length}\,(R/\mathfrak{q}) < \infty$ by the discussion preceding the lemma. Therefore the above inequality yields the desired upper bound. □

In Chapter 11, we have found that the Hilbert function of an ideal is essentially a polynomial function. Somewhat similarly, Lemma 12.4 relates the function $d \mapsto \text{length}\left(R/\mathfrak{m}^d\right)$ to a polynomial. Is there a connection? The answer is yes. In fact, we will interpret the function $d \mapsto \text{length}\left(R/\mathfrak{m}^{d+1}\right)$ as the Hilbert function of the associated graded ring, to be defined in the next section.

12.2 The Associated Graded Ring

Throughout this section, R will be a Noetherian local ring with maximal ideal $\mathfrak{m}$. We will write $K := R/\mathfrak{m}$ for the residue class field.

We first consider the subalgebra

$$R^* := R[\mathfrak{m} \cdot t] \subseteq R[t]$$

of the polynomial ring $R[t]$ generated by all at with $a \in \mathfrak{m}$. Eisenbud [17] calls R^* the *blowup algebra*, while Lang [33] calls it the *first associated graded ring*. A close relative of R^* is the *Rees ring* (see Matsumura [38, §15, Section 4]).

Recall that a ring S is called graded if it has a direct sum decomposition

$$S = S_0 \oplus S_1 \oplus S_2 \oplus \cdots = \bigoplus_{d \in \mathbb{N}_0} S_d$$

(as an abelian group) such that $S_i \cdot S_j \subseteq S_{i+j}$ for all $i, j \in \mathbb{N}_0$. Then elements of S_d are called *homogeneous* of degree d. The ring $R[t]$ is graded by $R[t]_d := R \cdot t^d$, and since R^* is generated by homogeneous elements, it is graded by

$$R_d^* := R \cdot t^d \cap R^* = \mathfrak{m}^d \cdot t^d \cong \mathfrak{m}^d$$

(where the second equality is obvious from the definition of R^*). Now we define the **associated graded ring** of R as the quotient ring $\mathrm{gr}(R) := R^*/(\mathfrak{m})_{R^*}$. Since $\mathrm{gr}(R)$ is formed by factoring out an ideal generated by homogeneous elements (of degree 0), $\mathrm{gr}(R)$ is graded, too. Moreover, $\mathrm{gr}(R)$ is annihilated by $\mathfrak{m}$, so it is a graded K-algebra. There is an important description of the graded components:

$$\mathrm{gr}(R)_d = R_d^*/(R_d^* \cdot \mathfrak{m}) \cong \mathfrak{m}^d/\mathfrak{m}^{d+1}. \tag{12.3}$$

In particular, $\mathrm{gr}(R)_0 \cong K$. If we view the graded components of $\mathrm{gr}(R)$ as $\mathfrak{m}^d/\mathfrak{m}^{d+1}$ according to the above isomorphism, multiplication of homogeneous elements works as follows: For $a \in \mathfrak{m}^i$ and $b \in \mathfrak{m}^j$, the product of the residue classes is

$$(a + \mathfrak{m}^{i+1}) \cdot (b + \mathfrak{m}^{j+1}) = ab + \mathfrak{m}^{i+j+1} \in \mathfrak{m}^{i+j}/\mathfrak{m}^{i+j+1}. \tag{12.4}$$

Some textbooks define $\mathrm{gr}(R)$ as a graded algebra with components $\mathfrak{m}^d/\mathfrak{m}^{d+1}$, and multiplication is given by (12.4). It should be pointed out that $\mathrm{gr}(R)$ and R^* can be (and often are) defined in greater generality, where R is any ring and $\mathfrak{m}$ is substituted by any ideal $I \subseteq R$. Exercise 12.5 gives a presentation of the associated graded ring of the coordinate ring of an affine variety, localized at a point. This is very interesting, since it provides a geometric interpretation in terms of the so-called tangent cone. Some explicit examples of associated graded rings are computed in Exercise 12.6.

Is $\mathrm{gr}(R)$ Noetherian? The answer is yes, for a very simple reason: Every ideal in R is finitely generated, so in particular $\mathfrak{m} = (c_1, \ldots, c_k)_R$. Therefore $R^* = R[c_1 t, \ldots, c_k t]$ is Noetherian by Corollary 2.12, and so $\mathrm{gr}(R)$ is Noetherian, too. In fact, it is generated as a K-algebra by the elements $a_i := c_i t + (\mathfrak{m})_{R^*}$. Notice that these generators lie in the graded component $\mathrm{gr}(R)_1$ of degree 1. Graded algebras that are generated by their degree-1 component are called *standard graded*. To make the connection to Hilbert functions, let $K[x_1, \ldots, x_k]$ be a polynomial ring and let $I \subseteq K[x_1, \ldots, x_k]$

be the kernel of the map $K[x_1, \dots, x_k] \to \operatorname{gr}(R)$, $x_i \to a_i$. It is clear from the definitions that $\operatorname{gr}(R)_d$ is generated as a K-vector space by the monomials of degree d in the a_i. Setting $A := K[x_1, \dots, x_k]/I \cong \operatorname{gr}(R)$, we get

$$A_{\leq d} \cong \bigoplus_{i=0}^{d} \operatorname{gr}(R)_i \cong \bigoplus_{i=0}^{d} \mathfrak{m}^i/\mathfrak{m}^{i+1}$$

for all d, so using (12.1) yields

$$\begin{aligned} \operatorname{length}\left(R/\mathfrak{m}^{d+1}\right) &= \sum_{i=0}^{d} \operatorname{length}\left(\mathfrak{m}^i/\mathfrak{m}^{i+1}\right) \\ &= \sum_{i=0}^{d} \dim_K\left(\mathfrak{m}^i/\mathfrak{m}^{i+1}\right) = \dim_K(A_{\leq d}) = h_I(d). \end{aligned} \tag{12.5}$$

With Corollary 11.10 and Theorem 11.13 we have proved the following:

Proposition 12.5 (The Hilbert–Samuel polynomial of R). *There exists a polynomial $p_R \in \mathbb{Q}[x]$ such that*

$$\operatorname{length}\left(R/\mathfrak{m}^{d+1}\right) = p_R(d)$$

for all sufficiently large d, and $\deg(p_R) = \dim(\operatorname{gr}(R))$.

The function h_R given by $h_R(d) := \operatorname{length}\left(R/\mathfrak{m}^{d+1}\right)$ is called the **Hilbert–Samuel function** of R, and the polynomial p_R from Proposition 12.5 is called the **Hilbert–Samuel polynomial**. With this notation, (12.5) becomes

$$h_R(d) = h_I(d) \quad \text{and} \quad p_R = p_I. \tag{12.6}$$

With Proposition 12.5, Lemma 12.4 reads as

$$\dim(\operatorname{gr}(R)) \leq \dim(R). \tag{12.7}$$

The next goal is to prove that this is actually an equality. For this we need two lemmas. The first one is a version of the Artin–Rees lemma, and its proof makes essential use of the Noether property of R^*.

Lemma 12.6 (Artin–Rees lemma). *Let $I \subseteq R$ be an ideal. Then there exists a nonnegative integer r such that*

$$I \cap \mathfrak{m}^n = \mathfrak{m}^{n-r} \cdot (I \cap \mathfrak{m}^r)$$

for all $n \geq r$.

Proof. Let $J_d := \sum_{i=0}^{d} R^*(I \cap \mathfrak{m}^i)t^i$ be the ideal in R^* generated by the $(I \cap \mathfrak{m}^i)t^i$ with $i \leq d$. Since R^* is Noetherian, there exists a nonnegative integer r such that $J_n = J_r$ for $n \geq r$. Let $n \geq r$. Observe that $(I \cap \mathfrak{m}^n)t^n$ is contained in the homogeneous part R_n^* of R^* of degree n. Therefore

$$\begin{aligned}(I \cap \mathfrak{m}^n)t^n &\subseteq R_n^* \cap J_n = R_n^* \cap \sum_{i=0}^r R^*(I \cap \mathfrak{m}^i)t^i = \sum_{i=0}^r R_{n-i}^*(I \cap \mathfrak{m}^i)t^i \\ &= \sum_{i=0}^r \mathfrak{m}^{n-i}(I \cap \mathfrak{m}^i)t^n = \sum_{i=0}^r \mathfrak{m}^{n-r}\mathfrak{m}^{r-i}(I \cap \mathfrak{m}^i)t^n \\ &\subseteq \mathfrak{m}^{n-r}(I \cap \mathfrak{m}^r)t^n.\end{aligned}$$

So $\mathfrak{m}^{n-r} \cdot (I \cap \mathfrak{m}^r) \supseteq I \cap \mathfrak{m}^n \supseteq \mathfrak{m}^{n-r} \cdot (I \cap \mathfrak{m}^r)$, and the lemma is proved. □

In most textbooks, the Artin–Rees lemma is presented in a more general setting. In fact, R may be replaced by any Noetherian ring, $\mathfrak{m}$ by any ideal J, and I by a submodule of a Noetherian R-module M. Then instead of $\mathfrak{m}^n$ one considers the filtration $J^n M$ of M. One can also generalize this filtration to what is called a J-stable filtration. The proofs of the more general versions of the Artin–Rees lemma work essentially as the above proof. For details, see Eisenbud [17, Lemma 5.1]. Now we come to the second lemma.

Lemma 12.7. *Let $a \in \mathfrak{m}$. If a is not a zero divisor, then*

$$\dim(\operatorname{gr}(R/Ra)) < \dim(\operatorname{gr}(R)).$$

Proof. By Proposition 12.5, we need to show that $\deg(p_{R/Ra}) < \deg(p_R)$. So we need to compare the Hilbert–Samuel functions $h_{R/Ra}$ and h_R. Since $\mathfrak{m}/Ra$ is the maximal ideal of R/Ra, $h_{R/Ra}(d)$ is the length of the module

$$M_d := (R/Ra) \Big/ (\mathfrak{m}/Ra)^{d+1}.$$

The natural epimorphism $R \to M_d$ has kernel $Ra + \mathfrak{m}^{d+1}$, so we obtain an epimorphism $R/\mathfrak{m}^{d+1} \to M_d$ with kernel $\left(Ra + \mathfrak{m}^{d+1}\right)/\mathfrak{m}^{d+1} \cong Ra/\left(Ra \cap \mathfrak{m}^{d+1}\right)$. Therefore

$$\operatorname{length}(M_d) = \operatorname{length}\left(R/\mathfrak{m}^{d+1}\right) - \operatorname{length}\left(Ra/\left(Ra \cap \mathfrak{m}^{d+1}\right)\right), \tag{12.8}$$

where Theorem 12.3(c) was used. Applying the Artin–Rees lemma (Lemma 12.6) to $I = Ra$ yields an r such that $Ra \cap \mathfrak{m}^{d+1} = \mathfrak{m}^{d+1-r}(Ra \cap \mathfrak{m}^r) \subseteq \mathfrak{m}^{d+1-r}a$ for $d + 1 \geq r$. So for sufficiently large d we have an epimorphism

$$Ra/\left(Ra \cap \mathfrak{m}^{d+1}\right) \twoheadrightarrow Ra/\mathfrak{m}^{d+1-r}a \cong R/\mathfrak{m}^{d+1-r},$$

where the isomorphism comes from the fact that multiplication by a induces isomorphisms $\mathfrak{m}^i \cong \mathfrak{m}^i a$ of R-modules for all i (including $i = 0$). By Theorem 12.3(c), this gives a lower bound for the length of $Ra/\left(Ra \cap \mathfrak{m}^{d+1}\right)$, and substituting this into (12.8) yields

$$h_{R/Ra}(d) = \text{length}(M_d) \leq \text{length}\left(R/\mathfrak{m}^{d+1}\right) - \text{length}\left(R/\mathfrak{m}^{d+1-r}\right) = h_R(d) - h_R(d-r)$$

for sufficiently large d. From this we conclude that $\deg\left(p_{R/Ra}\right) < \deg\left(p_R\right)$, as desired. □

We are now ready to prove that R and $\text{gr}(R)$ have the same dimension. This is a central result of what is referred to as dimension theory in many textbooks (i.e., Atiyah and Macdonald [2], Matsumura [38], Eisenbud [17]), and that is why this chapter is called dimension theory, too. However, the results on parameter systems from Section 7.1 are usually considered as part of dimension theory. In fact, the main assertion of dimension theory is often stated as the equality of the following three numbers: (1) the dimension of R, (2) the size of a system of parameters of R, and (3) the degree of the Hilbert–Samuel polynomial of R.

Theorem 12.8 (The dimensions of R and $\text{gr}(R)$). *Let R be a Noetherian local ring and let $\text{gr}(R)$ be its associated graded ring. Then*

$$\dim(R) = \dim\left(\text{gr}(R)\right).$$

Equivalently, the Hilbert–Samuel polynomial of R has degree equal to $\dim(R)$.

Proof. From (12.7) we know that $\dim(\text{gr}(R)) \leq \dim(R)$. For the reverse inequality we use induction on $\dim(\text{gr}(R))$. We first reduce to the case that R is an integral domain. We need to prove that $\dim(R/P) \leq \dim(\text{gr}(R))$ for every $P \in \text{Spec}(R)$. For every d, the natural epimorphism $R/\mathfrak{m}^{d+1} \twoheadrightarrow (R/P)\big/(\mathfrak{m}/P)^{d+1}$ shows that $h_{R/P}(d) \leq h_R(d)$, so $\dim(\text{gr}(R/P)) \leq \dim(\text{gr}(R))$ by Proposition 12.5. Therefore it suffices to show that $\dim(R/P) \leq \dim(\text{gr}(R/P))$ for every $P \in \text{Spec}(R)$. In other words, we may assume that R is an integral domain.

For the induction, we first treat the case that $\dim(\text{gr}(R)) = 0$. Since $\text{gr}(R)$ is an affine K-algebra, Theorem 5.11 yields $\dim_K(\text{gr}(R)) < \infty$. This means that only finitely many graded components of $\text{gr}(R)$ are nonzero, so $\mathfrak{m}^{d+1} = \mathfrak{m}^d$ for some d. By Nakayama's lemma (Theorem 7.3), this implies $\mathfrak{m}^d = \{0\}$, so $\mathfrak{m} = \{0\}$ since R is a domain. Therefore R is a field, and $\dim(R) = 0$.

Now assume that $\dim(\text{gr}(R)) > 0$ and let $P_0 \subsetneqq P_1 \subsetneqq \cdots \subsetneqq P_k$ be a chain of prime ideals in R of length $k > 0$. Choose $a \in P_1 \setminus \{0\}$. By Lemma 12.7, $\dim(\text{gr}(R/Ra)) < \dim(\text{gr}(R))$, so by induction $\dim(R/Ra) = \dim(\text{gr}(R/Ra)) < \dim(\text{gr}(R))$. But we have a chain $P_1/Ra \subsetneqq \cdots \subsetneqq P_k/Ra$ of prime ideals in R/Ra, so $\dim(R/Ra) \geq k-1$. Putting the inequalities together, we conclude that $k \leq \dim(\text{gr}(R))$ and are done. □

We will finish this chapter by exploring more connections between R and its associated graded ring $\text{gr}(R)$. We first need to prove the following theorem, which is a consequence of the Artin–Rees lemma and Nakayama's lemma.

Theorem 12.9 (Krull's intersection theorem). *If R is a Noetherian local ring with maximal ideal $\mathfrak{m}$ (as always in this section), then*

$$\bigcap_{n\in\mathbb{N}} \mathfrak{m}^n = \{0\}.$$

Proof. Set $I := \bigcap_{n\in\mathbb{N}} \mathfrak{m}^n$, and let r be the integer given by the Artin–Rees lemma (Lemma 12.6). Then $I \cap \mathfrak{m}^{r+1} = \mathfrak{m} \cdot (I \cap \mathfrak{m}^r)$. By the definition of I, this means that $I = \mathfrak{m} \cdot I$. Now Nakayama's lemma (Theorem 7.3) shows that $I = \{0\}$. □

A more general version of Krull's intersection theorem can be found in Eisenbud [17, Corollary 5.4]. In Exercise 12.7 we will see that Theorem 12.9 may fail if R is not Noetherian.

It is a consequence of Theorem 12.9 that for every nonzero $a \in R$ there exists a nonnegative integer d such that $a \in \mathfrak{m}^d$ but $a \notin \mathfrak{m}^{d+1}$. We write this d as $d =: \operatorname{ord}(a)$, the **order** of a. Using the description (12.3) of the graded components of $\operatorname{gr}(R)$, we also define $\operatorname{gr}(a) := a + \mathfrak{m}^{d+1} \in \operatorname{gr}(R)_d \setminus \{0\}$. Setting $\operatorname{gr}(0) := 0$, we obtain a map $R \to \operatorname{gr}(R)$. Unfortunately, this map is neither additive nor multiplicative in general. However, for two nonzero elements a and $b \in R$ of orders $\operatorname{ord}(a) = d$ and $\operatorname{ord}(b) = e$, the formula (12.4) yields $\operatorname{gr}(a) \cdot \operatorname{gr}(b) = ab + \mathfrak{m}^{d+e+1} \in \operatorname{gr}(R)_{d+e}$, so $\operatorname{gr}(a) \cdot \operatorname{gr}(b) \neq 0$ if and only if $ab \notin \mathfrak{m}^{d+e+1}$, and in this case multiplicativity holds, i.e.,

$$\operatorname{gr}(ab) = \operatorname{gr}(a)\operatorname{gr}(b). \tag{12.9}$$

We are now ready to prove the following theorem about heredity of some properties from $\operatorname{gr}(R)$ to R.

Theorem 12.10 (Properties passing from $\operatorname{gr}(R)$ to R). *Let R be a Noetherian local ring and let $\operatorname{gr}(R)$ be its associated graded ring.*

(a) *If $\operatorname{gr}(R)$ is an integral domain, then the same is true for R.*
(b) *If $\operatorname{gr}(R)$ is normal, then the same is true for R.*

Proof. (a) R is not the zero ring since it is local. Let $a, b \in R$ be nonzero elements of orders d and e, respectively. By hypothesis, $\operatorname{gr}(a) \cdot \operatorname{gr}(b) \neq 0$, so $ab \notin \mathfrak{m}^{d+e+1}$ by the discussion preceding (12.9). This implies $ab \neq 0$.

(b) By (a), R is an integral domain. Let $a, b \in R$ with $b \neq 0$ such that $a/b \in \operatorname{Quot}(R)$ is integral over R. We need to show that $a \in (b)_R$. Using induction on n, we will prove the (seemingly) weaker statement

$$a \in \mathfrak{m}^n + (b)_R \quad \text{for all nonnegative integers } n. \tag{12.10}$$

This is true for $n = 0$, so assume $n > 0$. By induction, there exist $\tilde{a} \in \mathfrak{m}^{n-1}$ and $r \in R$ with

$$a = \tilde{a} + rb. \tag{12.11}$$

We are done if $\tilde{a} \in \mathfrak{m}^n$, so assume $\operatorname{ord}(\tilde{a}) = n - 1$. From (12.11) we see that $\tilde{a}/b = a/b - r$ is integral over R. By Lemma 8.11, it is almost integral, so there exists $c \in R \setminus \{0\}$ such that $c\tilde{a}^n \in (b^n)_R$ for all $n \in \mathbb{N}_0$. Since $\operatorname{gr}(R)$ is an integral domain, (12.9) holds for all elements of R, so $\operatorname{gr}(c)\operatorname{gr}(\tilde{a})^n \in (\operatorname{gr}(b)^n)_{\operatorname{gr}(R)}$ for all n. This means that $\operatorname{gr}(\tilde{a})/\operatorname{gr}(b) \in \operatorname{Quot}(\operatorname{gr}(R))$ is almost integral over $\operatorname{gr}(R)$. Since $\operatorname{gr}(R)$ is Noetherian, we can use Lemma 8.11 to conclude that $\operatorname{gr}(\tilde{a})/\operatorname{gr}(b)$ lies in $\operatorname{gr}(R)$. Since $\operatorname{gr}(\tilde{a}) \in \operatorname{gr}(R)_{n-1}$ and $\operatorname{gr}(b) \in \operatorname{gr}(R)_{\operatorname{ord}(b)}$ are homogeneous, their fraction must be homogeneous of degree $n - 1 - \operatorname{ord}(b)$, so it can be written as $\operatorname{gr}(s)$ with $s \in \mathfrak{m}^{n-1-\operatorname{ord}(b)} \setminus \mathfrak{m}^{n-\operatorname{ord}(b)}$. Using (12.9) again, we obtain

$$0 = \operatorname{gr}(\tilde{a}) - \operatorname{gr}(s)\operatorname{gr}(b) = \operatorname{gr}(\tilde{a}) - \operatorname{gr}(sb) = \tilde{a} - sb + \mathfrak{m}^n,$$

so $\tilde{a} = (\tilde{a} - sb) + sb \in \mathfrak{m}^n + (b)_R$. With (12.11), we conclude that $a \in \mathfrak{m}^n + (b)_R$, so (12.10) is proved. We may assume $b \in \mathfrak{m}$, since otherwise b is invertible in R and so $a \in (b)_R$ is certainly true. Therefore $\overline{R} := R/(b)_R$ is a Noetherian local ring with maximal ideal $\overline{\mathfrak{m}} := \mathfrak{m}/(b)_R$. Applying the canonical map $R \to \overline{R}$, $x \mapsto \overline{x}$, to (12.10) yields $\overline{a} \in \overline{\mathfrak{m}}^n$ for all n. By Krull's intersection theorem (Theorem 12.9), this implies $\overline{a} = 0$, so $a \in (b)_R$ and we are done. □

Exercises for Chapter 12

12.1 (Dimension theory with minimal use of Part III). The goal of this exercise it to make it possible to skip the third part of the book almost entirely. Modify parts of Chapters 11 and 12 in such a way that the assertion $\dim(R) = \dim(\operatorname{gr}(R))$ from Theorem 12.8 can be proved with minimal dependence on the material from Part III. In particular, the proof should not depend on the material on monomial orderings and Gröbner bases.
Hint: Instead of showing that the Hilbert function becomes a polynomial eventually, consider the least integer δ such that the Hilbert function can be bounded above by a polynomial of degree δ. With this modification, the concepts of Hilbert polynomial and Hilbert–Samuel polynomial will be omitted.
(Solution on page 228)

12.2 (Length and exact sequences). Let

$$\{0\} \longrightarrow M_1 \longrightarrow M_2 \longrightarrow \cdots \longrightarrow M_{n-1} \longrightarrow M_n \longrightarrow \{0\}$$

be a finite exact sequence of modules over a ring R (see Exercise 6.3 for the definition of an exact sequence). Assume that at most one of the M_i has infinite length. Show that

$$\sum_{i=1}^{n}(-1)^i \operatorname{length}(M_i) = 0.$$

In particular, all M_i have finite length.

12.3 (Hilbert series of a graded module over a graded ring). In this exercise the Hilbert–Serre theorem is proved in a rather general situation. Let $R = R_0 \oplus R_1 \oplus R_2 \oplus \cdots$ be a graded ring. An R-module M is called **graded** if it has a direct sum decomposition

$$M = \bigoplus_{i \in \mathbb{Z}} M_i$$

(as an abelian group) such that $R_i M_j \subseteq M_{i+j}$ for all $i \in \mathbb{N}_0$ and $j \in \mathbb{Z}$. (So as an important special case, R itself is a graded R-module.) In particular, every M_i is an R_0-module. We make the following assumptions: R_0 is a Artinian, R is finitely generated as an R_0-algebra, and M is a finitely generated, graded R-module. First show that

$$R = R_0[a_1, \ldots, a_n]$$

with a_i homogeneous and $d_i := \deg(a_i)$ positive, and that there exists $m \in \mathbb{Z}$ such that $M_i = \{0\}$ for $i < m$. Show that every M_i has finite length as an R_0-module. So we may define the *Hilbert series* of M as a formal Laurent series by

$$H_M(t) := \sum_{i \in \mathbb{Z}} \operatorname{length}(M_i) t^i \in \mathbb{Z}((t)).$$

(Observe that in the important special case that R_0 is a field K, the length is equal to the K-dimension.) Show that the Hilbert series has the form

$$H_M(t) = \frac{c_m t^m + c_{m+1} t^{m+1} + \cdots + c_k t^k}{(1 - t^{d_1}) \cdots (1 - t^{d_n})} \tag{12.12}$$

with $k \geq m$ and $c_i \in \mathbb{Z}$.
Remark: If R is standard graded, i.e., if all a_i can be chosen of degree 1, then it follows as in Corollary 11.10 that there exists a polynomial $p_M \in \mathbb{Q}[x]$ such that $\operatorname{length}(M_i) = p_M(i)$ for all sufficiently large i.
Hint: For every i consider the map $M_{i-d_1} \to M_i$ given by multiplication by a_1. Complete this map to an exact sequence $\{0\} \to X \to M_{i-d_1} \to M_i \to Y \to \{0\}$. Use Exercise 12.2 and induction on n.

12.4 (Easier computation of the Hilbert–Samuel function). Let $\mathfrak{m} \subset R$ be a maximal ideal of a ring, and consider the localization $R_\mathfrak{m}$ with maximal ideal $\mathfrak{m}_\mathfrak{m}$. Show that for every nonnegative integer i there is an isomorphism

$$\mathfrak{m}_{\mathfrak{m}}^i/\mathfrak{m}_{\mathfrak{m}}^{i+1} \cong \mathfrak{m}^i/\mathfrak{m}^{i+1}$$

of R-modules. With $K := R/\mathfrak{m} \cong R_{\mathfrak{m}}/\mathfrak{m}_{\mathfrak{m}}$, show that the isomorphism is K-linear, so $\dim_K\left(\mathfrak{m}_{\mathfrak{m}}^i/\mathfrak{m}_{\mathfrak{m}}^{i+1}\right) = \dim_K\left(\mathfrak{m}^i/\mathfrak{m}^{i+1}\right)$.

***12.5 (The associated graded ring and the tangent cone).** Let $I \subseteq K[x_1,\ldots,x_n]$ be an ideal in a polynomial ring over a field. If $f \in I \setminus \{0\}$, write f_{in} for the **initial form** of f, defined to be the nonzero homogeneous part of f of least degree. Using this, define the **initial form ideal** as

$$I_{\text{in}} := (f_{\text{in}} \mid f \in I \setminus \{0\})_{K[x_1,\ldots,x_n]}.$$

Set $A := K[x_1,\ldots,x_n]/I$ and assume that $I \subseteq (x_1,\ldots,x_n) =: \mathfrak{n}$, so $\mathfrak{m} := \mathfrak{n}/I$ is a maximal ideal in A. Show that there is an isomorphism

$$\operatorname{gr}(A_{\mathfrak{m}}) \cong K[x_1,\ldots,x_n]/I_{\text{in}},$$

which sends homogeneous elements to homogeneous elements of the same degree.
Remark: Our assumption means that $(0,\ldots,0) \in \mathcal{V}(I)$. Since any point in $\mathcal{V}(I)$ can be shifted to $(0,\ldots,0)$ by changing coordinates, we obtain a presentation of the associated graded ring of the coordinate ring of an affine variety, localized at a point. The affine variety $\mathcal{V}(I_{\text{in}})$ is called the **tangent cone**. The geometric interpretation of the tangent cone is contained in its very name: It is the best approximation of $\mathcal{V}(I)$ by a *cone*, i.e., an affine variety made up of lines through the origin. I_{in} can be computed by Mora's tangent cone algorithm [39], which essentially is a subtle variant of Buchberger's algorithm applied to a "monomial ordering" that does not satisfy (2) from Definition 9.1(a). The book by Greuel and Pfister [22] has a systematic treatment of such orderings. *(Solution on page 230)*

12.6 (Examples of associated graded rings). For the following examples of local rings R, determine the Hilbert–Samuel function $h_R(d)$ and the associated graded ring $\operatorname{gr}(R)$. Here $\operatorname{gr}(R)$ should be determined only up to isomorphism.

(a) $R = K[[x_1,\ldots,x_n]]$, the formal power series ring in n indeterminates over a field. Why is R local? *Hint:* You may use Exercise 11.1.
(b) Let $X_1, X_2, X_3 \subseteq K^2$ be the cubic curves over an algebraically closed field K given by the equations $\xi_1^3 - \xi_2^2 = 0$, $\xi_2^2 - \xi_1^2(\xi_1+1) = 0$, and $\xi_2^2 - \xi_1(\xi_1^2+1) = 0$, respectively, as shown in Fig. 12.1. Let R_i be the localization of the coordinate ring of X_i at the point $(0,0)$. If you own this book, you can draw the tangent cones into the pictures in Fig. 12.1. *Hint:* You may use Exercise 12.5.
(c) This example is from algebraic number theory. Let $A = \mathbb{Z}[\sqrt{-3}] \subset \mathbb{C}$, $\mathfrak{m} = \left(2, 1+\sqrt{-3}\right) \subseteq A$, and $R = A_{\mathfrak{m}}$. Why is $\mathfrak{m}$ a maximal ideal?

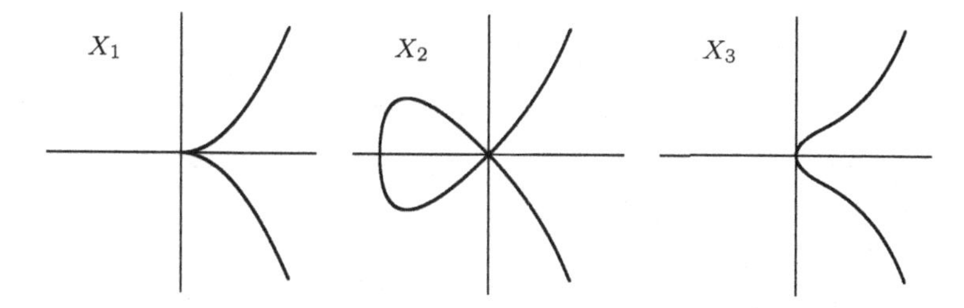

Fig. 12.1. Affine cubic curves, displaying a cusp, a double point, and smoothness

Hint: You may use Exercise 12.4. It may also be helpful to determine the powers of $\mathfrak{m}$ as $\mathbb{Z}$-modules.

12.7 (Hypotheses of Krull's intersection theorem). This exercise provides an example of a (non-Noetherian) local ring where Krull's intersection theorem (Theorem 12.9) fails. We will consider the ring R of *germs of continuous functions*, which we define as follows. Let I be the set of all continuous functions $\mathbb{R} \to \mathbb{R}$ that vanish on a neighborhood of 0. (Here $\mathbb{R}$ is equipped with the Euclidean topology.) Obviously I is an ideal in the ring $C^0(\mathbb{R}, \mathbb{R})$ of all continuous functions $\mathbb{R} \to \mathbb{R}$. The ring R is defined as $R := C^0(\mathbb{R}, \mathbb{R})/I$.

(a) Show that R is a local ring. *Hint:* You may use Exercise 6.7(b).
(b) Find a nonzero element of R that lies in every power $\mathfrak{m}^n$ of the maximal ideal of R. So Theorem 12.9 fails for R.

12.8 (Does Theorem 12.10 have a converse?). Construct an example of a normal local ring R such that the associated graded ring $\mathrm{gr}(R)$ is not an integral domain. This shows that the converse statements of Theorem 12.10(a) and (b) do not hold.
Hint: You may use Exercises 8.4 and 12.5.

Chapter 13
Regular Local Rings

As mentioned before, local rings serve for the study of the local behavior of a global object, such as an affine variety. In particular, notions of local "niceness" can be defined as properties of local rings. There is a range of much-studied properties of local rings. This includes the Cohen–Macaulay property, the Gorenstein property, normality, and regularity. In this book only normality and regularity are dealt with at some length, and one exercise, 13.3, is devoted to the Cohen–Macaulay property. It turns out that regularity is the nicest of these properties, meaning that it implies all others. After defining the notion of regularity of a Noetherian local ring R, we will see that this is equivalent to the condition that the associated graded ring $\mathrm{gr}(R)$ is isomorphic to a polynomial ring. If R is the coordinate ring of an affine variety, localized at a point, $\mathrm{gr}(R)$ can be interpreted as the coordinate ring of the tangent cone at that point (see Exercise 12.5). So in this situation regularity means that the tangent cone is (isomorphic to) affine n-space.

How can we determine the points x of an affine variety X where the localized coordinate ring $K[X]_x$ is regular? This is the topic of the second section of this chapter, where we prove the Jacobian criterion. A consequence is that an affine variety is "nice" at "most" of its points.

13.1 Basic Properties of Regular Local Rings

Throughout this section, let R be a Noetherian local ring with maximal ideal $\mathfrak{m}$ and residue class field $K := R/\mathfrak{m}$. If M is an R-module, then $M/\mathfrak{m}M$ is annihilated by $\mathfrak{m}$, so it is a K-vector space. The following lemma is a consequence of Nakayama's lemma. If you have done Exercise 7.3, you can skip the proof, since it is contained in the exercise.

Lemma 13.1 (Generating modules over a local ring). *In the above setting, assume M to be finitely generated. Let $m_1, \dots, m_n \in M$. Then the following statements are equivalent:*

G. Kemper, *A Course in Commutative Algebra*, Graduate Texts in Mathematics 256, DOI 10.1007/978-3-642-03545-6_14,

(a) M is generated by $m_1, \dots, m_n$ as an R-module.
(b) $M/\mathfrak{m}M$ is generated by $m_1 + \mathfrak{m}M, \dots, m_n + \mathfrak{m}M$ as a K-vector space.

In particular, all minimal generating systems of M have the same size, namely $\dim_K(M/\mathfrak{m}M)$.

Proof. It is clear that (a) implies (b). Conversely, assume (b) and set $N := (m_1, \dots, m_n) \subseteq M$. Then (b) implies $M \subseteq N + \mathfrak{m}M$, so $M/N \subseteq \mathfrak{m} \cdot M/N$. By Nakayama's lemma (Theorem 7.3), this implies $M/N = \{0\}$, so $M = N$. □

Applying Lemma 13.1 to $M = \mathfrak{m}$ shows that every minimal generating systems of $\mathfrak{m}$ has the size $\dim_K(\mathfrak{m}/\mathfrak{m}^2)$. Since we know from the principal ideal theorem (more precisely, from Corollary 7.6) that $\mathfrak{m}$ cannot be generated by fewer than $\dim(R)$ (the Krull dimension) elements, we obtain

$$\dim_K(\mathfrak{m}/\mathfrak{m}^2) \geq \dim(R). \tag{13.1}$$

This inequality prompts the definition of regularity.

Definition 13.2. *(a) The local ring R is called* **regular** *if*

$$\dim_K(\mathfrak{m}/\mathfrak{m}^2) = \dim(R).$$

(Here $\dim(R)$ signifies the Krull dimension.) So R is regular if and only if $\mathfrak{m}$ can be generated by $\dim(R)$ elements, which in turn is equivalent to the condition that R has a system of parameters that generates $\mathfrak{m}$. Such a system of parameters is called a **regular system of parameters***.*

(b) Let S be a Noetherian ring and $X := \operatorname{Spec}(S)$. An element $P \in X$ is called a **nonsingular point** *if the localization S_P is regular. Otherwise, P is called a* **singular point***. If X has no singular points, S is called a* **regular ring***.*

(c) A point $x \in X$ of an affine variety is called **nonsingular** *if the localization $K[X]_x$ of the coordinate ring at x is regular. Otherwise, x is called* **singular***. If every point is nonsingular, X is called* **nonsingular***.*

Remark. The above definition of a regular ring raises the following question: Is a regular *local* ring R also a regular ring in the sense of Definition 13.2(b)? In other words, is R_P regular for every $P \in \operatorname{Spec}(R)$? This is indeed true, but not at all easy to prove (see Matsumura [37, Corollary 18.G]). ◁

Example 13.3. By Definition 13.2(a), a zero-dimensional local ring is regular if and only if it is a field (equivalently, if and only if it is reduced). ◁

Before treating more examples, it is useful to establish the following regularity criterion in terms of the associated graded ring. It is a consequence of Theorem 12.8.

Theorem 13.4 (Associated graded ring and regularity)**.** *The local ring R is regular if and only if the associated graded ring $\operatorname{gr}(R)$ is isomorphic to a polynomial ring over K.*

Proof. Write $A := \operatorname{gr}(R)$. By Theorem 12.8 we have $\dim(A) = \dim(R) =: n$.

First assume that R is regular, so the maximal ideal $\mathfrak{m}$ is generated by n elements. By the discussion preceding (12.5) on page 172, it follows that A is generated by n elements as a K-algebra, so by Theorem 5.9 and Proposition 5.10 these elements must be algebraically independent. It follows that A is isomorphic to a polynomial ring.

Conversely, assume that A is isomorphic to a polynomial ring. By Corollary 5.7 it follows that A is generated by n elements $b_1, \dots, b_n$. By the discussion preceding (12.3) on page 171 we have a grading $A = \bigoplus_{d \in \mathbb{N}_0} A_d$ with $A_0 \cong K$ and $A_1 \cong \mathfrak{m}/\mathfrak{m}^2$. We may assume that the homogeneous component of degree 0 of every b_i is 0. Let $\pi\colon A \to A_1$ be the projection on the component of degree 1. Every $a \in A$ can be written as a polynomial over K in the b_i, and it follows that $\pi(a)$ is a K-linear combination of the $\pi(b_i)$. Therefore A_1 is generated by $\pi(b_1), \dots, \pi(b_n)$, and we get $\dim_K\left(\mathfrak{m}/\mathfrak{m}^2\right) = \dim_K(A_1) \le n$. With (13.1) this implies that R is regular. □

To get a geometric interpretation of regularity, consider the case that $R = K[X]_x$ is the coordinate ring of an affine variety, localized at a point $x \in X$. Exercise 12.5 gives a presentation $\operatorname{gr}(R) \cong K[x_1, \dots, x_n]/J$ of $\operatorname{gr}(R)$ in this case, where the variety $\mathcal{V}_{K^n}(J)$ can be interpreted as the tangent cone at x, i.e., the best approximation of X by an affine variety made up of lines passing through x. So roughly speaking, Theorem 13.4 tells us that x is a nonsingular point if and only if the tangent cone at x is an affine n-space. Geometrically, this makes a lot of sense, since nonsingularity should mean that the variety looks "nice" locally. However, there is a catch. Even if the tangent cone is some affine n-space, $\operatorname{gr}(R)$ need not necessarily be isomorphic to a polynomial ring, since J need not be a radical ideal. This happens, for example, if x is a cusp of X. In such a case, the geometric interpretation may be saved by viewing the affine variety of a nonradical ideal J has having "double components" or "hidden embedded components."

Example 13.5. (1) The formal power series ring $R := K[[x_1, \dots, x_n]]$ in n indeterminates over a field is a regular local ring. This can be seen by doing Exercise 12.6(a) (the result is $\operatorname{gr}(R) \cong K[x_1, \dots, x_n]$) and applying Theorem 13.4, or by observing that the maximal ideal is generated by $x_1, \dots, x_n$ and using Exercise 7.10 to conclude that $\dim(R) = n$.

(2) Let X_1, X_2, X_3 be the cubic curves from Exercise 12.6(b), shown in Fig. 12.1 on page 179. Let R_i be the localization of the coordinate ring of X_i at the point $x := (0, 0)$. In Exercise 12.6(b) the associated graded rings $\operatorname{gr}(R_i)$ are determined, and the result is isomorphic to a polynomial ring only for R_3. So x is a singular point of X_1 and X_2, but not of X_3. The curve X_1 is particularly interesting, since it has a cusp at x. Here the tangent cone may be viewed as a "double line." By changing coordinates one can also determine the associated graded ring of the localization at other points. The result is that all points other than the origin are nonsingular. This is what one expects from looking at Fig. 12.1. ◁

This may be a good place for a short digression on completions. By considering two elements from R as "near" if their difference lies in a high power of $\mathfrak{m}$, we get a concept of convergence. Krull's intersection theorem (Theorem 12.9) guarantees that with this concept, the limit of a convergent series is unique. Exercise 13.4 gives more background on this. We also have a concept of Cauchy sequences. Given this, R is called **complete** if every Cauchy sequence has a limit in R. Most of the local rings we have seen in this book are not complete. However, one can construct an extension $\widehat{R}$ of R that is a complete local ring; $\widehat{R}$ also is Noetherian, and it has the property that all of its elements are limits of R-valued sequences. The ring $\widehat{R}$ is called the **completion** of R. In a sense, the construction of completion mimics the passage from the rational numbers to the real numbers. From the construction of $\widehat{R}$ it can be shown that R and $\widehat{R}$ have the same associated graded ring. So the associated graded ring may serve to transport properties from R to $\widehat{R}$ and vice versa. For example, it follows from Theorem 12.8 that $\dim(\widehat{R}) = \dim(R)$, and it follows from Theorem 13.4 that $\widehat{R}$ is regular if and only if R is regular. A nice example of a complete local ring is the formal power series ring $K[[x_1, \ldots, x_n]]$ in n indeterminates over a field. In fact, it is the completion of $K[x_1, \ldots, x_n]_{(x_1,\ldots,x_n)}$ (see Exercise 13.5). Another well-known example of a complete ring is the ring $\mathbb{Z}_p$ of p-adic integers (with p a prime number), which plays an important role in algebraic number theory and computer algebra. In fact, $\mathbb{Z}_p$ is the completion of $\mathbb{Z}_{(p)}$, the ring of rational numbers with denominator not divisible by p.

Completion is an important tool in commutative algebra. Philosophically, the idea is that localization is not "local enough," but completion describes the behavior of a variety on a smaller scale. For example, the local ring at a point of an irreducible affine variety X contains all the information of the variety that is invariant under birational equivalence, since the field of fractions of the local ring is the function field $K(X) = \text{Quot}(K[X])$, and birational equivalence is just defined as isomorphy of the function fields. So the local ring still contains some sort of global information, even if it is regular. However, if we assume that the local ring is regular, it turns out that its completion is isomorphic to the formal power series ring $K[[x_1, \ldots, x_n]]$ with $n = \dim(X)$ (see Matsumura [37, Corollary 2 to Theorem 60, page 206]). So in this case completion eliminates global information. Another illustration of this philosophy is contained in Exercise 13.6. For more on completion, we refer to Eisenbud [17, Chapter 7].

By putting together Theorem 13.4, Proposition 8.8, and Theorem 12.10, we obtain the following:

Corollary 13.6. *(a) Every regular local ring is an integral domain.*
(b) Every regular local ring is normal.

In fact, a bit more is true: Every regular local ring is factorial. This is clear for zero-dimensional rings by Example 13.3, and will be proved for one-dimensional rings on page 197. In dimension > 1, the result is much harder to prove (see Eisenbud [17, Theorem 19.19]), and it will not be used in this book. By Exercise 12.8 there exist normal local rings that are not regular. So in general, the converse of Corollary 13.6 does not hold. However, the converse of Corollary 13.6(b) holds if R has dimension 1 (see Theorem 14.1).

Example 13.7. (1) Let S be a Noetherian ring, and suppose that $P \in \operatorname{Spec}(S)$ is contained in more than one irreducible component of $\operatorname{Spec}(S)$. This means that P contains more than one minimal prime ideal of S. By Theorem 6.5, it follows that the localization S_P has more than one minimal prime ideal. But since an integral domain has $\{0\}$ as the only minimal prime ideal, S_P is not an integral domain, and by Corollary 13.6(a) we conclude that S_P is not regular. So P is a singular point of $\operatorname{Spec}(S)$.

(2) A special case of (1) is the following: Every point of an affine variety that lies in the intersection of more than one irreducible component is a singular point.

(3) The last example is from algebraic number theory. Let $A = \mathbb{Z}[\sqrt{-3}] \subset \mathbb{C}$, $\mathfrak{m} = \left(2, 1 + \sqrt{-3}\right) \subseteq A$, and $R = A_\mathfrak{m}$ (see Exercise 12.6(c)). Then R is not normal, since $z := \frac{1+\sqrt{-3}}{2} \in \operatorname{Quot}(R) \setminus R$, but $z^2 - z + 1 = 0$. By Corollary 13.6(b), R is not regular. The result of Exercise 12.6(c) also shows that $\operatorname{gr}(R)$ is not isomorphic to a polynomial ring. ◁

13.2 The Jacobian Criterion

In this section we study the **singular locus** (which by definition is the set of all singular points) in the spectrum of an affine algebra. So as a special case we are also treating the singular locus of an affine variety.

We need some preparations from field theory, concerning separable field extensions. Recall that every algebraic field extension of a field of characteristic 0 is separable. In contrast, an algebraic element α of a field extension of a field K in characteristic $p > 0$ is separable if and only if its minimal polynomial $\operatorname{irr}(\alpha, K) \in K[x]$ cannot be written as a polynomial in x^p, i.e., $\operatorname{irr}(\alpha, K) \notin K[x^p]$. We say that a finitely generated (but not necessarily finite) field extension L of K is **separable** if there exists a transcendence basis T such that L is separable (as an algebraic extension) over $K(T)$, the subfield generated by T. In this case T is called a *separating transcendence basis.* We need the following proposition, which is not part of the standard curriculum of an abstract algebra course. Readers who are interested only in characteristic 0 can skip the proposition and Lemma 13.9.

Proposition 13.8 (Facts about separable field extensions).

(a) *Every finitely generated field extension of a perfect field is separable.*

(b) *If L is a finitely generated separable field extension of K, then every generating set of L over K contains a separating transcendence basis.*

Proof. (a) The proof follows Mac Lane [35]. Let K be a perfect field, which we may assume to have positive characteristic p. We will prove the following by induction on n: If L is a finitely generated extension of K with a transcendence basis T such that L has degree n over the separable closure of $K(T)$ in L, then L is separable over K. There is nothing to prove for $n = 1$, so assume $n > 1$. This means that there exists $\alpha \in L$ that is not separable over $K(T)$, so $g := \operatorname{irr}(\alpha, K(T)) \in K(T)[x^p]$. We write T^p for the set of all pth powers of elements of T. If all coefficients of g lay in $K(T^p)$, then g would be a pth power of a polynomial in $K(T)[x]$, since by hypothesis every element of K has a pth root in K. This would contradict the irreducibility of g, so $g \notin K(T^p)[x]$. Applying Lemma 13.9 below yields a new transcendence basis T' such that the separable closure of $K(T')$ in L contains that of $K(T)$. Since $\alpha \in T'$, the inclusion is strict, so the result follows by induction.

(b) If $\operatorname{char}(K) = 0$, every transcendence basis is separating. In the case $\operatorname{char}(K) = p > 0$ we proceed by induction on the transcendence degree $n := \operatorname{trdeg}(L/K)$. If $n = 0$, then $T = \emptyset$ is a separating transcendence basis by hypothesis. So we may assume $n > 0$. By hypothesis there exists a separating transcendence basis T. For every element $\alpha \in L$, the minimal polynomial $\operatorname{irr}(\alpha, K(T))$ is separable, so if all its coefficients lie in the subfield $K(T^p)$, then α is separable over $K(T^p)$. Assume that this happens for every element of a given generating set S of L over K. Then L would be separable over $K(T^p)$ (see Lang [33, Chapter VII, Theorem 4.8]). But every element t from T has $\operatorname{irr}(t, K(T^p)) = x^p - t^p$, which is inseparable. This contradiction shows that there exists $\alpha \in S$ such that $g := \operatorname{irr}(\alpha, K(T))$ does not lie in $K(T^p)[x]$. Applying Lemma 13.9 below yields a new transcendence basis T' such that L is separable over $K(T')$, and $\alpha \in T'$. So viewed as an extension of $K(\alpha)$, L has the separating transcendence basis $T' \setminus \{\alpha\}$, and it is still generated by S. By induction, L has a separating transcendence basis $T'' \subseteq S$ over $K(\alpha)$, so as an extension of K it has the separating transcendence basis $T'' \cup \{\alpha\} \subseteq S$. □

The following lemma was used in the proof.

Lemma 13.9. *Let L be an extension of a field K of characteristic $p > 0$. Let T be a finite transcendence basis, and write T^p for the set of all pth powers of elements of T. If the minimal polynomial $g := \operatorname{irr}(\alpha, K(T))$ of an $\alpha \in L$ does not lie in $K(T^p)[x]$, then there exists $t \in T$ such that $T' := (T \setminus \{t\}) \cup \{\alpha\}$ is a transcendence basis, and all elements from L that are separable over $K(T)$ are also separable over $K(T')$.*

Proof. Since $K[T]$ is factorial, there exists $0 \neq h \in K[T]$ such that $f := hg \in K[T][x]$ is a primitive polynomial, so by the Gauss lemma, f is irreducible (see Lang [33, Chapter V, Theorem 6.3]). Since h is the leading coefficient of f (as a polynomial in x), f does not lie in $K[T^p][x]$, so there exists $t \in T$ such that f, viewed as a polynomial in t, is separable. This shows that t

is separable over $K(T')$. Therefore T' is a new transcendence basis, and all elements of T are separable over $K(T')$. This implies that all elements from L that are separable over $K(T)$ are separable over $K(T')$ (see Lang [33, Chapter VII, Theorem 4.9]). □

We come back to the goal of calculating the singular locus in the spectrum of an affine algebra A. If $A = K[x_1, \ldots, x_n]/I$ is given as a quotient ring of a polynomial ring over a field, then an element of $X := \operatorname{Spec}(A)$ is given as P/I, where $P \subset K[x_1, \ldots, x_n]$ is a prime ideal with $I \subseteq P$. The main goal of this section is to prove the Jacobian criterion for the regularity of the local ring $A_{P/I}$. This criterion involves an irreducible component of X containing P/I. Such a component corresponds to a prime ideal $Q \subset K[x_1, \ldots, x_n]$ that is minimal over I, and contained in P. The criterion also involves the rank of a matrix of polynomials modulo P, defined as follows: If $(g_{i,j}) \in K[x_1, \ldots, x_n]^{m \times k}$, then $\operatorname{rank}(g_{i,j} \mod P)$ denotes the rank of the matrix $(g_{i,j} + P) \in \operatorname{Quot}(K[x_1, \ldots, x_n]/P)^{m \times k}$. The matrix that appears in the Jacobian criterion is made up of the (formal) partial derivatives of polynomials generating I. This is often called the *Jacobian matrix*.

Theorem 13.10 (Jacobian criterion). *Let $I = (f_1, \ldots, f_m) \subseteq K[x_1, \ldots, x_n]$ be an ideal in a polynomial ring over a field, and let $P \subset K[x_1, \ldots, x_n]$ be a prime ideal containing I. Furthermore, let $Q \subset K[x_1, \ldots, x_n]$ be a prime ideal that is minimal over I and is contained in P. Then*

(a)

$$\operatorname{rank}\left(\frac{\partial f_i}{\partial x_j} \mod P\right) \leq \operatorname{ht}(Q).$$

(b) If equality holds in (a)*, then the local ring $(K[x_1, \ldots, x_n]/I)_{P/I}$ is regular.*

(c) If $\operatorname{Quot}(K[x_1, \ldots, x_n]/P)$ is a (not necessarily finite) separable field extension of K, then the converse of (b) *holds. The separability hypothesis is automatically satisfied if K is a perfect field or if $P = (x_1 - \xi_1, \ldots, x_n - \xi_n)$ corresponds to a point $(\xi_1, \ldots, \xi_n) \in \mathcal{V}_{K^n}(I)$.*

If the separability hypothesis of Theorem 13.10(c) is not satisfied, then it can happen that the converse of (b) fails (see Exercise 13.7).

Before we turn to the proof of Theorem 13.10, let us note that it can be reformulated as follows: The nullity of the Jacobian matrix modulo P is greater than or equal to the dimension of every irreducible component of $\operatorname{Spec}(K[x_1, \ldots, x_n]/I)$ that contains P, with equality if and only if the local ring is regular (provided the hypothesis of part (c) holds). Moreover, notice that if $P = (x_1 - \xi_1, \ldots, x_n - \xi_n)$ corresponds to a point $(\xi_1, \ldots, \xi_n) \in \mathcal{V}_{K^n}(I)$, then the Jacobian matrix modulo P is $\left(\frac{\partial f_i}{\partial x_j}(\xi_1, \ldots, \xi_n)\right) \in K^{m \times n}$. The kernel of the Jacobian matrix modulo P can be interpreted as the tangent space at the point P. So for an affine variety X we have a tangent space attached to every point, and its dimension is greater than or equal to the dimension of an irreducible component on which the point lies. The singular points are those where the dimension of the tangent space exceeds that lower bound.

We need two lemmas for proving Theorem 13.10.

Lemma 13.11. *Let $P \subset K[x_1, \ldots, x_n]$ be a prime ideal of height m in a polynomial ring over a field.*

(a) There exist $f_1, \ldots, f_m \in P$ generating the localized ideal $P_P \subseteq K[x_1, \ldots, x_n]_P$.
(b) If Quot $(K[x_1, \ldots, x_n]/P)$ *is a separable field extension of K, then*

$$\operatorname{rank}\left(\frac{\partial f_i}{\partial x_j} \mod P\right) = m.$$

Proof. As a field extension of K, $L := \operatorname{Quot}(K[x_1, \ldots, x_n]/P)$ is generated by $\alpha_1, \ldots, \alpha_n$ with $\alpha_i := x_i + P$. We express this by writing $L = K(\alpha_1, \ldots, \alpha_n)$. By Corollaries 5.7 and 8.23, $K[x_1, \ldots, x_n]/P$ has dimension $k := n - m$, so by Theorem 5.9, L has transcendence degree k. Since every generating set of a field extension contains a transcendence basis, we may assume that $\alpha_1, \ldots, \alpha_k$ form a transcendence basis, so they are algebraically independent, and L is a finite extension of $L_0 := K(\alpha_1, \ldots, \alpha_k)$. If L is separable over K, we may use Proposition 13.8(b) and additionally assume that L is separable over L_0. For every $l \in \{0, \ldots, m\}$, consider the map

$$\varphi_l \colon K(x_1, \ldots, x_k)[x_{k+1}, \ldots, x_{k+l}] \to L, \ x_i \mapsto \alpha_i.$$

We claim that $\operatorname{im}(\varphi_l) = K(\alpha_1, \ldots, \alpha_{k+l}) =: L_l$ and $\ker(\varphi_l) = (f_1, \ldots, f_l)$ with $f_i \in K[x_1, \ldots, x_{k+i}] \cap P$. Additionally, if L is separable over L_0, then $\partial f_i / \partial x_{k+i} \notin P$. All this is true for $l = 0$. Using induction on l, we may assume that $l > 0$ and that $f_1, \ldots, f_{l-1}$ have already been found. Since α_{k+l} is algebraic over L_{l-1}, it follows that $L_l = L_{l-1}[\alpha_{k+l}]$ (see Lang [33, Chapter VII, Proposition 1.4]), and so $L_l = L_0[\alpha_{k+1}, \ldots, \alpha_{k+l}]$ by induction. This shows that $\operatorname{im}(\varphi_l) = L_l$. Set $g := \operatorname{irr}(\alpha_{k+l}, L_{l-1}) \in L_{l-1}[x_{k+l}]$. Since $L_{l-1} = L_0[\alpha_{k+1}, \ldots, \alpha_{k+l-1}]$, there exist $f_l \in K[x_1, \ldots, x_{k+l}]$ and $h \in K[x_1, \ldots, x_k] \setminus P$ such that $g = \frac{f_l(\alpha_1, \ldots, \alpha_{k+l-1}, x_{k+l})}{h(\alpha_1, \ldots, \alpha_k)}$. It follows that $f_l \in K[x_1, \ldots, x_{k+l}] \cap P$. Moreover, if L is separable over L_0, then L_l is also separable over L_{l-1} (see Lang [33, Chapter VII, Theorem 4.9]), and it follows that g has no multiple roots, so $\frac{\partial g}{\partial x_{k+l}}(\alpha_{k+l}) \neq 0$. This implies $\partial f_l / \partial x_{k+l} \notin P$. Clearly $f_l \in \ker(\varphi_l)$. For proving $\ker(\varphi_l) = (f_1, \ldots, f_l)$, take $f \in \ker(\varphi_l)$. Then g divides $f(\alpha_1, \ldots, \alpha_{k+l-1}, x_{k+l})$, so there exist $r \in K[x_1, \ldots, x_{k+l}]$ and $s \in K[x_1, \ldots, x_k] \setminus P$ with

$$\frac{f(\alpha_1, \ldots, \alpha_{k+l-1}, x_{k+l}) \cdot h(\alpha_1, \ldots, \alpha_k)}{f_l(\alpha_1, \ldots, \alpha_{k+l-1}, x_{k+l})} = \frac{r(\alpha_1, \ldots, \alpha_{k+l-1}, x_{k+l})}{s(\alpha_1, \ldots, \alpha_k)}.$$

Therefore all coefficients of $hsf - rf_l$ (as a polynomial in x_{k+l}) lie in $\ker(\varphi_{l-1})$. So by induction $hsf - rf_l \in (f_1, \ldots, f_{l-1})_{K(x_1,\ldots,x_k)[x_{k+1},\ldots,x_{k+l}]}$. Since $h, s \neq 0$, this implies $f \in (f_1, \ldots, f_l)$, and the claim is proved.

For $l = m$ we get $\ker(\varphi_m) = (f_1, \ldots, f_m)_{K(x_1,\ldots,x_k)[x_{k+1},\ldots,x_n]}$. The algebraic independence of $\alpha_1, \ldots, \alpha_k$ implies that $K(x_1, \ldots, x_k)[x_{k+1}, \ldots, x_n] \subseteq K[x_1, \ldots, x_n]_P$, so $(\ker(\varphi_m))_{K[x_1,\ldots,x_n]_P} = (f_1, \ldots, f_m)_{K[x_1,\ldots,x_n]_P}$ follows. We obtain

$$P_P \subseteq (\ker(\varphi_m))_{K[x_1,\ldots,x_n]_P} = (f_1, \ldots, f_m)_{K[x_1,\ldots,x_n]_P} \subseteq P_P,$$

proving (a).

For proving (b), consider the last m columns of the Jacobian matrix, which form the square matrix $A := (\partial f_i / \partial x_{k+j}) \in K[x_1, \ldots, x_n]^{m \times m}$. Since $f_i \in K[x_1, \ldots, x_{k+i}]$, A is a lower triangular matrix with diagonal entries $\partial f_i / \partial x_{k+i}$. Since these entries do not lie in P under the hypothesis of (b), $\det(A) \notin P$ follows, proving (b). □

Notice that in Lemma 13.11, $K[x_1, \ldots, x_n]_P$ is a local ring of dimension m with maximal ideal P_P. Therefore part (a) says that $K[x_1, \ldots, x_n]_P$ is regular, so $K[x_1, \ldots, x_n]$ is a regular ring. This is generalized in Exercise 13.1.

The following lemma gives an interpretation of the rank of the Jacobian determinant of $f_1, \ldots, f_m$ modulo a prime ideal in terms of the ideal generated by the f_i.

Lemma 13.12. *Let $I = (f_1, \ldots, f_m) \subseteq K[x_1, \ldots, x_n]$ be an ideal in a polynomial ring over a field, and let $P \subset K[x_1, \ldots, x_n]$ be a prime ideal containing I. If $L := K[x_1, \ldots, x_n]_P / P_P$ (which is a field isomorphic to* $\operatorname{Quot}(K[x_1, \ldots, x_n]/P)$*), then*

(a)

$$\operatorname{rank}\left(\frac{\partial f_i}{\partial x_j} \mod P\right) \leq \dim_L\left(\left(I_P + P_P^2\right) / P_P^2\right).$$

(b) If L is a separable field extension of K, then equality holds in (a).

Proof. We will construct linear maps $\varphi\colon L^m \to P_P/P_P^2$ and $\psi\colon P_P/P_P^2 \to L^n$. First

$$\varphi\colon L^m \to P_P/P_P^2,\ (g_1 + P_P, \ldots, g_m + P_P) \mapsto \sum_{j=1}^{m} g_j f_j + P_P^2$$

gives a well-defined, L-linear map with image $\operatorname{im}(\varphi) = \left(I_P + P_P^2\right) / P_P^2$. To define ψ, consider $f \in P_P^2$. This means that $hf = \sum_{j=1}^{r} g_j h_j$ with $h \in K[x_1, \ldots, x_n] \setminus P$ and $g_j, h_j \in P$. For $1 \leq i \leq n$ we get

$$h \frac{\partial f}{\partial x_i} + f \frac{\partial h}{\partial x_i} = \sum_{j=1}^{r} \left(g_j \frac{\partial h_j}{\partial x_i} + h_j \frac{\partial g_j}{\partial x_i}\right) \in P,$$

so $\partial f/\partial x_i \in P_P$. Therefore the map

$$\psi\colon P_P/P_P^2 \to L^n,\ f + P_P^2 \mapsto \left(\frac{\partial f}{\partial x_1} + P_P, \ldots, \frac{\partial f}{\partial x_n} + P_P\right),$$

is well defined. An easy calculation shows that ψ is L-linear. By considering the images of the standard basis vectors of L^m under the composition $\psi \circ \varphi\colon L^m \to L^n$, we see that $\psi \circ \varphi$ is given by the Jacobian matrix modulo P_P. It follows that

$$\begin{aligned}\operatorname{rank}\left(\frac{\partial f_i}{\partial x_j} \mod P\right) &= \dim_L\left(\operatorname{im}(\psi \circ \varphi)\right)\\ &\le \dim_L\left(\operatorname{im}(\varphi)\right) = \dim_L\left((I_P + P_P^2)/\,P_P^2\right),\end{aligned}$$

with equality if ψ is injective. So we have proved (a), and for proving (b) it suffices to show that ψ is injective under the hypothesis of (b). Under this hypothesis, Lemma 13.11(b) is applicable. Notice that ψ is independent of $f_1, \ldots, f_m$. So for showing the injectivity of ψ we may assume that $f_1, \ldots, f_m$ are the polynomials given by Lemma 13.11. Then $I_P = P_P$ by Lemma 13.11(a), so φ is surjective. Moreover, $\dim_L(\operatorname{im}(\psi \circ \varphi)) = \operatorname{rank}(\partial f_i/\partial x_j \mod P) = m$ by Lemma 13.11(b), so $\psi \circ \varphi\colon L^m \to L^n$ is injective. Therefore ψ has to be injective, and the proof is complete. □

We are now ready for the proof of the Jacobian criterion.

Proof of Theorem 13.10. Let $Q_0 \in \operatorname{Spec}(K[x_1, \ldots, x_n])$ be a prime ideal with $I \subseteq Q_0 \subseteq P$ having minimal height among all prime ideals between I and P. Assume that the theorem has been proved for Q_0 in the place of Q. Then part (a) also follows for Q, since $\operatorname{ht}(Q_0) \le \operatorname{ht}(Q)$. Moreover, if $Q \ne Q_0$, then $(K[x_1, \ldots, x_n]/I)_{P/I}$ is not regular by Example 13.7(1). So (b) tells us that the inequality in (a) is strict for Q_0 and therefore also for Q. Moreover, the converse of (b) is trivially true since $(K[x_1, \ldots, x_n]/I)_{P/I}$ is not regular. So we may assume that Q has minimal height among the prime ideals between I and P. By Lemma 1.22 and Theorems 6.5 and 8.22, this implies

$$\dim\left(K[x_1, \ldots, x_n]_P/I_P\right) = \operatorname{ht}(P) - \operatorname{ht}(Q).$$

The quotient ring $R := K[x_1, \ldots, x_n]_P/I_P$ is a Noetherian local ring with maximal ideal P_P/I_P and residue class field $L := K[x_1, \ldots, x_n]_P/P_P$. Since $R \cong (K[x_1, \ldots, x_n]/I)_{P/I}$, we are interested in whether R is regular. From (13.1) we obtain

$$\dim_L\left((P_P/I_P)\big/\,(P_P/I_P)^2\right) \ge \dim(R) = \operatorname{ht}(P) - \operatorname{ht}(Q),$$

with equality if and only if R is regular. We have L-linear isomorphisms

$$(P_P/I_P)\big/ (P_P/I_P)^2 \cong P_P \,/ (I_P + P_P^2) \cong (P_P/P_P^2) \,\big/ \,\Big((I_P + P_P^2)/\, P_P^2\Big),$$

so

$$\dim_L (P_P/P_P^2) - \dim_L \Big((I_P + P_P^2)/\, P_P^2\Big) \geq \operatorname{ht}(P) - \operatorname{ht}(Q),$$

with equality if and only if R is regular. Since Lemma 13.11(a) shows that $K[x_1, \ldots, x_n]_P$ is regular, $\dim_L (P_P/P_P^2)$ is equal to $\operatorname{ht}(P)$, so combining the above inequality with Lemma 13.12(a) yields

$$\operatorname{rank}\left(\frac{\partial f_i}{\partial x_j} \mod P\right) \leq \dim_L \Big((I_P + P_P^2)/\, P_P^2\Big) \leq \operatorname{ht}(Q), \tag{13.2}$$

and R is regular if and only if the second inequality is in fact an equality. Parts (a) and (b) follow immediately from this. Moreover, if the hypothesis of part (c) is satisfied, then Lemma 13.12(b) shows that the first equality of (13.2) is in fact an equality, so (c) also follows.

The remark that the hypothesis of (c) is satisfied if K is a perfect field follows from Proposition 13.8(a). □

The Jacobian criterion gives a straightforward procedure for determining the singular locus of an affine variety $X \subseteq K^n$. Assume for simplicity that X is equidimensional of dimension m. Then a point $x \in X$ is singular if and only if the Jacobian matrix of the polynomials defining X, evaluated at x, has rank less than $h := n - m$. This is equivalent to the condition that all $h \times h$ minors of the Jacobian matrix vanish at x. So the singular locus is given by the polynomials defining X together with the $h \times h$ minors of the Jacobian matrix. In particular, we see that the singular locus is Zariski closed. But we do not (yet) know whether it may happen that X consists entirely of singular points.

Also notice that in the case of a hypersurface $X = \mathcal{V}(f)$, the singular locus consists of the points in X where all partial derivatives $\partial f/\partial x_i$ vanish. Exercises 13.9 and 13.10 have some explicit examples.

The following result states the closedness of the singular locus in a slightly more general situation. To address the question whether all points may be singular, we consider the **nonsingular locus**, the set of all nonsingular points, and show that under reasonable hypotheses this set is dense. The upshot is that in most situations singular points may be regarded as "rare." For example, it follows that an affine curve (= an affine variety over an algebraically closed field that is equidimensional of dimension 1) has only finitely many singular points.

Corollary 13.13 (The singular locus and the nonsingular locus).

(a) If A is an affine algebra over a perfect field, then the singular locus X_{sing} in $X :=$ Spec(A) is closed.
(b) If R is a reduced Noetherian ring, then the nonsingular locus in Spec(R) *is dense.*

(c) *The singular locus X_{sing} in an affine variety X is closed.*
(d) *The nonsingular locus in an affine variety over an algebraically closed field is open and dense.*

Proof. (a) Write $A = K[x_1, \ldots, x_n]/I$ with $I = (f_1, \ldots, f_m)$, and let $Q_1, \ldots, Q_k \in \operatorname{Spec}(K[x_1, \ldots, x_n])$ be the prime ideals that are minimal over I. For a positive integer h, let $J_h \subseteq K[x_1, \ldots, x_n]$ be the ideal generated by the $h \times h$ minors of the Jacobian matrix of the f_i, and set $J_0 := K[x_1, \ldots, x_n]$. With this, Theorem 13.10 yields

$$X_{\text{sing}} = \bigcup_{i=1}^{k} \mathcal{V}_X\left(\left(J_{\operatorname{ht}(Q_i)} + Q_i\right)/I\right),$$

which is closed.

(b) Since the Zariski closure of the nonsingular locus consists of those prime ideals containing the intersection of all prime ideals $P \in \operatorname{Spec}(R)$ such that R_P is regular, it suffices to show that R_P is regular for all minimal prime ideals P of R. An easy calculation shows that every localization of a reduced ring is reduced, so in particular $\operatorname{nil}(R_P) = \{0\}$ for P a minimal prime ideal. By Theorem 6.5, R_P has precisely one prime ideal. By Corollary 3.14(c), this prime ideal is equal to $\operatorname{nil}(R_P) = \{0\}$. It follows that R_P is a field and therefore a regular local ring.

(c) Let $I \subseteq K[x_1, \ldots, x_n]$ be the ideal corresponding to the affine variety X. Let $Q_1, \ldots, Q_k$ and J_h be as in the proof of part (a). With this notation, Theorem 13.10 yields

$$X_{\text{sing}} = \bigcup_{i=1}^{k} \mathcal{V}_{K^n}\left(J_{\operatorname{ht}(Q_i)} + Q_i\right),$$

which is closed.

(d) With I and $Q_1, \ldots, Q_k$ as above, we have $K[X] = K[x_1, \ldots, x_n]/I$, and the Q_i/I are the minimal prime ideals of $K[X]$. By (a) there exists an ideal $J \subseteq K[x_1, \ldots, x_n]$ such that $K[X]_{P/I}$ (with $P \subset K[x_1, \ldots, x_n]$ a prime ideal containing I) is regular if and only if $J \not\subseteq P$. Part (b) applies to $K[X]$ by Theorem 1.25(a), so by the proof of (b), J is not contained in any of the Q_i. So $X_i := \mathcal{V}(Q_i) \not\subseteq \mathcal{V}(J) = X_{\text{sing}}$ holds for all i. In other words, $X_i \cap X_{\text{sing}}$ is properly contained in X_i. If $Y \subseteq X$ is the Zariski closure of the nonsingular locus, we obtain $X_i = (X_i \cap X_{\text{sing}}) \cup (X_i \cap Y)$. Since K is algebraically closed we have $\mathcal{I}(X_i) = Q_i$, so X_i is irreducible by Theorem 3.10(a), and it follows that $X_i = X_i \cap Y$. Therefore $X = X_1 \cup \cdots \cup X_k \subseteq Y$. This shows that the nonsingular locus is dense. $\square$

Example 13.14. Consider the algebra $A = K[x]/(x^2)$ with K a field. Then $P = (x)$ is the only prime ideal in $K[x]$ containing $I := (x^2)$. The Jacobian matrix reduced modulo P is zero. So by the Jacobian criterion (Theorem 13.10), the local ring $R := A_{P/I}$ is not regular. (This is also clear since R is zero-dimensional but not a field.) This example shows that for nonreduced rings, the nonsingular locus may be empty. ◁

It holds in much greater generality that the singular locus is a closed subset of the spectrum of a ring. For example, all so-called *excellent rings* satisfy this. In fact, excellence is defined by a list of properties, among them the closedness of the singular locus and the Noether property. The notion of an excellent ring is due to Grothendieck. His objective was to capture some properties that many Noetherian rings share with affine algebras, but to exclude other, more pathological, rings. The notion has led to some fruitful research. Important results are that every localization of an excellent ring and every finitely generated algebra over an excellent ring are excellent again. Moreover, all fields and the ring $\mathbb{Z}$ are excellent, and so are formal power series rings over fields. So informally speaking, all Noetherian rings that mathematicians usually deal with are excellent. More on excellent rings can be found in Matsumura [37, Chapter 13]. But there do exist Noetherian rings for which the singular locus is not closed. One such example was given by Nagata [40, §5].

Exercises for Chapter 13

13.1 (Regular rings). Prove the following.

(a) If R is a regular local ring with maximal ideal $\mathfrak{m}$ and $P \in \operatorname{Spec}(R[x])$ is a prime ideal with $\mathfrak{m} \subseteq P$, then $R[x]_P$ is regular.
(b) If S is a regular Noetherian ring, then so is $S[x]$.
(c) $\mathbb{Z}$ and all polynomial rings $\mathbb{Z}[x_1, \ldots, x_n]$ are regular rings.

***13.2 (Quotient rings of regular local rings).** Let R be a regular local ring and $I \subseteq R$ an ideal. Show that the following statements are equivalent:

(a) R/I is a regular local ring.
(b) $I = (a_1, \ldots, a_k)$, where the a_i are taken from a regular system of parameters of R.

Show that if (b) is satisfied, then $\dim(R/I) = \dim(R) - k$. Give an example in which I is proper but R/I is not regular.

13.3 (Cohen–Macaulay rings). In this exercise we look at Cohen–Macaulay rings. A **regular sequence** of length n is a sequence $a_1, \ldots, a_n$ of elements of a ring R such that $(a_1, \ldots, a_n) \neq R$ and such that for every $i \in \{1, \ldots, n\}$, multiplication with a_i induces an injective map on $R/(a_1, \ldots, a_{i-1})$. If R is a Noetherian local ring, then R is called **Cohen–Macaulay** if R has a regular sequence of length equal to $\dim(R)$. In general, a Noetherian ring is called **Cohen–Macaulay** if the localization at every prime ideal is Cohen–Macaulay.

(a) Show that every regular local ring is Cohen–Macaulay.
(b) Give an example of a Cohen–Macaulay local ring that is not regular.
(c) Show that the affine algebra $A := K[x_1, x_2]/(x_1^2, x_1x_2)$ is not Cohen–Macaulay.

Cohen–Macaulay rings are an important topic in commutative algebra. In this exercise we have barely scratched the surface. Readers can find more on the subject in Eisenbud [17, Chapters 17 and 18] and Bruns and Herzog [8].

13.4 (The Krull topology). Let R be a ring and let $I_0 \supseteq I_1 \supseteq I_2 \supseteq \cdots$ be a descending chain (also known as a filtration) of ideals in R. An important special case is $I_n = I^n$ with $I \subseteq R$ an ideal. We define the **Krull topology** on R (with respect to the chain (I_n)) by calling a subset $U \subseteq R$ open if for every $a \in U$ there exists a nonnegative integer n such that the residue class $a + I_n$ is contained in U.

(a) Show that this defines a topology on R.
(b) We say that an R-valued sequence $(a_k)_{k \in \mathbb{N}_0}$ converges to an $a \in R$ if for every neighborhood U of a there are at most finitely many k with $a_k \notin U$. Prove the equivalence of the following three statements:

    (1) R is a Hausdorff space.
    (2) Every R-valued sequence converges to at most one $a \in R$.
    (3) $\bigcap_{n \in \mathbb{N}_0} I_n = \{0\}$.

    So if R is a Noetherian local ring with maximal ideal $\mathfrak{m}$ and $I_n = \mathfrak{m}^n$, these statements hold by Krull's intersection theorem (Theorem 12.9).
(c) Let $(a_k)_{k \in \mathbb{N}_0}$ and $(b_k)_{k \in \mathbb{N}_0}$ be two R-valued sequences converging to a and b, respectively. Show that $(a_k + b_k)$ and $(a_k \cdot b_k)$ converge to $a + b$ and $a \cdot b$, respectively.

13.5 (The formal power series ring is complete). Suppose that $S = K[[x_1, \dots, x_n]]$ is the formal power series ring in n indeterminates over a field. Recall that S is a local ring with maximal ideal $\mathfrak{m} = (x_1, \dots, x_n)_S$ (see Exercise 12.6(a)).

(a) Let $(f_k)_{k \in \mathbb{N}}$ be a sequence with $f_k \in S$ such that for every nonnegative integer m there exists k_m such that $f_{k'} - f_k \in \mathfrak{m}^m$ for all $k, k' \geq k_m$. In other words, assume that (f_k) is a Cauchy sequence. Show that there exists $f \in S$ such that $\lim_{k \to \infty} f_k = f$. So S is a complete local ring.
(b) Show that every polynomial $f \in K[x_1, \dots, x_n] \setminus (x_1, \dots, x_n)_{K[x_1, \dots, x_n]}$ is invertible in S. So the local ring $R := K[x_1, \dots, x_n]_{(x_1, \dots, x_n)}$ is embedded in S.
(c) Show that every $f \in S$ is the limit of a convergent sequence $(f_k)_{k \in \mathbb{N}_0}$ with $f_k \in K[x_1, \dots, x_n]$. So S is the completion of R.

(d) Assume that $\mathrm{char}(K) \neq 2$. Show that the polynomial $1 + x_1$ has a square root in S. Conclude that $R \subsetneq S$, so R is not complete.

Remark: By the result of Exercise 12.6(a), we have $\mathrm{gr}(S) \cong K[x_1, \ldots, x_n]$. From Exercise 12.5 it follows that $\mathrm{gr}(R) \cong K[x_1, \ldots, x_n]$, too. This exemplifies that a local ring and its completion have the same associated graded ring.

13.6 (Completion is more local than localization). The goal of this exercise is to illustrate the idea that completion provides a look at a smaller scale than localization. The exercise, and in particular Fig. 13.1, were inspired by Eisenbud [17, Section 7.2 and Figure 7.3]. Consider the cubic curve $X \subseteq K^2$ given by the equation $\xi_2^2 - \xi_1^2(\xi_1 + 1) = 0$, where K is an algebraically closed field of characteristic not equal to 2. Let R be the localization of the coordinate ring $K[X]$ at the (singular) point $(0,0)$.

(a) Show that R is an integral domain, so that $\mathrm{Spec}(R)$ is irreducible.
(b) Show that the completion $\widehat{R}$ is not an integral domain. So $\mathrm{Spec}(\widehat{R})$ decomposes into components. *Hint:* Inspired by Exercise 13.5(d), construct two R-valued Cauchy sequences whose product converges to 0. Then use Exercise 13.4(c).

Remark: In fact, it is not hard to prove that $\mathrm{Spec}(\widehat{R})$ has two irreducible components. Since X has a double point at $(0,0)$, we should expect to see two components on a small scale. So completion meets this expectation, but localization does not. See Fig. 13.1 for an illustration. *(Solution on page 230)*

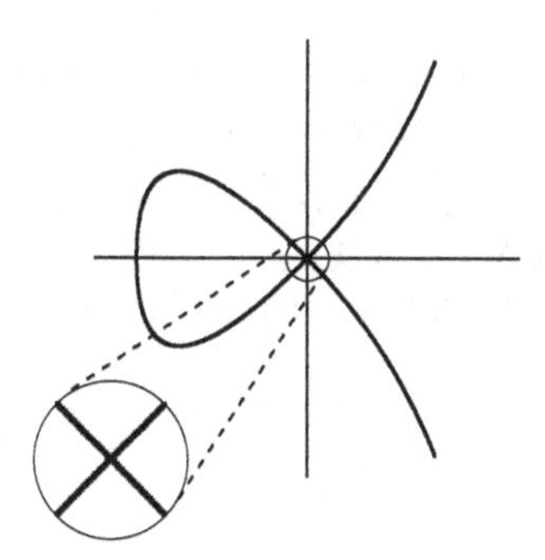

Fig. 13.1. Local and "more local": enlarged area represents completion

13.7 (Hypotheses of the Jacobian criterion). Give an example in which the converse of Theorem 13.10(b) fails.

13.8 (A nicer version of the Jacobian criterion?). It would be nice to have the following unified version of parts (b) and (c) of Theorem 13.10:

(b') Equality holds in (a) *if and only if* $(K[x_1,\dots,x_n]/I)_{P/I}$ *is regular and* $\operatorname{Quot}(K[x_1,\dots,x_n]/P)$ *is a separable extension of* K.

Is this true? Give a proof or a counterexample.

13.9 (The singular locus). Determine the singular locus of the affine varieties X over $K = \mathbb{C}$ given by the following equations:

(a) $x_1^3 - x_2^2 = 0$ (see Fig. 12.1 on page 179)
(b) $x_2^2 - x_1^2(x_1 + 1) = 0$ (see Fig. 12.1 on page 179)
(c) $x_2^2 - x_1(x_1^2 + 1) = 0$ (see Fig. 12.1 on page 179)
(d) $\left((x_1 - 3)^2 + x_2^2 - 25\right) \cdot \left((x_1 + 3)^2 + x_2^2 - 25\right)$
(e) $x_1^2 - x_2^2 x_3 = 0$ (this surface is called the *Whitney umbrella*)
(f) $(x_1^3 - x_2^2) \cdot (x_1^2 + x_2^2 - 2) = 0$ and $(x_1^3 - x_2^2) \cdot x_3 = 0$

Visualize the results by a drawing or in your imagination.

13.10 (Elliptic curves). Let K be an algebraically closed field of characteristic not equal to 2. Let $a, b \in K$. Show that the cubic curve E in K^2 given by the equation

$$x_2^2 = x_1^3 + ax_1 + b$$

is nonsingular if and only if $4a^3 + 27b^2 \neq 0$.
Remark: The above equation is the *Weierstrass normal form* of a cubic curve. If E is nonsingular, it is called an **elliptic curve**.

13.11 (The singular locus of $\mathbb{Z}[\sqrt{-3}]$). This is an example from algebraic number theory. Consider the ring $R := \mathbb{Z}[\sqrt{-3}] \subset \mathbb{C}$.

(a) Show that the ring $S = \mathbb{Z}\left[(1 + \sqrt{-3})/2\right]$ is Euclidean. *Hint:* You may use the *norm function* $N\colon \mathbb{C} \to \mathbb{R},\ z \mapsto |z|^2$.
(b) Use (a) and Example 13.7(3) to determine the singular locus $X_{\text{sing}} \subseteq X := \operatorname{Spec}(R)$.

Chapter 14
Rings of Dimension One

Noetherian rings of dimension 0 are rather well understood: They are semilocal, and a Noetherian local ring of dimension 0 is regular if and only if it is a field. The next step is to study one-dimensional rings. In geometry, one-dimensional rings occur as coordinate rings of affine curves. In algebraic number theory, they occur as rings of algebraic integers. The final chapter of this book is devoted to rings of dimension one. We first show that a Noetherian local ring of dimension one is regular if and only if it is normal. As a consequence, we see that the process of normalization, when applied to an affine curve, amounts to desingularization.

In the second section of this chapter we look at the multiplicative theory of ideals. We extend the notion of ideals by including so-called *fractional ideals*, and ask which ideals are invertible as fractional ideals. This is closely linked having height one.

The last section is about *Dedekind domains*. These can be characterized as normal Noetherian domains of dimension ≤ 1. It turns out that this is equivalent to the condition that all nonzero ideals are invertible (as fractional ideals). Yet another equivalent condition is that every ideal can be written as a product of prime ideals. If this is satisfied, then the factorization of an ideal as a product of prime ideals is unique. So ideals in Dedekind domains enjoy the unique factorization property, while elements in general do not. The extent to which a Dedekind domain fails to be factorial is measured by the *ideal class group*, which we introduce. As an application, we will see that the group law on an elliptic curve can be defined by a correspondence between points and elements of the ideal class group of the coordinate ring.

14.1 Regular Rings and Normal Rings

We start by taking a closer look at one-dimensional regular local rings. By definition, the maximal ideal of a one-dimensional regular local ring R is a principal ideal $\mathfrak{m} = (\pi)$. A generator π is often called a **uniformizing parameter**. It follows that $\mathfrak{m}^n = (\pi^n)$ for all nonnegative integers n. Krull's

G. Kemper, *A Course in Commutative Algebra*, Graduate Texts in Mathematics 256, DOI 10.1007/978-3-642-03545-6_15,

intersection theorem (Theorem 12.9) shows that for every nonzero $a \in R$ there exists a maximal integer n such that $a \in \mathfrak{m}^n$, so $a = u \cdot \pi^n$ with $u \in R^\times$ an invertible element. Since R is an integral domain by Corollary 13.6(a), we can form $K := \operatorname{Quot}(R)$ and write every $a \in K^\times$ (:= the multiplicative group $K \setminus \{0\}$) as $a = u \cdot \pi^n$ with $n \in \mathbb{Z}$ and $u \in R^\times$. It is easy to see that n and u are unique (and n does not depend on the choice of the uniformizing parameter). A consequence is that R is factorial with exactly one prime element, up to invertible elements. (As mentioned before, it is true but much harder to show that regular local rings of any dimension are factorial.) Mapping a to n defines a map ν: $K^\times \to \mathbb{Z}$. This map is a group homomorphism, and if we set $\nu(0) := \infty$, then ν satisfies $\nu(a+b) \geq \min\{\nu(a), \nu(b)\}$ for all $a, b \in K$, and $\nu(a) = \infty$ if and only if $a = 0$. A map with these properties is called a **discrete valuation** on K. We can retrieve R from K by means of ν, since

$$R = \{a \in K \mid \nu(a) \geq 0\}.$$

This is usually expressed by saying that R is the **valuation ring** belonging to the valuation ν. One also says that R is a **discrete valuation ring** (abbreviated DVR). Viewing regular local rings of dimension one as discrete valuation rings has become so common that these rings are often just referred to as DVRs. This is justified since as a converse of what we have just found, all DVRs are one-dimensional regular local rings (see Exercise 14.1).

Theorem 14.1. *A Noetherian local ring of dimension one is regular if and only if it is normal.*

Proof. Regularity implies normality by Corollary 13.6(b).

For the converse, assume that R is a one-dimensional normal Noetherian local domain with maximal ideal $\mathfrak{m}$. By Corollary 7.9 there exists $a \in \mathfrak{m}$ with $\sqrt{(a)} = \mathfrak{m}$. By the Noether property there exists an ideal P that is maximal among all colon ideals $(a) : (y) := \{x \in R \mid xy \in (a)\} \subseteq R$ with $y \in R \setminus (a)$. So $P := (a) : (b)$ with $b \in R \setminus (a)$. We claim that P is a prime ideal. Indeed, $P \neq R$ since $b \notin (a)$, and if $x, y \in R \setminus P$, then $xb \notin (a)$ and $(a) : (b) \subseteq (a) : (xb)$, so $(a) : (xb) = P$ by the maximality. Therefore $y \notin (a) : (xb)$, so $xy \notin P$. We have $(a) \subseteq P$, and since $\mathfrak{m}$ is the only prime ideal of R that contains (a), we conclude that $\mathfrak{m} = P = (a) : (b)$. Clearly $a \neq 0$, so we may consider the R-submodule

$$I := \frac{b}{a} \cdot \mathfrak{m} \subseteq \operatorname{Quot}(R).$$

From $\mathfrak{m} = (a) : (b)$ we get $I \subseteq R$, so I is an ideal. By way of contradiction assume that $I \subseteq \mathfrak{m}$. Then $\mathfrak{m}$ is an $R\left[\frac{b}{a}\right]$-module, so by Lemma 8.3, b/a is integral over R. By hypothesis, this implies $b/a \in R$, so $b \in (a)$, a contradiction. We conclude that $I = R$. Multiplying this equation by a/b yields $\mathfrak{m} = R \cdot \frac{a}{b}$, so $\mathfrak{m}$ is a principal ideal. Therefore R is regular. □

Exercise 12.8 shows that this result does not extend to higher dimensions. In fact, there are examples of nonregular normal Noetherian local rings of all dimensions ≥ 2.

Theorem 14.1 has some nice consequences. For example, if R is a Noetherian normal ring, then R_P is normal for all $P \in \operatorname{Spec}(R)$ by Proposition 8.10, so Theorem 14.1 says that R_P is regular for all P with $\operatorname{ht}(P) \leq 1$. Geometrically, this means that if X is a normal variety over an algebraically closed field, then the singular locus has codimension at least 2 in X. Both these statements are referred to as *regularity in codimension 1*. However, regularity in codimension 1 does not imply normality; a second condition, usually called "S2," is required (see Eisenbud [17, Theorem 11.15], and Exercise 14.3 for an explicit example). The situation is better for rings of dimension 1. In fact, it follows from Proposition 8.10 and Theorem 14.1 that a one-dimensional Noetherian domain is normal if and only if it is regular, and an irreducible affine curve is normal if and only if it is nonsingular. An important point is that normality is a property that can be achieved by normalization (whereas there is no such process as "regularization" in general). So in particular, by combining Corollary 8.28 with Theorem 14.1 we get the following result.

Corollary 14.2 (Desingularization of affine curves). *Let X be an irreducible affine curve. Then there exists an affine curve $\widetilde{X}$ with a surjective morphism $f: \widetilde{X} \to X$ such that:*

(a) $\widetilde{X}$ is nonsingular.
(b) All fibers of f are finite, and if $x \in X$ is a nonsingular point, then the fiber of x consists of one point.

Generalizing Corollary 14.2, we could speak of "desingularization in codimension 1" of a higher-dimensional irreducible affine variety. Moreover, in Exercise 14.4, the corollary is generalized to arbitrary affine curves. What Corollary 14.2 does can best be pictured in the situation of a double point: The two branches of the curve that cross are taken apart by raising one to a higher dimension, thereby deleting the double point. Sometimes one also speaks of *blowing up* a singularity. Example 8.9(4) illustrates this. The example also shows that the "higher" dimension can in fact be smaller. The following is an example in which the dimension does go up.

Example 14.3. We wish to desingularize the plane complex curve $X \subseteq \mathbb{C}^2$ given by the equation $x_1^4 + x_2^4 - x_1^2 = 0$, which is irreducible by the Eisenstein criterion (see Lang [33, Chapter V, Theorem 7.1]). The curve X is shown in Fig. 14.1. The idea is to desingularize X by forming the normalization of the coordinate ring $A := \mathbb{C}[X]$. How can we find quotients of elements of A that are integral over A? The Jacobian criterion (Theorem 13.10) yields $(0, 0)$ as the only singular point. By Theorem 14.1, the localization of the coordinate ring $A = \mathbb{C}[X]$ is normal at all points except $(0, 0)$. So the normalization $\widetilde{A}$ is contained in all A_x with $x \neq (0, 0)$. This means that an $f/g \in \widetilde{A}$ satisfies $g(x) \neq 0$ for $x \neq (0, 0)$. From this is it straightforward to try the residue class

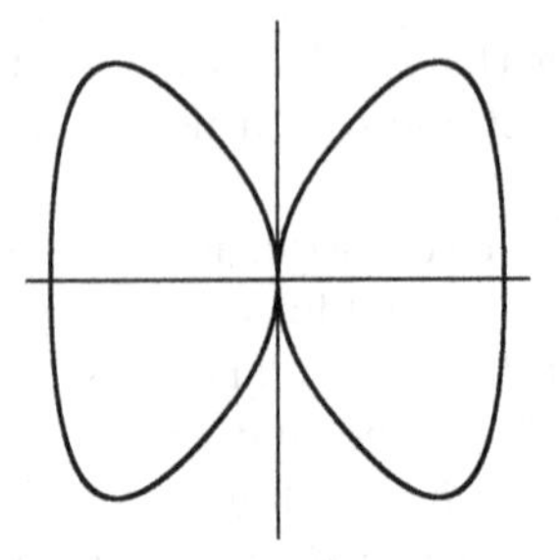

Fig. 14.1. A "butterfly" curve

of x_1 as the denominator g. By trial and error, we find that $a := \overline{x}_2^2/\overline{x}_1$ (with $\overline{x}_i := x_i + (x_1^4 + x_2^4 - x_1^2) \in A$) is integral over A, since dividing the defining equation by x_1^2 yields $\overline{x}_1^2 + a^2 - 1 = 0$. Putting this equation together with the defining equation for a, we consider the variety

$$\widetilde{X} := \left\{(\xi_1, \xi_2, \xi_3) \in \mathbb{C}^3 \mid \xi_1^2 + \xi_3^2 - 1 = \xi_1\xi_3 - \xi_2^2 = 0\right\} \subset \mathbb{C}^3.$$

We hope and guess that $\widetilde{X}$ is the desired desingularization. To verify this, we first check that

$$f\colon \widetilde{X} \to X,\ (\xi_1, \xi_2, \xi_3) \mapsto (\xi_1, \xi_2),$$

is a morphism, since $(\xi_1, \xi_2, \xi_3) \in \widetilde{X}$ obviously implies $\xi_1^4 + \xi_2^4 - \xi_1^2 = 0$. Secondly, every point $(\xi_1, \xi_2) \in X \setminus \{(0,0)\}$ has the unique preimage $(\xi_1, \xi_2, \xi_2^2/\xi_1)$, and the singular point $(0,0)$ has two preimages: $(0,0,1)$ and $(0,0,-1)$. Finally, the Jacobian matrix of $\widetilde{X}$ is

$$J = \begin{pmatrix} 2x_1 & 0 & 2x_3 \\ x_3 & -2x_2 & x_1 \end{pmatrix}.$$

For points $(\xi_1, \xi_2, \xi_3) \in \widetilde{X}$ with $\xi_2 \neq 0$, also ξ_1 and ξ_3 are nonzero, so $J(\xi_1, \xi_2, \xi_3)$ has rank 2. On the other hand, if $\xi_2 = 0$, then ξ_1 or ξ_3, but not both, are zero, and again $\operatorname{rank}(J(\xi_1, \xi_2, \xi_3)) = 2$. By the Jacobian criterion, this shows that $\widetilde{X}$ is nonsingular. So we have indeed found a desingularization. With a bit more work (i.e., by verifying that the equations for $\widetilde{X}$ define a prime ideal) we could also establish that $\widetilde{X}$ is exactly the normalization of X.

This example shows very nicely what happens: The original plane curve is wound around the cylinder given by the equation $\xi_1^1 + \xi_3^2 - 1 = 0$ in such a way that the branches of the curve are on different sides of the cylinder. In this way the double point is blown up. ◁

More examples are contained in Exercise 14.5.

In dimension greater than one, the existence and calculation of a desingularization is a much harder problem. In fact, in positive characteristic the existence problem is still open. For a good overview and an in-depth treatment, readers should turn to Cutkosky [14].

14.2 Multiplicative Ideal Theory

For any ring R, the set of ideals together with the ideal product forms an abelian monoid with R as neutral element. The only invertible element in this monoid is R itself. The situation becomes more interesting if we enlarge our view by including *fractional ideals*, according to the following definition.

Definition 14.4. *Let R be an integral domain and $K := \operatorname{Quot}(R)$ its field of fractions.*

(a) A **fractional ideal** *is an R-submodule $I \subseteq K$. The product of two fractional ideals is defined as the product of ordinary ideals (see Definition 2.5), making the set of fractional ideals into an abelian monoid with neutral element R. (It should be noted that some authors require fractional ideals to be nonzero, and/or impose the additional condition that there exist a nonzero $a \in R$ with $aI \subseteq R$.)*

(b) A fractional ideal is called **invertible** *if there exists a fractional ideal J with $I \cdot J = R$. So the invertible fractional ideals form an abelian group, which we write as $C(R)$. (We will give an explanation for the choice of the letter C on page 205.)*

It is possible to generalize the above definition to rings that need not be integral domains by considering the total ring of fractions instead of the field of fractions. However, almost none of the theory that we will develop here carries over to this case. So we continue to assume that R is an integral domain.

If a product $I \cdot J$ of fractional ideals is invertible then so are I and J (multiply the inverse of $I \cdot J$ by J and by I), and conversely. For every nonzero $a \in K$, the principal fractional ideal $(a)_R$ is invertible (with inverse $(a^{-1})_R$). This gives a homomorphism from $K^\times$ into $C(R)$ with kernel $R^\times$. In general, this is not surjective, i.e., there may exist nonprincipal invertible ideals, as the following example shows.

Example 14.5. In the ring $R := \mathbb{Z}\left[\sqrt{-5}\right] \subseteq \mathbb{C}$, consider the ideal $I = \left(2, 1+\sqrt{-5}\right)_R \subseteq R$. If $J := \left(1, \frac{1-\sqrt{-5}}{2}\right)_R \subseteq \operatorname{Quot}(R)$, then

$$I \cdot J = \left(2, 1-\sqrt{-5}, 1+\sqrt{-5}, 3\right)_R = R,$$

so I is invertible. But I is not a principal ideal. Indeed, from the assumption $I = (z)_R$ with $z = a + b\sqrt{-5}$, $a, b \in \mathbb{Z}$, we deduce that $a^2 + 5b^2$ (the norm of z, which by definition is the product of z and its complex conjugate) divides 4 and 6, the norms of 2 and of $1 + \sqrt{-5}$. This implies $a = \pm 1$ and $b = 0$, so $I = R$. But $I = \{x + y\sqrt{-5} \mid x, y \in \mathbb{Z},\ x \equiv y \mod 2\} \neq R$.

We have already studied this ring R in Example 8.9(3), and seen that it is normal but not factorial. ◁

So invertible ideals generalize principal ideals. But they are not very far away from being principal, as the following result shows.

Proposition 14.6 (Invertible ideals are locally principal). *Let R be an integral domain and $I \subseteq K :=$ Quot(R) a fractional ideal. Then the following statements are equivalent:*

(a) I is invertible.
(b) If $I' := \{a \in K \mid aI \subseteq R\}$, then $I \cdot I' = R$.
(c) I is nonzero, finitely generated, and for every prime ideal $P \in$ Spec(R) there exists $a \in I$ such that the localization of I satisfies

$$I_P = (a)_{R_P}.$$

We describe the latter property of I by saying that I is locally principal.

Proof. We start by showing that (a) implies (c). So we assume that there exists a fractional ideal $J \subseteq K$ with $I \cdot J = R$. In particular, we have $1 = \sum_{i=1}^n a_i b_i$ with $a_i \in I$ and $b_i \in J$. So every $x \in I$ satisfies $x = \sum_{i=1}^n x b_i a_i$, and $x b_i \in I \cdot J = R$. Therefore I is generated by $a_1, \dots, a_n$. Clearly I is nonzero. Moreover, for every $P \in$ Spec(R) there exist $a \in I$ and $y \in J$ with $ay \in R \setminus P$ (otherwise, $I \cdot J$ would be contained in P). So for a general element $b/u \in I_P$ (with $b \in I$ and $u \in R \setminus P$) we have

$$\frac{b}{u} = \frac{by}{uay} \cdot a \in (a)_{R_P},$$

since $by \in I \cdot J = R$ and $uay \in R \setminus P$. So I is locally principal.

Now we assume (c) and wish to deduce (b). By the definition of I', $I \cdot I' \subseteq R$ is an ideal. By way of contradiction, assume that it is proper. Then there exists a maximal ideal $P \in$ Spec(R) with $I \cdot I' \subseteq P$. (This conclusion requires Zorn's lemma.) By hypothesis we have $a \in I$ with $I_P = (a)_{R_P}$, and $I = (a_1, \dots, a_n)$. It follows that there exists $u \in R \setminus P$ with $u a_i \in (a)_R$ for all i, so $uI \subseteq (a)_R$. Since $I \neq \{0\}$, a is nonzero, and it follows that $u/a \in I'$, so $u = a \cdot u/a \in I \cdot I'$, in contradiction to $I \cdot I' \subseteq P$. Therefore (b) holds.

Finally, (b) implies (a) since I' is a fractional ideal, and we are done. Let us add that (b) can easily be deduced directly from (a). □

In view of part (b) of the above proposition, we define

$$I^{-1} := \{a \in \text{Quot}(R) \mid aI \subseteq R\}$$

for any fractional ideal of an integral domain.

The finiteness condition in part (c) cannot be omitted: Although it may seem unlikely, there are examples of Noetherian domains with fractional ideals that are locally principal but not finitely generated (see Exercise 14.6).

We draw a few consequences from Proposition 14.6.

Corollary 14.7 (Properties of invertible ideals). *Let $I \in C(R)$ be an invertible fractional ideal of a Noetherian domain R.*

(a) There exist invertible ideals $J_1, J_2 \subseteq R$ with $I = J_1 \cdot J_2^{-1}$.
(b) If $I \subseteq R$, then every prime ideal $P \in \operatorname{Spec}(R)$ that is minimal over I has height 1.
(c) If $I =: P$ is a prime ideal of R, then P has height 1 and R_P is regular.

Proof. (a) By Proposition 14.6, I is finitely generated. If $a \in R \setminus \{0\}$ is a common denominator of all elements in a generating set, then $J_1 := I \cdot (a) \subseteq R$ and $I = J_1 \cdot (a)^{-1}$. Since $J_2 := (a)$ and I are invertible, the same holds for J_1.

(b) Let $P \in \operatorname{Spec}(R)$ be minimal over I. Then P_P is minimal over I_P, which by Proposition 14.6 is a principal ideal. So $\operatorname{ht}(P_P) \leq 1$ by the principal ideal theorem (Theorem 7.4). Since $\{0\} \neq I \subseteq P_P$, the height must be equal to 1. So $\operatorname{ht}(P) = \operatorname{ht}(P_P) = 1$.

(c) By part (b), P has height 1, so $\dim(R_P) = 1$. By Proposition 14.6, the maximal ideal of R_P is principal, so R_P is regular. □

So we cannot hope that prime ideals of height other than 1 are invertible. But when are all height-one prime ideals invertible? By the corollary, a necessary condition for this is regularity in codimension 1. So a normal Noetherian domain would be a good candidate. However, in Exercise 14.7 we find an example of a normal Noetherian domain with a prime ideal of height 1 that is *not* invertible. So more is required. Recall that by Proposition 8.8, factoriality is a stronger condition than normality, and by Proposition 8.10, the condition that every localization at a prime ideal is factorial lies between the two. We call an integral domain R **locally factorial** if R_P is factorial for every $P \in \operatorname{Spec}(R)$.

Theorem 14.8 (Invertible ideals in a locally factorial ring). *Let R be a Noetherian domain.*

(a) If R is locally factorial, then every height-one prime ideal of R is invertible.
(b) If every height-one prime ideal of R is invertible, then an ideal $I \subseteq R$ is invertible if and only if it is a finite product of prime ideals of height 1 (where $I = R$ occurs as the empty product).

Remark. As mentioned before, every regular ring is locally factorial. (We have proved this only for rings of dimension at most 1; see page 197.) So all regular domains lie within the scope of the theorem. ◁

Proof of Theorem 14.8. (a) Let $Q \subset R$ be a prime ideal of height 1. We use Proposition 14.6. Clearly Q is finitely generated and nonzero, so we need to show only that $Q_P \subseteq R_P$ is a principal ideal for every $P \in \operatorname{Spec}(R)$. If $Q \nsubseteq P$, then Q contains elements that are invertible in R_P, so $Q_P = (1)_{R_P}$ is a principal ideal. On the other hand, if $Q \subseteq P$, then by Theorem 6.5, Q_P is a prime ideal of R_P of height 1. Since R_P is factorial, it follows by Lemma 5.14 that Q_P is a principal ideal in this case, too.

(b) It follows from the hypothesis that every product of height-one prime ideals is also invertible. We prove the converse by Noetherian induction. So assume that there exists an invertible ideal that is *not* a product of height-one prime ideals. By the Noether property we can choose I be maximal among all counterexamples. Since R is not a counterexample, $I \neq R$, and therefore there exists a prime ideal $P \in \operatorname{Spec}(R)$ that is minimal over I. By Corollary 14.7(b), P has height 1, so it is invertible. Using Lemma 14.9 below, we obtain $I \subsetneqq J := I \cdot P^{-1} \subseteq R$. Since I is invertible, so is J. With the maximality of I, this implies that J is a product of height-one prime ideals. So the same holds for I, and we are done. □

In the proof we have used the following lemma.

Lemma 14.9. *Let R be a Noetherian domain and let $I \subseteq R$ be a nonzero ideal that is contained in an invertible prime ideal P. Then $I \subsetneqq I \cdot P^{-1} \subseteq R$.*

Proof. From $I \subseteq P$ it follows that $J := I \cdot P^{-1} \subseteq P \cdot P^{-1} = R$. Moreover, $I = J \cdot P \subseteq J$. Assume that $I = J$. Then $I = P \cdot I$. This localizes to $I_P = P_P \cdot I_P$, which by Nakayama's lemma (Theorem 7.3) gives $I_P = \{0\}$. Since there are no zero divisors, we obtain $I = \{0\}$, contradicting the hypothesis. □

Theorem 14.8(b) becomes even more interesting if we combine it with the following unique factorization result.

Proposition 14.10 (Unique factorization of invertible ideals). *Let R be an integral domain and let $I \subseteq R$ be an invertible ideal that has a factorization*

$$I = P_1 \cdots P_n$$

with P_i prime ideals (where $n = 0$ occurs if $I = R$). Then this factorization is unique up to the order of the factors.

Proof. We use induction on n. Let $I = Q_1 \cdots Q_m$ be another factorization with $Q_i \in \operatorname{Spec}(R)$. If $n = 0$ then $m = 0$, since otherwise $I \subseteq Q_1 \subsetneqq R{=}I$. Consider the case $n > 0$. By renumbering, we may assume that P_1 is minimal among the P_i. Since $Q_1 \cdots Q_m \subseteq P_1$, there exists i with $Q_i \subseteq P_1$. By renumbering, we may assume $i = 1$. Since $P_1 \cdots P_n \subseteq Q_1$, there exists j with $P_j \subseteq Q_1$, so $P_j \subseteq Q_1 \subseteq P_1$. With the minimality of P_1, this implies $P_1 = Q_1$. Since I is invertible, so are all P_i. Multiplying by $P_1^{-1} = Q_1^{-1}$ gives $P_2 \cdots P_n = Q_2 \cdots Q_m$, and the result follows by induction. □

Assume that R is a locally factorial Noetherian domain, or more generally a Noetherian domain in which all height-one prime ideals are invertible. We can extend Theorem 14.8(b) and Proposition 14.10 to invertible *fractional* ideals. In fact, if $I \subseteq \text{Quot}(R)$ is an invertible fractional ideal, then it follows by Corollary 14.7(a) and Theorem 14.8(b) that I can be written as a product of height-one prime ideals and inverses of height-one prime ideals. Conversely, it follows from the group property of $C(R)$ that every such product is invertible. More formally, let $\mathcal{M} \subseteq \text{Spec}(R)$ be the set of all prime ideals of height 1. Then a fractional ideal I is invertible if and only if it can be written as

$$I = \prod_{Q \in \mathcal{M}} Q^{e_{I,Q}} \tag{14.1}$$

with $e_{I,Q} \in \mathbb{Z}$, and all but finitely many $e_{I,Q}$ equal to 0. It follows from Proposition 14.10 that the $e_{I,Q}$ are unique. In fact, if there existed two different factorizations, we could multiply both by height-one prime ideals until we obtained two different factorizations of a nonfractional ideal, contradicting Proposition 14.10. It also follows that $I \subseteq R$ if and only if all $e_{I,Q}$ are nonnegative.

If we multiply two invertible ideals, the corresponding exponents $e_{I,Q}$ in (14.1) get added. So our results can be expressed by saying that the group $C(R)$ of invertible fractional ideals is isomorphic to the free abelian group generated by the height-one prime ideals. This motivates the following definition. For any ring R, the group $\text{Div}(R)$ of **Weil divisors** is defined to be the free abelian group generated by the height-one prime ideals of R. In contrast to $C(R)$, the group of Weil divisors is usually written additively, so a Weil divisor is a "formal" $\mathbb{Z}$-linear combination of height-one prime ideals. In particular, if R is the coordinate ring of an affine curve, a Weil divisor can be written as a formal $\mathbb{Z}$-linear combination of points.

In this context, an invertible ideal of an integral domain R is called a **Cartier divisor**, and $C(R)$ is the group of Cartier divisors. This explains the use of the letter C. (It should be noted that the standard definition of Cartier divisors in algebraic geometry is different; see Hartshorne [26, page 141].) So if R is a locally factorial Noetherian domain (or, more generally, a Noetherian domain in which all height-one prime ideals are invertible), we have $C(R) \cong \text{Div}(R)$. Using the isomorphism, we can speak of the Weil divisor associated to an invertible ideal or to a nonzero element $a \in R$: The latter is $\sum_{i=1}^n e_i \cdot P_i$ if $(a) = \prod_{i=1}^n P_i^{e_i}$. The situation becomes less nice when we relax the conditions on R. Exercise 14.8 deals with the case that R is a normal Noetherian domain.

14.3 Dedekind Domains

In this theory, the best-behaved domains are those in which every nonzero ideal is invertible. We will study these rings now, and see that the invertibility of nonzero ideals is equivalent to various other interesting conditions.

Theorem 14.11 (Rings with a perfect multiplicative ideal theory). *For an integral domain R, the following statements are equivalent:*

(a) Every nonzero ideal of R is invertible.
(b) R is Noetherian and every ideal of R is locally principal.
(c) R is Noetherian and normal and has dimension at most 1.
(d) Every ideal $I \subseteq R$ is a finite product of prime ideals (where $I = R$ occurs as the empty product).

If these conditions are satisfied, then the factorization of a nonzero ideal as a product of prime ideals is unique up to the order of the factors. Moreover, every finitely generated, nonzero fractional ideal has a unique factorization as (14.1).

Proof. It follows from Proposition 14.6 that (a) implies (b).

We now assume (b) and wish to deduce (c). It follows that for every $P \in \text{Spec}(R)$, $P_P \subseteq R_P$ is a principal ideal. If $\text{ht}(P) = 0$, then $P = \{0\}$ and $R_P = \text{Quot}(R)$ is regular. Otherwise, it follows that R_P is one-dimensional and regular. Therefore R is regular (and hence normal by Corollary 13.6(b) and Proposition 8.10) and of dimension at most 1. So we have deduced (c).

Next we assume (c) and wish to prove (d). By (c), R is locally factorial since for every $P \in \text{Spec}(R)$, the local ring R_P is a field (in the case $P = \{0\}$) or a discrete valuation ring (by Theorem 14.1 and the discussion preceding it). So by Theorem 14.8(a), every height-one prime ideal of R is invertible. By way of contradiction, assume that there exists an ideal $I \subseteq R$ that is not a finite product of prime ideals. Since R is Noetherian, we may assume I to be maximal with this property. We have $\{0\} \neq I \subsetneqq R$, so there exists a prime ideal P that contains I. Since $\dim(R) \leq 1$ and $P \neq \{0\}$, P must have height 1, so it is invertible. Lemma 14.9 yields $I \subsetneqq I \cdot P^{-1} \subseteq R$, so by the maximality of I, $I \cdot P^{-1}$ is a finite product of prime ideals. Therefore the same is true for I.

The most work is required for deducing (a) from (d). We will first show that (under the assumption (d)) every invertible prime ideal is maximal. From this we will draw the (at first sight surprising) consequence that every nonzero prime ideal is invertible, which together with the hypothesis (d) implies (a) directly. So let $P \subseteq R$ be an invertible prime ideal. To show that P is maximal, we need to prove that $P + (a) = R$ for every $a \in R \setminus P$. We have factorizations

$$P + (a) = P_1 \cdots P_n \quad \text{and} \quad P + (a^2) = P_1' \cdots P_m'$$

as products of prime ideals. Computing modulo P and writing $\overline{a} := a + P \in \overline{R} := R/P$, we get

$$(\overline{a})_{\overline{R}} = \overline{P_1} \cdots \overline{P_n} \quad \text{and} \quad (\overline{a}^2)_{\overline{R}} = \overline{P_1'} \cdots \overline{P_m'}.$$

This gives two factorizations of $(\overline{a}^2)$, which is an invertible ideal of $\overline{R}$. By Proposition 14.10 it follows that $m = 2n$ and, after renumbering, $\overline{P_i} = \overline{P_{2i-1}'} = \overline{P_{2i}'}$ for $i = 1, \ldots, n$. By Lemma 1.22, the same holds for the original P_i and P_j', and we conclude that $P + (a^2) = (P + (a))^2$. In particular, every $x \in P$ can be written as $x = y + az + a^2 w$ with $y \in P^2$, $z \in P$, and $w \in R$. But then $w \in P$ since $a^2 w = x - y - az \in P$ and $a^2 \notin P$. So $x \in P^2 + a \cdot P$, and we obtain

$$P \subseteq P \cdot (P + (a)) \subseteq P.$$

Multiplying by P^{-1} yields $P + (a) = R$, as claimed.

The second (and final) step is to show that every nonzero prime ideal is invertible. So assume that $\{0\} \neq Q \in \operatorname{Spec}(R)$. Choose a nonzero $b \in Q$. By hypothesis, we have $(b) = Q_1 \cdots Q_r$ with $Q_i \in \operatorname{Spec}(R)$. Since (b) is invertible, the Q_i are invertible, too, so by what we have shown they are maximal. Since $Q_1 \cdots Q_r \subseteq Q$, there exists an i with $Q_i \subseteq Q$, so $Q = Q_i$ by the maximality of Q_i. Therefore Q is invertible, and the proof of the equivalence of (a) through (d) is complete.

The uniqueness of a factorization of a nonzero ideal follows from (a), (d), and Proposition 14.10. If I is a finitely generated, nonzero fractional ideal, there exists a nonzero $a \in R$ such that $J := aI \subseteq R$. Since J and (a) are products of prime ideals, I has a factorization as (14.1). If there are two such factorizations, we can multiply both by prime ideals until we obtain two factorizations of a nonfractional ideal. So the factorizations are unique after all. □

An integral domain that satisfies the equivalent conditions from Theorem 14.11 is called a **Dedekind domain**. Of these conditions, (c) is the one that tends to be easiest to verify. The condition (b) shows that Dedekind domains are not too far away from principal ideal domains. Although our investigation originated from studying condition (a), condition (d) and the unique factorization statement may be the most interesting. Notice that elements of a Dedekind domain do not always enjoy the unique factorization property that holds for ideals: consider Example 8.9(3). So ideals are "idealized" elements. Many more properties that are equivalent to R being a Dedekind domain can be found in the literature. For instance, Larsen and McCarthy [34, Theorem 6.20] list 16.

An important class of Dedekind domains comes from algebraic geometry: If X is an irreducible, nonsingular affine curve, then the coordinate ring $K[X]$ is a Dedekind domain since it satisfies (c) from Theorem 14.11.

Another class of arguably even more importance comes from number theory: Let K be a **number field**, i.e., a finite field extension of $\mathbb{Q}$. Then the ring of **algebraic integers** in K is defined as the integral closure of $\mathbb{Z}$ in K, and is written as $\mathcal{O}_K$. It follows from Lemma 8.27 that $\mathcal{O}_K$ is Noetherian. Being an integral closure of a ring in a field, it is also normal. Since $\mathcal{O}_K$ is an integral extension of $\mathbb{Z}$, it has dimension 1 by Corollary 8.13. So $\mathcal{O}_K$ satisfies condition (c) and is therefore a Dedekind domain. Rings of algebraic integers are the central object of study in the field of *algebraic number theory*. Historically, much of the interest in rings of algebraic integers originated from the study of Diophantine problems. For instance, the question which integers can be represented as $x^2 + dy^2$ (with $x, y, d \in \mathbb{Z}$, but d fixed) can be translated into a question about algebraic integers using the factorization $x^2 + dy^2 = \left(x + y\sqrt{-d}\right)\left(x - y\sqrt{-d}\right)$. So one is led to calculations in the ring $\mathcal{O}_K$ of algebraic integers in the number field $K = \mathbb{Q}(\sqrt{-d})$. Clearly the question whether $\mathcal{O}_K$ is factorial plays a central role in this game. The answer is yes for some d (e.g., $d = 1$), but no for most (e.g., $d = 5$; see Example 8.9(3)). Another extremely well-known Diophantine equation is the Fermat equation $x^n + y^n = z^n$. With ζ_{2n} a primitive $(2n)$th root of unity, this translates to

$$\prod_{i=1}^{n} \left(x - \zeta_{2n}^{2i-1} y\right) = z^n,$$

an equation in the ring $\mathcal{O}_K$ of algebraic integers in the cyclotomic field $K = \mathbb{Q}(\zeta_{2n})$. Again, the question whether $\mathcal{O}_K$ is factorial arises naturally. In fact, there were attempts at proving Fermat's last theorem that hinged on the assumption that $\mathcal{O}_K$ is factorial. Again, this is false for most n. The following example illustrates how the nonuniqueness of factorization in a ring of algebraic integers is resolved by turning to ideals.

Example 14.12. Consider the ring $R = \mathbb{Z}\left[\sqrt{-5}\right]$. In Example 8.9(3) we have seen that R is normal, so R is the ring of algebraic integers in $\mathbb{Q}(\sqrt{-5})$. There we have also exhibited an example of a nonunique factorization:

$$6 = 2 \cdot 3 = \left(1 + \sqrt{-5}\right)\left(1 - \sqrt{-5}\right). \tag{14.2}$$

How do the corresponding principal ideals $(2)_R$, $(3)_R$, etc. factorize? In Exercise 14.9 it is shown that every ideal of a Dedekind domain is generated by two elements. With this in mind, it is not too hard to find the following factorizations, which are easy to verify:

$$\begin{aligned}
(2)_R &= \left(2, 1 + \sqrt{-5}\right)_R^2, \\
(3)_R &= \left(3, 1 + \sqrt{-5}\right)_R \left(3, 1 - \sqrt{-5}\right)_R, \\
\left(1 + \sqrt{-5}\right)_R &= \left(2, 1 + \sqrt{-5}\right)_R \left(3, 1 + \sqrt{-5}\right)_R, \\
\left(1 - \sqrt{-5}\right)_R &= \left(2, 1 + \sqrt{-5}\right)_R \left(3, 1 - \sqrt{-5}\right)_R.
\end{aligned}$$

To see that $(2, 1+\sqrt{-5})_R$ and $(3, 1 \pm \sqrt{-5})_R$ are prime ideals, observe that they are the kernels of the ring homomorphisms $R \to \mathbb{F}_2$, $a + b\sqrt{-5} \mapsto a+b \mod 2$, and $R \to \mathbb{F}_3$, $a + b\sqrt{-5} \mapsto a \mp b \mod 3$, respectively. So we get the unique factorization

$$(6)_R = (2, 1+\sqrt{-5})_R^2 \, (3, 1+\sqrt{-5})_R \, (3, 1-\sqrt{-5})_R$$

of ideals, and the nonuniqueness in (14.2) is explained by regroupings of the above factors. ◁

We have mentioned before that for any integral domain R, the principal ideals (a) with $a \in \operatorname{Quot}(R) \setminus \{0\}$ form a subgroup of $C(R)$. The quotient group

$$\operatorname{Cl}(R) := C(R) \Big/ \Big\{(a) \mid a \in \operatorname{Quot}(R) \setminus \{0\}\Big\}$$

is called the **ideal class group** of R. This name is most intuitive in the case that R is a Dedekind domain, and some authors restrict the definition to that case. Since $C(R)$ and $\operatorname{Div}(R)$ are isomorphic if R is a Dedekind domain, $\operatorname{Cl}(R)$ is isomorphic to the group of equivalence classes of Weil divisors, where two Weil divisors are called **linearly equivalent** if they map to a principal fractional ideal in $C(R)$. For a Dedekind domain R, the ideal class group is trivial if and only if R is a principal ideal domain (which by the following theorem is equivalent to R being factorial). So $\operatorname{Cl}(R)$ can be viewed as quantifying the extent to which a Dedekind domain fails to be factorial.

Theorem 14.13 (Factorial Dedekind domains). *For a Dedekind domain R, the following statements are equivalent:*

(a) R is factorial;
(b) R is a principal ideal domain.

Proof. First assume that R is factorial. By Lemma 5.14, it follows that every prime ideal of height 1 is principal. Since every nonzero ideal is a product of height-one prime ideals, this implies (b).

The fact that every principal ideal domain is factorial is usually part of an abstract algebra course (see Lang [33, Chapter II, Theorem 4.2]). We give a (shorter) proof for the case of Dedekind domains here. Let $a \in R$ be a nonzero, noninvertible element. Then $(a) = P_1 \cdots P_n$ with P_i prime ideals. By assumption, we have $P_i = (p_i)$ with $p_i \in R$ prime elements. Multiplying p_1 by an invertible element if necessary, we can achieve that $a = p_1 \cdots p_n$. Now suppose that we have another factorization $a = q_1 \cdots q_m$ with $q_j \in R$ irreducible. Since p_1 is a prime element that divides the product of the q_j, it divides one of the q_j, say q_1. Therefore we can achieve $q_1 = p_1$ by multiplying q_1 by an invertible element if necessary. Continuing in this way, we end up with $p_i = q_i$ for $i = 1, \ldots, n$ and $1 = q_{n+1} \cdots q_m$, so $m = n$. This proves the uniqueness of factorization. □

A generalization of Theorem 14.13 is given in Exercise 14.10.

How large can ideal class groups become? For rings $\mathcal{O}_K$ of algebraic integers in a number field, the answer is that the ideal class group is finite. Its order is called the *class number*. This is one of the central results of algebraic number theory. For a proof, see Neukirch [42, Chapter I, Theorem 6.3]. This is in sharp contrast to the behavior in more general cases. In fact, we will see in the following example that for a nonsingular, irreducible affine curve X, $\mathrm{Cl}(K[X])$ can become infinite. (In fact, it is finite only in exceptional cases.) Moreover, Claborn [10] proved that any abelian group whatsoever is isomorphic to the ideal class group of a suitable Dedekind domain.

We finish this chapter with an example that shows how the ideal class group can be used to give an elliptic curve the structure of an abelian group.

Example 14.14 (The group law on an elliptic curve). Let $E \subseteq K^2$ be an elliptic curve over an algebraically closed field K of characteristic not equal to 2, given by the equation

$$x_2^2 = x_1^3 + ax_1 + b$$

with $a, b \in K$ such that $4a^3 + 27b^2 \neq 0$ (see Exercise 13.10). The goal of this example is to give E (enriched by a point at infinity) the structure of an abelian group, which is isomorphic to the ideal class group of the coordinate ring $R := K[E]$. For some details and proofs we will refer to the exercises. By Exercise 13.10, E is nonsingular, so R is a Dedekind domain. For two points $P_1, P_2 \in E$, let L be the line passing through P_1 and P_2. If $P_1 = P_2$, take the tangent line to E through P_1. (The remark at the end of Exercise 14.11 says exactly how this is done.) If L is not parallel to the x_2-axis, then L meets E at another point P_3. This is shown in Fig. 14.2, and proved in Exercise 14.11. Notice that P_3 may be equal to P_1 or to P_2 if $P_1 \neq P_2$ and L is tangent to E at this point, or if $P_1 = P_2$ is an inflection point of E. If $l \in K[x_1, x_2]$ is a polynomial of degree 1 defining L and $\overline{l} \in R$ is the corresponding regular function on X, then $\overline{l}$ vanishes at the points P_i, so it lies in

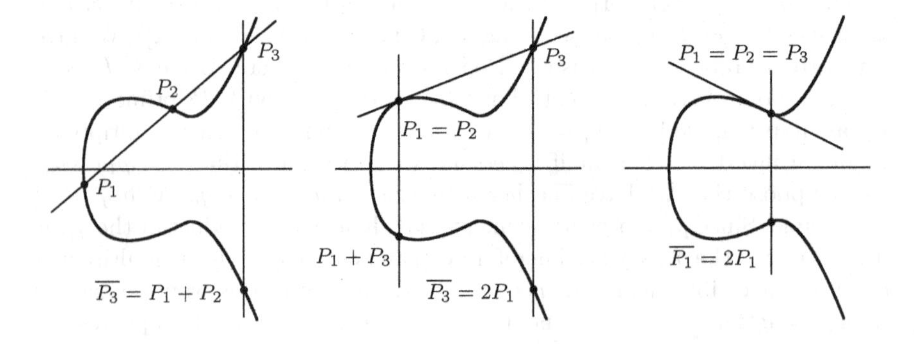

Fig. 14.2. The group law on the elliptic curve $y^2 = x^3 - x + 1$

the corresponding maximal ideals $\mathfrak{m}_{P_i} \in \operatorname{Spec}_{\max}(R)$. It is very plausible that $(\overline{l})_R = \mathfrak{m}_{P_1}\mathfrak{m}_{P_2}\mathfrak{m}_{P_3}$. Exercise 14.11 gives an exact proof. (The subtlety lies in the multiplicities in the case that some P_i coincide.) So the Weil divisor $P_1 + P_2 + P_3$ is linearly equivalent to 0. We write this as

$$P_1 + P_2 + P_3 \sim 0. \tag{14.3}$$

Next we consider the case that L is parallel to the x_2-axis. This happens if and only if $P_2 = \overline{P_1}$, where for any point $P = (\xi_1, \xi_2)$ we write $\overline{P} := (\xi_1, -\xi_2)$. In this case, P_1 and $\overline{P_1}$ are the only intersections of L and E. So for every $P \in E$ we obtain

$$P + \overline{P} \sim 0. \tag{14.4}$$

Putting this together with (14.3) yields

$$P_1 + P_2 \sim \overline{P}_3. \tag{14.5}$$

This already looks like an addition on E. To show that it really defines a group law, consider the map

$$\varphi\colon E \to \operatorname{Cl}(R),\ P \mapsto [\mathfrak{m}_P] \quad (\text{the class of } \mathfrak{m}_P \text{ in } \operatorname{Cl}(R)).$$

So in terms of Weil divisors, φ maps every point to its equivalence class. Let $d = \sum_{i=1}^m n_i P_i \in \operatorname{Div}(R)$ (with coefficients $n_i \in \mathbb{Z}$ and $P_i \in E$) be a Weil divisor. We obtain another Weil divisor $\overline{d} = \sum_{i=1}^m k_i Q_i$ by substituting every P_i with $n_i < 0$ in d by $-\overline{P_i}$. Then $d \sim \overline{d}$ by (14.4), and all coefficients k_i in $\overline{d}$ are nonnegative. If the coefficient sum of $\overline{d}$ is greater than 2, we can use (14.5) to find a Weil divisor that is linearly equivalent to $\overline{d}$, but has coefficient sum one smaller than that of $\overline{d}$. So by induction on the coefficient sum, we see that every Weil divisor is linearly equivalent to a point $P \in E$ or to 0. We conclude that every nontrivial element of $\operatorname{Cl}(R)$ lies in the image of φ.

The most difficult part of this discussion is to prove that φ is injective, i.e., that for two distinct points $P, Q \in E$, there exists no $f \in \operatorname{Quot}(R)$ with $(f)_R = \mathfrak{m}_P \cdot \mathfrak{m}_Q^{-1}$. This is the content of Exercise 14.12. In this exercise, it is also shown that the trivial class is not in the image of φ, i.e., there exists no $f \in \operatorname{Quot}(R)$ such that $(f)_R = \mathfrak{m}_P$ with $P \in E$. With this, we can extend φ to a bijection between $\widehat{E} := E \cup \{\infty\}$ and $\operatorname{Cl}(R)$ by mapping ∞ to the trivial class. The geometric interpretation of the additional point ∞ is that it is the point at infinity. This makes sense since we can think of the line through P and $\overline{P}$ as meeting E at infinity. Having a bijection between $\widehat{E}$ and the abelian group $\operatorname{Cl}(R)$, we can use this to transfer the group law from $\operatorname{Cl}(R)$ to $\widehat{E}$. With this, (14.5) indeed defines the sum of two points $P_1, P_2 \in E$ as given by the following recipe: Draw the line through P_1 and P_2 and take the third point P_3 of E meeting this line (always counting intersections with multiplicities). Then reflect P_3 in the x_1-axis to obtain the desired point $\overline{P_3} = P_1 + P_2$. Special cases apply: $P + \infty := P$, and $P + \overline{P} := \infty$.

It is of course possible to define the addition on $\widehat{E}$ directly by this recipe. Then the main difficulty is to verify the associative law (e.g., this takes 12 pages in the book of Washington [52]). But by using the bijection with $\mathrm{Cl}(R)$, we get the associative law automatically. This approach also gives a conceptual explanation of why the group law is defined in such a seemingly arbitrary way. On the other hand, it provides the ideal class group $\mathrm{Cl}(R)$ with the structure of a projective variety. In this way, elliptic curves act as the first significant example for the theories of Jacobian varieties and abelian varieties, which are deep and fascinating subjects in algebraic geometry.

Another important aspect is rational points. Suppose that $k \subseteq K$ is a subfield with $a, b \in k$ (i.e., the equation defining E lies in $k[x_1, x_2]$). A point $P \in E(k) := k^2 \cap E$ is called (k-)**rational**. If P is a rational point, then clearly the same is true for $-P = \overline{P}$. Moreover, if $P_1, P_2 \in E(k)$ with $P_1 \neq -P_2$, then substituting a parametrization of the line through P_1 and P_2 into the equation defining E gives a polynomial of degree 3 with coefficients in k. (Exercise 14.11 has more details.) Since this polynomial has two zeros in k, corresponding to the points P_1 and P_2 (or a double zero if $P_1 = P_2$), its third zero lies in k, too. This means that $P_1 + P_2$ is also a rational point. So we have seen that $\widehat{E}(k) := E(k) \cup \{\infty\}$ is a subgroup of $\widehat{E}$.

This has applications in cryptography. In fact, if k is a (large) finite field, then $\widehat{E}(k)$ provides a finite group G in which the *discrete logarithm problem* (i.e., determining n from the given data g and g^n, with $g \in G$, written multiplicatively) is supposedly very hard. This gives rise to public-key cryptosystems. In this business, the choice of the elliptic curve and of a "base point" $P \in E$ with large order are crucial for the security of the cryptosystem. Applications to cryptography are among the reasons why elliptic curves have become very fashionable (and useful) in recent years. See Washington [52] for a good introduction to elliptic curves and their use in cryptography. ◁

Exercises for Chapter 14

14.1 (Discrete valuation rings). Let K be a field and let $\nu\colon K \to \mathbb{Z} \cup \{\infty\}$ be a discrete valuation. Assume that ν is *nontrivial*, i.e., $\operatorname{im}(\nu) \neq \{0, \infty\}$. Show that the valuation ring $R := \{a \in K \mid \nu(a) \geq 0\}$ is a one-dimensional regular local ring.

14.2 (Discrete valuations on the rational function field). Let $K(x)$ be the rational function field over a field. Classify all nontrivial discrete valuations on $K(x)$ that vanish on $K^\times$.
Hint: You will find that the valuation rings are in bijective correspondence with the set of all monic irreducible polynomials in $K[x]$ together with one extra element, usually written as ∞ (why?).

14.3 (Regular in codimension 1 does not imply normal). This exercise deals with an example of an affine domain that is regular in codimension 1 but not normal. The example is drawn from Shafarevich [46, Chapter II, §5.1], where is appears in geometric terms. The example is the subalgebra

$$A := K[f_1, f_2, f_3, f_4] \subseteq K[x_1, x_2]$$

with

$$f_1 = x_1,\ f_2 = x_1x_2,\ f_3 = x_2(x_2 - 1),\ f_4 = x_2^2(x_2 - 1),$$

where $K[x_1, x_2]$ is the polynomial algebra in two indeterminates over a field.

(a) Show that $K[x_1, x_2]$ is the normalization of A.
(b) Show that there exist two maximal ideals $\mathfrak{n}_1, \mathfrak{n}_2 \in \operatorname{Spec}_{\max}(K[x_1, x_2])$ with $A \cap \mathfrak{n}_i = (f_1, f_2, f_3, f_4)_A =: \mathfrak{m}$.
*(c) Show that $K[x_1, x_2] \subseteq A_P$ for all $P \in \operatorname{Spec}(A) \setminus \{\mathfrak{m}\}$, and conclude that there exists $Q \in \operatorname{Spec}(K[x_1, x_2])$ with $A_P = K[x_1, x_2]_Q$. *Hint:* Two of the relations of the f_i are $f_1^2 f_3 + f_2(f_1 - f_2) = 0$ and $f_3^3 + f_4(f_3 - f_4) = 0$.
(d) Conclude that A is a two-dimensional nonnormal domain such that the singular locus in $\operatorname{Spec}(A)$ is $\{\mathfrak{m}\}$, so regularity in codimension 1 holds.

14.4 (Desingularization of nonirreducible curves). Show that Corollary 14.2 holds for all (not necessarily irreducible) affine curves.
Hint: Use Exercises 4.3 and 6.6.

14.5 (Examples of desingularization). Find desingularizations of the plane complex curves given by the following equations.

(a) $x_1^3 - x_2^2 = 0$ (the cubic curve with a cusp shown in Fig. 12.1)
(b) $x_1^4 - x_1^2 + x_2^2 = 0$ (lemniscate of Gerono, an ∞-shaped curve)
(c) $x_1^6 + x_2^6 - x_1^2$ (butterfly-shaped, similar to Fig. 14.1)
(d) $x_1^4 + x_2^4 - x_1x_2$ (another figure-eight curve, but tilted by 45° and with perpendicular crossing)

Hint: It may be hard to do (d) by hand. If you have access to MAGMA [5] you can use the function `Normalization`.

14.6 (Finite generation of fractional ideals). (a) Give an example of a fractional $\mathbb{Z}$-ideal $I \subseteq \mathbb{Q}$ that is locally principal but not finitely generated.
(b) Show that for a nonzero fractional ideal $I \subseteq \operatorname{Quot}(R)$ of a Noetherian domain R, I^{-1} is finitely generated.
(c) For your example in (a), what are I^{-1}, $I \cdot I^{-1}$, and $(I^{-1})^{-1}$?

14.7 (A noninvertible prime ideal of height 1). This example is taken from Hutchins [28, Example 47] (with a slight modification), and due to Gilmer [20, page 554, Exercise 2]. Consider the ring $R = \mathbb{Z}[x, x^2/2] \subset \mathbb{Q}[x]$.

(a) Show that R is a normal Noetherian domain. *Hint:* For this part, it may lead to a nicer notation to consider the isomorphic ring $S := \mathbb{Z}\left[x, \sqrt{2x}\right]$. You may look at Example 8.9(3) for inspiration.
(b) Show that the ideal $P := \left(x, x^2/2\right)_R$ is a prime ideal of height 1.
(c) Show that P is not invertible.

14.8 (Cartier divisors and Weil divisors). Let R be a normal Noetherian domain. The goal of this exercise is to construct an injective homomorphism $C(R) \to \text{Div}(R)$. Write $\mathcal{M}$ for the set of height-one prime ideals of R, and write $\mathcal{F}$ for the set of all finitely generated nonzero fractional ideals. For each $Q \in \mathcal{M}$, R_Q is a Dedekind domain, so for $I \in \mathcal{F}$ there exists a unique $e_{I,Q} \in \mathbb{Z}$ with $I_Q = Q_Q^{e_{I,Q}}$.

(a) Show that
$$\Phi\colon \mathcal{F} \to \text{Div}(R),\ I \mapsto \sum_{Q \in \mathcal{M}} e_{I,Q} \cdot Q,$$
defines a homomorphism of monoids. *Hint:* The hardest part is to show that $e_{I,Q} = 0$ for all but finitely many Q.
(b) Show that the restriction
$$\Psi := \Phi|_{C(R)}\colon C(R) \to \text{Div}(R)$$
of Φ to $C(R)$ is an injective group homomorphism. *Hint:* Use Exercise 8.3.
*(c) Show that Ψ is surjective if and only if every $P \in \mathcal{M}$ is invertible. In this case, Ψ coincides with the isomorphism described on page 205. *Hint:* If $\Psi(I) = P \in \mathcal{M}$, consider $P \cdot I^{-1}$.

Remark: It follows that Exercise 14.7 gives an example in which Ψ is not surjective.

***14.9 (Properties of Dedekind domains).** Let R be a Dedekind domain. Prove the following.

(a) If $P_1, \ldots, P_n \in \text{Spec}(R)$ are pairwise distinct nonzero prime ideals and $e_1, \ldots, e_n$ are nonnegative integers, there exists $a \in R \setminus \{0\}$ such that
$$(a) = P_1^{e_1} \cdots P_n^{e_n} \cdot J$$
with $J \subseteq R$ an ideal in whose factorization none of the P_i appear.
(b) Every ideal of R is generated by at most two elements.

14.10 (Factorial rings). Show that for an integral domain R, the following statements are equivalent:

(a) R is factorial of dimension ≤ 1.
(b) R is a principal ideal domain.

If these conditions are satisfied, then R is Noetherian. Is it true that every factorial ring is Noetherian?

Exercises 14.11 and 14.12 fill the gaps in Example 14.14. Together with the example, they form a nice application project of our methods.

14.11 (Divisor of a line intersecting a curve). In this exercise we study a situation that seems rather special, but is general enough to handle elliptic curves, for example. Let K be an algebraically closed field and let $X \subset K^2$ be a nonsingular, irreducible affine curve. So $\mathcal{I}(X) = (g)$ with $g \in K[x_1, x_2]$ irreducible (see Theorem 5.13). Consider a line

$$L = \{(a\xi + b, c\xi + d) \mid \xi \in K\} \subset K^2 \quad \text{(with } a, b, c, d \in K, \ a \text{ or } c \text{ nonzero)},$$

and assume $L \neq X$. With t a new indeterminate, set $f := g(at + b, ct + d) \in K[t]$ and let $f = a_n \cdot \prod_{i=1}^{n}(t - \xi_i)$ with $a_n \in K \setminus \{0\}$ and $\xi_i \in K$, not necessarily distinct. So the $P_i := (a\xi_i + b, c\xi_i + d)$ are the points of the intersection $L \cap X$, counted with "multiplicities." Multiplicity greater than one means that L is "tangent" to X in P_i. Let $\mathfrak{m}_i \in \operatorname{Spec}_{\max}(K[X])$ be the maximal ideal belonging to P_i. Furthermore, let $l := cx_1 - ax_2 + ad - bc$ (which defines L), and let $\bar{l} := l + (g) \in K[X]$ be the corresponding regular function on X. Show that

$$(\bar{l}) = \mathfrak{m}_1 \cdots \mathfrak{m}_n.$$

So the Weil divisor $P_1 + \cdots + P_n$ is linearly equivalent to 0.
Remark: If X is an elliptic curve defined as in Example 14.14, then f has degree 3 if $a \neq 0$, i.e., if L is not parallel to the x_2-axis. So in this case we get three points whose sum is linearly equivalent to 0. Otherwise, f has degree 2, so the sum of two points is linearly equivalent to 0. It is also clear that (for general X) if P is a point of X, then by setting $a := \frac{\partial g}{\partial x_2}(P)$, $b := -\frac{\partial g}{\partial x_1}(P)$, and $(b, d) := P$, one achieves that the polynomial f will become divisible by t^2, which geometrically means that L is tangent to X in P. *(Solution on page 231)*

***14.12 (Rational functions on an elliptic curve).** Let K be an algebraically closed field of characteristic not equal to 2, and let $E \subset K^2$ be an elliptic curve given by the equation

$$x_2^2 = x_1^3 + ax_1 + b$$

with $a, b \in K$, $4a^3 + 27b^2 \neq 0$ (see Exercise 13.10). Let $R := K[E]$ be the coordinate ring and $L := \operatorname{Quot}(R)$ the *field of rational functions* on E. By a **place** of L we mean a discrete valuation ring $\mathcal{O}$ such that $K \subset \mathcal{O} \subset L$ and $\operatorname{Quot}(\mathcal{O}) = L$. So giving a place of L is the same as giving a nontrivial discrete valuation on L that vanishes on $K^\times$ (see Exercise 14.1).

(a) Show that L has the following places: (1) the localizations $R_P =: \mathcal{O}_P$ of R at points $P \in E$, and (2) one further place, which we will write as $\mathcal{O}_\infty$. We will write the maximal ideals of the places as $\mathfrak{p}_P$ and $\mathfrak{p}_\infty$. Also show that $R \cap \mathfrak{p}_\infty = \{0\}$. *Hint:* The last statement can be proved by using a suitable K-automorphism $\varphi\colon L \to L$.

(b) Show that L is not isomorphic (as a K-algebra) to the rational function field $K(x)$. This result is usually expressed by saying that E is not a *rational curve*. *Hint:* This can be done by giving a K-automorphism $\varphi\colon L \to L$ that fixes four places of L (in the sense that $\varphi(\mathcal{O}) = \mathcal{O}$), and showing that $K(x)$ has no such automorphism.

(c) Assume that there exists $f \in L$ such that $(f)_R = \mathfrak{m}_P \cdot \mathfrak{m}_Q^{-1}$ with $P, Q \in E$ distinct points, or $(f)_R = \mathfrak{m}_P$ (:= the maximal ideal of R belonging to P). In other words, assume that as a Weil divisor, P is linearly equivalent to Q or to 0. Show that this implies $L \cong K(f)$, contradicting (b). *Hint:* Consider the integral closure A of $K[f]$ in L. Apply the structure theorem for finitely generated modules over a principal ideal domain (see Lang [33, Chapter XV, Theorem 2.2]) to A.

Remark: Part (c) shows that for a nonrational, nonsingular, irreducible affine curve that has only one point at infinity, no point is linearly equivalent to another point or to 0. In this context, it would be more natural to consider *projective curves*. Then zeros and poles at infinity would be included in the divisor of a rational function, and the hypothesis on the number of points at infinity would vanish. *(Solution on page 232)*

Solutions of Some Exercises

1.3. Let $\mathfrak{n} \in \operatorname{Spec}_{\max}(R)$ and consider the homomorphism

$$\varphi: R[x] \to R/\mathfrak{n}, \ f \mapsto f(0) + \mathfrak{n}.$$

The kernel $\mathfrak{m}$ of φ is a maximal ideal of $R[x]$, and $R \cap \mathfrak{m} = \mathfrak{n}$, so $\mathfrak{n} \in \operatorname{Spec}_{\mathrm{rab}}(R)$.

2.5.

(a) That S generates A means that for every element $f \in A$ there exist finitely many elements $f_1, \dots, f_m \in S$ and a polynomial $F \in K[T_1, \dots, T_m]$ in m indeterminates such that $f = F(f_1, \dots, f_m)$. Let $P_1, P_2 \in K^n$ be points with $f(P_1) \neq f(P_2)$. Then

$$F(f_1(P_1), \dots, f_m(P_1)) \neq F(f_1(P_2), \dots, f_n(P_2)),$$

so $f_i(P_1) \neq f_i(P_2)$ for at least one i. This yields part (a).

(b) Consider the polynomial ring $B := K[x_1, \dots, x_n, y_1, \dots, y_n]$ in $2n$ indeterminates. Polynomials from B define functions $K^n \times K^n \to K$. For $f \in K[x_1, \dots, x_n]$, define

$$\Delta f := f(x_1, \dots, x_n) - f(y_1, \dots, y_n) \in B.$$

So for $P_1, P_2 \in K^n$ we have $\Delta f(P_1, P_2) = f(P_1) - f(P_2)$. Consider the ideal

$$I := (\Delta f \mid f \in A)_B \subseteq B.$$

G. Kemper, *A Course in Commutative Algebra*, Graduate Texts in Mathematics 256, DOI 10.1007/978-3-642-03545-6,

By Hilbert's basis theorem (Corollary 2.13), B is Noetherian, so by Theorem 2.9 there exist $f_1, \dots, f_m \in A$ such that

$$I = (\Delta f_1, \dots, \Delta f_m)_B.$$

We claim that $S := \{f_1, \dots, f_m\}$ is A-separating. For showing this, take two points P_1 and P_2 in K^n and assume that there exists $f \in A$ with $f(P_1) \neq f(P_2)$. Since $\Delta f \in I$, there exist $g_1, \dots, g_m \in B$ with

$$\Delta f = \sum_{i=1}^{m} g_i \Delta f_i,$$

so

$$\sum_{i=1}^{m} g_i(P_1, P_2) \Delta f_i(P_1, P_2) = \Delta f(P_1, P_2) \neq 0.$$

Therefore we must have $\Delta f_i(P_1, P_2) \neq 0$ for some i, so $f_i(P_1) \neq f_i(P_2)$.

(c) $S = \{x, xy\}$ is R-separating.

3.6. Define a partial ordering "$\leq$" on set $\mathcal{M} := \{P \in \operatorname{Spec}(R) \mid P \subseteq Q\}$ by

$$P \leq P' \iff P' \subseteq P$$

for $P, P' \in \mathcal{M}$. Let $\mathcal{C} \subseteq \mathcal{M}$ be a chain (=totally ordered subset) in $\mathcal{M}$. Set $\mathcal{C}' := \mathcal{C} \cup \{Q\}$ and $P := \bigcap_{P' \in \mathcal{C}'} P'$. Clearly P is an ideal of R, and $P \subseteq Q$. For showing that P is a prime ideal, take $a, b \in R$ with $ab \in P$ but $b \notin P$. There exists $P_0 \in \mathcal{C}'$ with $b \notin P_0$. Let $P' \in \mathcal{C}'$. Since $\mathcal{C}'$ is a chain, we have $P' \subseteq P_0$ or $P_0 \subseteq P'$. In the first case, $b \notin P'$ but $ab \in P'$, so $a \in P'$. In particular, $a \in P_0$. From this, $a \in P'$ follows in the case that $P_0 \subseteq P'$. We have shown that $a \in P$, so $P \in \mathcal{M}$. By the definition of the ordering, P is an upper bound for $\mathcal{C}$. Now Zorn's lemma yields a maximal element of $\mathcal{M}$, which is a minimal prime ideal contained in Q.

If $R \neq \{0\}$, there exists a maximal ideal $\mathfrak{m}$ of R (by Zorn's lemma applied to $\{I \subsetneqq R \mid I \text{ ideal}\}$ with the usual ordering), and by the above, $\mathfrak{m}$ contains a minimal prime ideal.

5.3. As in the proof of Theorem 5.9 and Proposition 5.10, we only have to show that $\operatorname{trdeg}(A) \leq \dim(A)$. By hypothesis, $A \subseteq B$ with B an affine K-algebra. By induction on n, we will show the following, stronger claim:

Claim. If $\operatorname{trdeg}(A) \geq n$, then there exists a chain

$$Q_0 \subseteq Q_1 \subseteq \cdots \subseteq Q_n$$

in $\operatorname{Spec}(B)$ such that with $P_i := A \cap Q_i \in \operatorname{Spec}(A)$ there are strict inclusions $P_{i-1} \subsetneqq P_i$ for $i = 1, \ldots, n$.

The claim is correct for $n = 0$. To prove it for $n > 0$, let $a_1, \ldots, a_n \in A$ be algebraically independent. As in the proof of Theorem 5.9, we see that there exists a minimal prime ideal M_i of B (not A!) such that the a_i are algebraically independent modulo M_i. Replacing B by B/M_i and A by $A/A\cap M_i$, we may assume that B is an affine domain. Set $L := \operatorname{Quot}(K[a_1])$, $A' := L \cdot A$ and $B' := L \cdot B$, which are all contained in $\operatorname{Quot}(B)$. A' has transcendence degree at least $n-1$ over L. By induction, there is a chain

$$Q_0' \subseteq Q_1' \subseteq \cdots \subseteq Q_{n-1}'$$

in $\operatorname{Spec}(B')$ such that with $P_i' := A' \cap Q_i' \in \operatorname{Spec}(A')$ there are strict inclusions $P_{i-1}' \subsetneqq P_i'$ for $i = 1, \ldots, n-1$. Set $Q_i := B \cap Q_i' \in \operatorname{Spec}(B)$ and $P_i := A \cap Q_i = A \cap P_i' \in \operatorname{Spec}(A)$. For $i = 1, \ldots, n-1$, we have $P_{i-1} \subsetneqq P_i$, since $P_{i-1} = P_i$ would imply

$$P_i' \subseteq (L \cdot A) \cap P_i' \subseteq L \cdot P_i = L \cdot P_{i-1} \subseteq L \cdot P_{i-1}' = P_{i-1}' \subseteq P_i'.$$

As in the proof of Theorem 5.9, we see that A/P_{n-1} is not algebraic over K. Since A/P_{n-1} is contained in B/Q_{n-1}, it follows from Lemma 1.1(b) that A/P_{n-1} is not a field. Choose a maximal ideal $Q_n \subset B$ which contains Q_{n-1}. By Proposition 1.2, $P_n := A \cap Q_n$ is a maximal ideal of A. Clearly $P_{n-1} \subseteq P_n$. Since A/P_{n-1} is not a field, the inclusion is strict. So we have shown the claim, and the result follows.

6.8.

(a) We prove that the negations of both statements are equivalent. First, if $a \in P$, then $U_a \cap P \neq \emptyset$ since $a \in U_a$. If $P + (a)_R = R$, then $1 = b + xa$ with $b \in P$ and $x \in R$, so $b = 1 - xa \in U_a \cap P$. Conversely, if $U_a \cap P \neq \emptyset$, then $a^m(1 + xa) \in P$ with $m \in \mathbb{N}_0$ and $x \in R$. This implies $a \in P$ or $1 + xa \in P$. In the second case we obtain $P + (a)_R = R$.

(b) Assume that $\dim(R) \leq n$, and let $Q_0 \subsetneqq \cdots \subsetneqq Q_k$ be a chain of prime ideals in $U_a^{-1}R$, with $a \in R$. By Theorem 6.5, setting $P_i := \varepsilon^{-1}(Q_i)$ (with $\varepsilon\colon R \to U_a^{-1}R$ the canonical map) yields a chain of length k in $\operatorname{Spec}(R)$, and we have $U_a \cap P_i = \emptyset$. By part (a), this implies that P_i is not a maximal ideal (otherwise, $P_i + (a)_R$ would be R), so we can append a maximal ideal to this chain. Therefore $k + 1 \leq \dim(R) \leq n$, and we conclude $\dim\left(U_a^{-1}R\right) \leq n - 1$.

Conversely, assume $\dim\left(U_a^{-1}R\right) \leq n-1$ for all $a \in R$. Let $P_0 \subsetneqq \cdots \subsetneqq P_k$ be a chain in $\mathrm{Spec}(R)$ of length $k > 0$. Choose $a \in P_k \setminus P_{k-1}$. Then $P_{k-1} + (a)_R \neq R$ (both ideals are contained in P_k), so $U_a \cap P_{k-1} = \emptyset$ by part (a). By Theorem 6.5, setting $Q_i := U_a^{-1}P_i$ $(i = 0, \ldots, k-1)$ yields a chain of length $k-1$ in $\mathrm{Spec}\left(U_a^{-1}R\right)$. Therefore $k - 1 \leq \dim\left(U_a^{-1}R\right) \leq n-1$. We conclude $\dim(R) \leq n$ if $n > 0$. If $n = 0$, the above argument shows that there cannot exist a chain of prime ideals in R of positive length, so $\dim(R) \leq 0$.

(c) We use induction on n, starting with the case $n = 0$. By part (b), $\dim(R) \leq 0$ is equivalent to $U_a^{-1}R = \{0\}$ for all $a \in R$. This condition is equivalent to $0 \in U_a$, which means that there exist $m \in \mathbb{N}_0$ and $x \in R$ with $a^m(1 - xa) = 0$. This is equivalent to $a^m \in (a^{m+1})_R$, which is (6.5) for $n = 0$.
Now assume $n > 0$. By part (b), $\dim(R) \leq n$ is equivalent to $\dim\left(U_a^{-1}R\right) \leq n-1$ for all $a \in R$. By induction, this is equivalent to the following: For all $a_0, \ldots, a_{n-1} \in R$ and all $u_0, \ldots, u_{n-1} \in U_a$, there exist $m_0, \ldots, m_{n-1} \in \mathbb{N}_0$ such that

$$\prod_{i=0}^{n-1} \left(\frac{a_i}{u_i}\right)^{m_i} \in \left(\frac{a_j}{u_j} \cdot \prod_{i=0}^{j} \left(\frac{a_i}{u_i}\right)^{m_i} \,\middle|\, j = 0, \ldots, n-1\right)_{U_a^{-1}R}.$$

Multiplying generators of an ideal by invertible ring elements does not change the ideal. Since the $\varepsilon(u_i)$ are invertible in $U_a^{-1}R$, it follows that the above condition is independent of the u_i. In particular, the condition is equivalent to

$$\prod_{i=0}^{n-1} \varepsilon(a_i)^{m_i} \in \left(\varepsilon(a_j) \cdot \prod_{i=0}^{j} \varepsilon(a_i)^{m_i} \,\middle|\, j = 0, \ldots, n-1\right)_{U_a^{-1}R}.$$

By the definition of localization, this is equivalent to the existence of $m \in \mathbb{N}_0$ and $x \in R$ with

$$a^m(1 + xa) \cdot \prod_{i=0}^{n-1} a_i^{m_i} \in \left(a_j \cdot \prod_{i=0}^{j} a_i^{m_i} \,\middle|\, j = 0, \ldots, n-1\right)_R.$$

Writing a_n and m_n instead of a and m, we see that this condition is equivalent to (6.5).

7.4.

(a) Let $P \in \mathrm{Spec}(R)$ be a prime ideal containing $I := (x)_R$. For all nonnegative integers i we have $(xy^i)^2 = x \cdot xy^{2i} \in I$, so $xy^i \in P$. Therefore P contains the ideal $(x, xy, xy^2, \ldots)_R$, which is maximal. So

$$P = (x, xy, xy^2, \dots)_R.$$

(b) The ideal

$$Q := (xy, xy^2, xy^3, \dots)_R$$

is properly contained in P, and $R/Q \cong K[x]$, so Q is a prime ideal. The chain

$$\{0\} \subsetneqq Q \subsetneqq P$$

shows that $\text{ht}(P) \geq 2$. But $\dim(R) \leq \text{trdeg}(R) = 2$ by Theorem 5.5, so $\text{ht}(P) = 2$.

(c) Let $S_n = K[x, y_1, \dots, y_{n-1}]$ be a polynomial ring in n indeterminates (countably many for $n = \infty$), and set $R_n := K + S_n \cdot x$. As in (a), we see that $P = S_n \cdot x$ is the unique prime ideal of R_n containing $(x)_{R_n}$. For $0 \leq k < n$, we have prime ideals

$$Q_k := x \cdot (y_1, \dots, y_k)_{S_n} = R \cap (y_1, \dots, y_k)_{S_n} \in \text{Spec}(R)$$

forming a strictly ascending chain. Since all Q_k are properly contained in P, we obtain $\text{ht}(P) \geq n$, and equality follows by Theorem 5.5.

7.7. We first show that S is infinite-dimensional. For $i \in \mathbb{N}_0$, we have strictly ascending chains of prime ideals

$$Q_{i,j} = \left(x_{i^2+1}, \dots, x_{i^2+j}\right)_R \subset R \quad (1 \leq j \leq 2i+1)$$

with $Q_{i,j} \cap U = \emptyset$. By Theorem 6.5, this corresponds to a chain of length $2i$ in $\text{Spec}(S)$. It follows that $\dim(S) = \infty$.

For showing that S is Noetherian, we first remark that R_{P_i} is Noetherian for all $i \in \mathbb{N}_0$. Indeed, with $R_i := K[x_{(i+1)^2+1}, x_{(i+1)^2+2}, x_{(i+1)^2+3}, \dots] \subseteq R$ we have $R_i \setminus \{0\} \subset R \setminus P_i$, so R_{P_i} is a localization of $\text{Quot}(R_i)[x_1, \dots, x_{(i+1)^2}]$. Therefore R_{P_i} is Noetherian by Corollaries 2.13 and 6.4. Now let $I \subseteq R$ be a nonzero ideal. Take $f \in I \setminus \{0\}$, and choose $n \in \mathbb{N}_0$ such that all indeterminates x_j occurring in f satisfy $j \leq (n+1)^2$. Since R_{P_i} is Noetherian, there exist $f_1, \dots, f_m \in I$ such that

$$(I)_{R_{P_i}} = (f_1, \dots, f_m)_{R_{P_i}} \quad \text{for} \quad 0 \leq i \leq n. \tag{S.7.1}$$

Take $g \in I$ and consider the ideal

$$J := \{h \in R \mid h \cdot g \in (f_1, \dots, f_m, f)_R\} \subseteq R.$$

Clearly $f \in J$. By (S.7.1), for $0 \le i \le n$ there exists $h_i \in R \setminus P_i$ with $h_i \in J$. By Lemma 7.7, there exists $h \in J \setminus \bigcup_{i=0}^n P_i$. Assume that $J \subseteq \bigcup_{i \in \mathbb{N}_0} P_i$. Then there exists $i > n$ with $h \in P_i$. With $\varphi_i \colon R \to R$ the homomorphism sending $x_{i^2+1}, x_{i^2+2}, \ldots, x_{(i+1)^2}$ to 0 and fixing all other indeterminates, this means $\varphi_i(h) = 0$. The choice of n implies that $\varphi_i(f) = f$. Since $f + h \in J$, there exists $j \in \mathbb{N}_0$ with $f + h \in P_j$, so $\varphi_j(f + h) = 0$. We obtain

$$\varphi_j(h) = \varphi_j\left(f + h - \varphi_i(f + h)\right) = \varphi_j(f + h) - \varphi_i\left(\varphi_j(f + h)\right) = 0, \quad \text{(S.7.2)}$$

so $\varphi_j(f) = \varphi_j(f + h) - \varphi_j(h) = 0$. This implies $j \le n$. Since $h \in P_j$ by (S.7.2), this is a contradiction to the choice of h. We conclude that there exists $u \in J \setminus \bigcup_{i \in \mathbb{N}_0} P_i$. In other words, $u \in U$ and $ug \in (f_1, \ldots, f_m, f)_R$, so $g \in (f_1, \ldots, f_m, f)_S$. It follows that

$$(I)_S = (f_1, \ldots, f_m, f)_S.$$

Since every ideal $I' \subseteq S$ in S can be written as $I' = (I)_S$ with $I = R \cap I' \subseteq R$, we conclude that every ideal in S is finitely generated, so S is Noetherian.

8.7. Set $K := \operatorname{Quot}(R)$. For showing that $\widetilde{R[x]} \subseteq \widetilde{R}[x]$, let $f \in \operatorname{Quot}(R[x]) = K(x)$ be integral over $R[x]$, so

$$f^m = \sum_{i=1}^{m-1} g_i f^i \quad \text{with} \quad g_i \in R[x]. \quad \text{(S.8.1)}$$

Then f is integral over $K[x]$, so $f \in K[x]$ by Example 8.9(1). Therefore there exists $u \in R \setminus \{0\}$ with $uf^k \in R[x]$ for all $0 \le k < m$. In order to reduce to the case that R is Noetherian, we may substitute R by the subring generated by the coefficients of all uf^k ($0 \le k < m$) and of all g_i from (S.8.1). By (S.8.1), $uf^k \in R[x]$ holds for all $k \ge 0$. If $a_n \in K$ is the highest coefficient of f, this implies $ua_n^k \in R$ for all k, so $R[a_n] \subseteq u^{-1}R$. By Theorem 2.10 (and using that R is Noetherian), this implies that $R[a_n]$ is finitely generated as an R-module, so $a_n \in \widetilde{R}$ by Lemma 8.3. This implies that $\widehat{f} := f - a_n x^n$ is integral over $R[x]$, so by induction on n we obtain $\widehat{f} \in \widetilde{R}[x]$. This completes the proof of $\widetilde{R[x]} \subseteq \widetilde{R}[x]$.

Conversely, let $f \in \widetilde{R}[x]$. Then all coefficients of f are integral over R and therefore also over $R[x]$, so f itself is integral over $R[x]$. This implies $f \in \widetilde{R[x]}$. The equivalence $R[x]$ normal $\Longleftrightarrow$ R normal is now clear.

8.11. Clearly $c_i - c_i(x) \in \mathfrak{m}$ for all i, so

$$I := (c_1 - c_1(x), \ldots, c_n - c_n(x))_A \subseteq \mathfrak{m}.$$

By Corollary 8.24, $\operatorname{ht}(\mathfrak{m}) = \dim(A) = n$. So all we need to show is that $\mathfrak{m}_\mathfrak{m} \subseteq \sqrt{I_\mathfrak{m}}$.

A is integral over $K[c_1, \ldots, c_n]$, so for every $a \in A$ there exist polynomials $g_1, \ldots, g_m \in K[x_1, \ldots, x_n]$ such that

$$a^m + g_1(c_1, \ldots, c_n)\, a^{m-1} + \cdots + g_{m-1}(c_1, \ldots, c_n)\, a + g_m(c_1, \ldots, c_n) = 0.$$

Computing modulo I and setting $\gamma_i := c_i(x) \in K$, this yields

$$a^m + g_1(\gamma_1, \ldots, \gamma_n)\, a^{m-1} + \cdots + g_{m-1}(\gamma_1, \ldots, \gamma_n)\, a + g_m(\gamma_1, \ldots, \gamma_n) \in I,$$

so A/I is algebraic. By Theorem 5.11, it follows that it is Artinian. The ideals $(\mathfrak{m}/I)^k \subseteq A/I$ form a descending chain, so there exists $k \in \mathbb{N}$ with $(\mathfrak{m}/I)^k = (\mathfrak{m}/I)^{k+1}$. Localizing at $\mathfrak{m}$, we obtain $M := (\mathfrak{m}_\mathfrak{m}/I_\mathfrak{m})^k = (\mathfrak{m}_\mathfrak{m}/I_\mathfrak{m})^{k+1}$. So M is a finitely generated $R_\mathfrak{m}$-module satisfying $\mathfrak{m}_\mathfrak{m} M = M$. Nakayama's lemma (Theorem 7.3) yields $M = \{0\}$, so $\mathfrak{m}_\mathfrak{m}^k \subseteq I_\mathfrak{m}$. This implies $\mathfrak{m}_\mathfrak{m} \subseteq \sqrt{I_\mathfrak{m}}$.

9.2.

(a) It is clear from the definition that $\mathcal{C}$ is closed under addition. From this, the result follows for $\alpha_i \in \mathbb{N}_{>0}$. Take $\mathbf{c} \in \mathbb{Z}^n$ such that $k\mathbf{c} = \mathbf{e} - \mathbf{f}$ with $k \in \mathbb{N}_{>0}$ and $\mathbf{e}, \mathbf{f} \in \mathbb{N}_0^n$ such that $\mathbf{f} < \mathbf{e}$. There exists $\mathbf{x} \in \mathbb{N}_0^n$ with $\mathbf{x} \equiv -\mathbf{e} \mod k$ (componentwise congruence), so also $\mathbf{x} \equiv -\mathbf{f} \mod k$ since $\mathbf{f} \equiv \mathbf{e} \mod k$. Set $\mathbf{e}' := (\mathbf{e}+\mathbf{x})/k$ and $\mathbf{f}' := (\mathbf{f}+\mathbf{x})/k$. Then $\mathbf{e}', \mathbf{f}' \in \mathbb{N}_0^n$, $\mathbf{e}' - \mathbf{f}' = \mathbf{c}$, and $k\mathbf{f}' < k\mathbf{e}'$ (where we used (3) from Definition 9.1(a)). If $\mathbf{e}' \le \mathbf{f}'$, then also $k\mathbf{e}' \le k\mathbf{f}'$ by induction on k (using (3) from Definition 9.1(a) again), a contradiction. By (1) from Definition 9.1(a), we conclude $\mathbf{f}' < \mathbf{e}'$ and so $\mathbf{c} \in \mathcal{C}$.
Since we already have the result for $\alpha_i \in \mathbb{N}_{>0}$, it follows for $\alpha_i \in \mathbb{Q}_{>0}$ from the above.
Now assume $\alpha_i \in \mathbb{R}_{>0}$ and $\mathbf{c}_i \in \mathcal{C}$ such that $\mathbf{c} := \sum_{i=1}^m \alpha_i \mathbf{c}_i \in \mathbb{Z}^n$. We will see that the α_i can be modified in such a way to make them rational. The set

$$L := \Big\{(\beta_1, \ldots, \beta_m) \in \mathbb{R}^m \mid \sum_{i=1}^m \beta_i \mathbf{c}_i = \mathbf{c}\Big\} \subseteq \mathbb{R}^m$$

is the solution set of an inhomogeneous system of linear equations with coefficients in $\mathbb{Q}$, so L is theimage of a map $\varphi\colon \mathbb{R}^l \to \mathbb{R}^m$, $(\gamma_1, \ldots, \gamma_l) \mapsto$

$v_0 + \sum_{j=1}^{l} \gamma_j v_j$ with $v_0, \ldots, v_l \in \mathbb{Q}^m$. By hypothesis $(\alpha_1, \ldots, \alpha_m) \in \operatorname{im}(\varphi) \cap \mathbb{R}^m_{>0}$, so the preimage $U := \varphi^{-1}(\mathbb{R}^m_{>0}) \subseteq \mathbb{R}^l$ is nonempty. Since φ is continuous, U is open. It follows that there is a point $(\gamma_1, \ldots, \gamma_l) \in U \cap \mathbb{Q}^l$. So $(\alpha'_1, \ldots, \alpha'_m) := \varphi(\gamma_1, \ldots, \gamma_l) \in \mathbb{Q}^m \cap L \cap \mathbb{R}^m_{>0} = \mathbb{Q}^m_{>0} \cap L$, and therefore $\sum_{i=1}^{m} \alpha'_i \mathbf{c}_i = \mathbf{c}$. By what we have shown already, it follows that $\mathbf{c} \in \mathcal{C}$.

(b) It follows from (2) in Definition 9.1(a) that the standard basis vectors $\mathbf{e}_j \in \mathbb{R}^n$ lie in $\mathcal{C}$, so we may include them into the given list of $\mathbf{c}_i$. By definition, $\mathbf{0} \notin \mathcal{C}$, and so $\mathbf{0} \notin \mathcal{H}$ by part (a). (Notice that if some α_i are zero, this means that we are just considering fewer vectors $\mathbf{c}_i$.)
$\mathcal{H}$ is the image of the compact set

$$\mathcal{D} := \left\{(\alpha_1, \ldots, \alpha_m) \in \mathbb{R}^m_{\geq 0} \mid \alpha_1 + \cdots + \alpha_m = 1\right\}$$

under the map $\psi\colon \mathbb{R}^m \to \mathbb{R}^n$, $(\alpha_1, \ldots, \alpha_m) \mapsto \sum_{i=1}^{m} \alpha_i \mathbf{c}_i$. Also consider the map $\delta\colon \mathcal{D} \to \mathbb{R}_{\geq 0}$, $x \mapsto \langle \psi(x), \psi(x) \rangle$, where $\langle \cdot, \cdot \rangle$ denotes the Euclidean scalar product. With $d := \inf(\operatorname{im}(\delta))$, there exists a $\mathcal{D}$-valued sequence (x_k) such that $\delta(x_k)$ converges to d. By the Bolzano–Weierstrass theorem we may substitute (x_k) by a convergent subsequence. With $x = \lim_{k\to\infty} x_k \in \mathcal{D}$, the continuity of δ implies $\delta(x) = \lim_{k\to\infty} \delta(x_k) = d$. Setting, $\mathbf{w}' := \psi(x) \in \mathcal{H}$, we get $d = \langle \mathbf{w}', \mathbf{w}' \rangle$. Since $0 \notin \mathcal{H}$, this implies $d > 0$. We claim that $\langle \mathbf{w}', \mathbf{c} \rangle \geq d$ for all $\mathbf{c} \in \mathcal{H}$. Indeed, for all $\alpha \in \mathbb{R}$ with $0 \leq \alpha \leq 1$ we have $\mathbf{w}' + \alpha(\mathbf{c} - \mathbf{w}') \in \mathcal{H}$, so the definition of d implies

$$d \leq \langle \mathbf{w}' + \alpha(\mathbf{c} - \mathbf{w}'), \mathbf{w}' + \alpha(\mathbf{c} - \mathbf{w}') \rangle = \\ d + 2\left(\langle \mathbf{w}', \mathbf{c} \rangle - d\right)\alpha + \langle \mathbf{c} - \mathbf{w}', \mathbf{c} - \mathbf{w}' \rangle \alpha^2.$$

Applying this with α small yields $\langle \mathbf{w}', \mathbf{c} \rangle \geq d$, so in particular $\langle \mathbf{w}', \mathbf{c}_i \rangle > 0$ for all i. So the preimage of $\mathbb{R}^m_{>0}$ under the map $\mathbb{R}^n \to \mathbb{R}^m$, $\mathbf{w} \to (\langle \mathbf{w}, \mathbf{c}_1 \rangle, \ldots, \langle \mathbf{w}, \mathbf{c}_m \rangle)$ is nonempty. Since the map is continuous, the preimage is open, and it follows that it contains points in $\mathbb{Q}^n$. So there exists $\mathbf{w} \in \mathbb{Q}^n$ with $\langle \mathbf{w}, \mathbf{c}_i \rangle > 0$ for all i. Multiplying $\mathbf{w}$ by a common denominator of the components, we may assume $\mathbf{w} \in \mathbb{Z}^n$. Since the standard basis vectors $\mathbf{e}_j$ are contained among the $\mathbf{c}_i$, it follows that $\mathbf{w} \in \mathbb{N}^n_{>0}$.

(c) Let $G = \{g_1, \ldots, g_r\}$. For $1 \leq i < j \leq r$ set $g_{i,j} := \operatorname{spol}(g_i, g_j)$. By Buchberger's criterion (Theorem 9.12), we have $g_{i,j} = \sum_{k=1}^{r} g_{i,j,k} \cdot g_k$ with $g_{i,j,k} \in K[x_1, \ldots, x_n]$ such that $\operatorname{LM}(g_{i,j,k} \cdot g_k) \leq \operatorname{LM}(g_{i,j})$. Let $M \subset K[x_1, \ldots, x_n]$ be the set of all g_i, $g_{i,j}$, and $g_{i,j,k}$. For a monomial $t = x_1^{e_1} \cdots x_n^{e_n}$, write $\mathbf{e}(t) := (e_1, \ldots, e_n)$. Observe that for $g \in K[x_1, \ldots, x_n]$ and $t \in \operatorname{Mon}(g)$ with $t \neq \operatorname{LM}(g)$, we have $\mathbf{e}(\operatorname{LM}(g)) - \mathbf{e}(t) \in \mathcal{C}$. Form the finite set

$$D := \left\{\mathbf{e}\left(\operatorname{LM}(g)\right) - \mathbf{e}(t) \mid g \in M \text{ and } \operatorname{LM}(g) \neq t \in \operatorname{Mon}(g)\right\} \subset \mathcal{C}.$$

By part (b) there exists $\mathbf{w} \in \mathbb{N}_{>0}^n$ such that $\langle \mathbf{w}, \mathbf{c} \rangle > 0$ for all $\mathbf{c} \in D$. By the definition of "$\leq_\mathbf{w}$" it follows that $\mathrm{LM}_{\leq_\mathbf{w}}(g) = \mathrm{LM}_{\leq}(g)$ for all $g \in M$. Here the subscripts indicate the monomial ordering that is used. This implies

$$\mathrm{spol}_{\leq_\mathbf{w}}(g_i, g_j) = \mathrm{spol}_{\leq}(g_i, g_j) = g_{i,j} = \sum_{k=1}^{r} g_{i,j,k} \cdot g_k$$

and $\mathrm{LM}_{\leq_\mathbf{w}}(g_{i,j,k} \cdot g_k) = \mathrm{LM}_{\leq}(g_{i,j,k} \cdot g_k) \leq \mathrm{LM}_{\leq}(g_{i,j}) = \mathrm{LM}_{\leq_\mathbf{w}}(g_{i,j})$. Applying Buchberger's criterion (Theorem 9.12) again yields that G is a Gröbner basis with respect to "$\leq_\mathbf{w}$". Moreover, we obtain

$$L_{\leq_\mathbf{w}}(I) = (\mathrm{LM}_{\leq_\mathbf{w}}(g_1), \ldots, \mathrm{LM}_{\leq_\mathbf{w}}(g_r)) = (\mathrm{LM}_{\leq}(g_1), \ldots, \mathrm{LM}_{\leq}(g_r)) = L_{\leq}(I).$$

10.3. By substituting R by its image in A, we may assume that $R \subseteq A$ is a subring. $\mathrm{Quot}(A)$ is finitely generated as a field extension of $\mathrm{Quot}(R)$, so the same is true for $\mathrm{Quot}(B)$. It follows that there exists a subalgebra $C \subseteq B$ such that $\mathrm{Quot}(C) = \mathrm{Quot}(B)$, and C is finitely generated. Since A is finitely generated as a C-algebra, Corollary 10.2 applies and yields an $a \in C \setminus \{0\}$ such that A_a is free as a module over C_a, and there exists a basis $\mathcal{M}$ with $1 \in \mathcal{M}$. We claim that $B_a = C_a$. The inclusion $C_a \subseteq B_a$ is clear. Conversely, for every $x \in B_a$ we have

$$x = \sum_{b \in \mathcal{M}} c_b \cdot b$$

with $c_b \in C_a$, and only finitely many c_b are nonzero. Since $\mathrm{Quot}(B) = \mathrm{Quot}(C)$, there exists $y \in C \setminus \{0\}$ such that $yx \in C$, so

$$yx \cdot 1 = \sum_{b \in \mathcal{M}} yc_b \cdot b.$$

The linear independence of $\mathcal{M}$ yields $c_b = 0$ for $b \neq 1$, so $x = c_1 \cdot 1 \in C_a$. We have shown that $B_a = C_a$. This completes the proof, since C_a is clearly finitely generated.

10.7. According to the hypothesis, we have $Y = \bigcup_{i=1}^m L_i$ with $L_i \subseteq X$ locally closed. Being a subset of a Noetherian space, the closure $\overline{Y}$ is Noetherian, too, so Theorem 3.11 yields

$$\overline{Y} = \bigcup_{j=1}^{n} Z_j$$

with Z_j the irreducible components, which are closed in X. Pick a Z_j and let Z_j^* be the union of all other components. Since

$$Z_j = Z_j \cap \overline{Y} = \bigcup_{i=1}^{m} \left(Z_j \cap \overline{L_i}\right),$$

there exists i with $Z_j \subseteq \overline{L_i}$. L_i is not a subset of Z_j^*, since otherwise $Z_j \subseteq Z_j^*$, so Z_j would be contained in a component other than itself. Write $L_i = C_i \cap U_i$ with C_i closed and U_i open, and form $U_j' := U_i \setminus Z_j^*$, which is also open. Then $L_i \not\subseteq Z_j^*$ and $L_i \subseteq \overline{Y} = Z_j \cup Z_j^*$ imply $U_j' \cap Z_j \neq \emptyset$. We have $Z_j = \overline{(U_j' \cap Z_j)} \cup (Z_j \setminus U_j')$. With the irreducibility of Z_j, this yields

$$Z_j = \overline{U_j' \cap Z_j}.$$

Moreover,

$$U_j' \cap Z_j \subseteq U_i \cap Z_j \subseteq U_i \cap \overline{L_i} = L_i \subseteq Y.$$

Form the open set $U' := \bigcup_{j=1}^{n} U_j'$. Then

$$U := U' \cap \overline{Y} = \bigcup_{j=1}^{n} \left(U_j' \cap (Z_j \cup Z_j^*)\right) = \bigcup_{j=1}^{n} \left(U_j' \cap Z_j\right) \subseteq Y,$$

and

$$\overline{U} = \bigcup_{j=1}^{n} \overline{U_j' \cap Z_j} = \bigcup_{j=1}^{n} Z_j = \overline{Y}.$$

So U is a subset of Y that is open and dense in $\overline{Y}$.

10.9. Since X is a union of finitely many locally closed sets, it suffices to prove the result for the case that X itself is locally closed. So $X = \mathcal{V}_{\operatorname{Spec}(S)}(I) \setminus \mathcal{V}_{\operatorname{Spec}(S)}(J)$ with $I, J \subseteq S$ ideals. If $J = (a_1, \ldots, a_n)_S$, then X is the union of all $\mathcal{V}_{\operatorname{Spec}(S)}(I) \setminus \mathcal{V}_{\operatorname{Spec}(S)}(a_i)$. So we may assume

$$X := \mathcal{V}_{\operatorname{Spec}(S)}(I) \setminus \mathcal{V}_{\operatorname{Spec}(S)}(a) = \{Q \in \operatorname{Spec}(S) \mid I \subseteq Q \text{ and } a \notin Q\}$$

with $a \in S$. With ψ: $S \to S_a/I_a$ the canonical map, Lemma 1.22 and Theorem 6.5 yield $X = \psi^*\left(\operatorname{Spec}(S_a/I_a)\right)$. So

$$\varphi^*(X) = (\psi \circ \varphi)^*\left(\operatorname{Spec}(S_a/I_a)\right).$$

Observe that S_a is generated as an R-algebra by $\frac{1}{a}$ and the images of the generators of S, so S_a/I_a is finitely generated as an R-algebra, too. Applying Corollary 10.8 to $\psi \circ \varphi \colon R \to S_a/I_a$ shows that $\varphi^*(X)$ is constructible.

11.7. In order to avoid introducing a lot of additional notation, it is useful to choose and fix the weight vector $\mathbf{w} = (w_1, \ldots, w_n) \in \mathbb{N}_{>0}^n$ throughout, and from now on write deg for $\deg_{\mathbf{w}}$. Everything in Definition 11.1 carries over to the weighted situation. (Notice that $\dim_K(A_{\leq d}) < \infty$ since all w_i are positive.) The formulas in Proposition 11.4 need to be modified as follows:

$$H_I(t) = \frac{1 - t^{\deg(f)}}{\prod_{i=0}^n (1 - t^{w_i})} \quad \text{if} \quad f \neq 0, \quad H_I(t) = \frac{1}{\prod_{i=0}^n (1 - t^{w_i})} \quad \text{if} \quad f = 0,$$

where we set $w_0 := 1$. The induction step in the proof works by using the direct sum decomposition

$$K[x_1, \ldots, x_n]_{\leq d} = \bigoplus_{\substack{i,j \in \mathbb{N}_0, \\ i + w_n j = d}} K[x_1, \ldots, x_{n-1}]_{\leq i} \cdot x_n^j,$$

which implies

$$H_n(t) = H_{n-1}(t) \cdot \left(\sum_{j=0}^{\infty} t^{w_n j} \right) = H_{n-1}(t) \cdot \frac{1}{1 - t^{w_n}} = \frac{1}{\prod_{i=0}^n (1 - t^{w_i})}.$$

The definition of a *weighted degree ordering* is straightforward, and Theorem 11.6 and its proof carry over word by word to the weighted situation. Ditto for the concept of homogeneity and Lemma 11.7. In Algorithm 11.8, the proof of Theorem 11.9, and Corollary 11.10, every occurrence of the denominator $(1 - t)^{n+1}$ should be replaced by $\prod_{i=0}^n (1 - t^{w_i})$. Obtaining an analogue of the Hilbert polynomial is a bit less straightforward. Write $w := \operatorname{lcm}\{w_1, \ldots, w_n\}$. Since $\frac{1}{1 - t^{w_i}} = \frac{1 + t^{w_i} + t^{2w_i} + \cdots t^{w - w_i}}{1 - t^w}$, the formula from the first part of Corollary 11.10 can be rewritten as

$$H_I(t) = \frac{a_0 + a_1 t + \cdots + a_k t^k}{(1 - t^w)^{n+1}}.$$

Since

$$\frac{1}{(1 - t^w)^{n+1}} = \sum_{d=0}^{\infty} \binom{d + n}{n} t^{wd} = \sum_{\substack{d \in \mathbb{N}_0, \\ d \equiv 0 \bmod w}} \binom{d/w + n}{n} t^d$$

we get

$$H_I(t) = \sum_{d=0}^{\infty} \sum_{\substack{0 \le i \le \min\{k,d\}, \\ i \equiv d \bmod w}} a_i \binom{(d-i)/w + n}{n} t^d.$$

So if we define

$$p_{I,j} := \sum_{\substack{0 \le i \le k, \\ i \equiv j \bmod w}} a_i \binom{(x-i)/w + n}{n} \in \mathbb{Q}[x] \quad (j = 0, \ldots, w-1),$$

we get $h_I(d) = p_{I,j}(d)$ for $d \ge k$ with $d \equiv j \mod w$. So instead of one Hilbert polynomial we obtain w polynomials to choose from according to the congruence class modulo w. We could substitute the degree of the Hilbert polynomial by the maximal degree of the $p_{I,j}$. Equivalently (and more conveniently), we define $\deg(h_I)$ to be the minimal k such that the Hilbert function is bounded above by a polynomial of degree k. With this, we get an analogue of Lemma 11.12, where $K[y_1, \ldots, y_m]$ may be equipped with another weighted degree. The proof remains unchanged. Now consider the proof of Theorem 11.13. By the freedom of the choice of the weight vector in Lemma 11.12, we may equip $K[y_1, \ldots, y_m, z_1, \ldots, z_r]$ with the "standard" weight vector $(1, 1, \ldots, 1)$. Therefore the proof of $\deg(p_J) = m$ remains valid, and we obtain the generalized form $\deg(h_I) = \dim(A)$ of Theorem 11.13. So Corollary 11.14 follows for "$\le$" a weighted degree ordering.

Finally, let "$\le$" be an arbitrary monomial ordering and $I \subseteq K[x_1, \ldots, x_n]$ an ideal. By Exercise 9.2(c) there exists a weight vector $\mathbf{w} \in \mathbb{N}_{>0}^n$ such that $L_\le(I) = L_{\le_\mathbf{w}}(I)$. Clearly "$\le_\mathbf{w}$" is a weighted degree ordering, so with the generalized version of Corollary 11.14 we get

$$\dim\left(K[x_1, \ldots, x_n]/I\right) = \dim\left(K[x_1, \ldots, x_n]/L_\le(I)\right).$$

12.1. We keep Definition 11.1 except for the definition of the Hilbert series, which we omit. We omit Example 11.2 and Remark 11.3(a). The other parts of Remark 11.3 are optional. Remark 11.5 is replaced by the following

Lemma. *For the zero ideal* $\{0\} \subset K[x_1, \ldots, x_n]$ *the formula*

$$h_{\{0\}}(d) = \binom{d+n}{n}$$

holds.

Proof. Since the Hilbert function of the zero ideal depends on the number n of indeterminates, we will write it in this proof as $h_n(d)$. We proceed by induction on n. For $n = 0$, $h_0(d) = 1$, so the formula is correct. For $n > 0$, we use the direct sum decomposition (11.1) on page 153, which implies

$$h_n(d) = \sum_{i=0}^{d} h_{n-1}(i) = \sum_{i=0}^{d} \binom{i+n-1}{n-1},$$

where induction was used for the second equality. We now show by induction on d that the latter sum equals $\binom{d+n}{n}$. This is correct for $d = 0$. For $d > 0$, we obtain

$$h_n(d) = \sum_{i=0}^{d} \binom{i+n-1}{n-1} = \binom{d+n-1}{n} + \binom{d+n-1}{n-1} = \binom{d+n}{n},$$

using a well-known identity of binomial coefficients in the last step. □

The Lemma implies that the Hilbert function of an ideal $I \subseteq K[x_1, \ldots, x_n]$ is bounded above by a polynomial. So we can define $\delta(I) \in \mathbb{N}_0 \cup \{-1\}$ to be the smallest integer δ such that h_I can be bounded above by a polynomial in $\mathbb{Q}[x]$ of degree δ. We skip the rest of Section 11.1. We modify the assertion of Lemma 11.12 to $\delta(I) = \delta(J)$. The proof works for the modified assertion with a slight change of last two sentences. The assertion of Theorem 11.13 becomes $\delta(I) = \dim(A)$. In the proof of Theorem 11.13, we replace $\deg(p_I)$ and $\deg(p_J)$ by $\delta(I)$ and $\delta(J)$, and use the above Lemma instead of Remark 11.5. Otherwise, the proof needs no modification. We skip everything else from Section 11.2. So only the following material is required from Part III: The shortened Definition 11.1, the above Lemma, the definition of $\delta(I)$, and the modified versions of Lemma 11.12 and Theorem 11.13.

We make no change to Section 12.1, except using the above Lemma instead of Remark 11.5 in the proof of Lemma 12.4. In Section 12.2, we modify the assertion of Proposition 12.5 to: $\dim(\text{gr}(R))$ *is the least degree of a polynomial providing an upper bound for* $\text{length}\left(R/\mathfrak{m}^{d+1}\right)$. This follows from (12.5) and the modified Theorem 11.13. We omit the definition of the Hilbert–Samuel polynomial. The modified version of Proposition 12.5 and Lemma 12.4 yield (12.7). The next modification is to the proof of Lemma 12.7. We start with: "In order to use Proposition 12.5, we compare the Hilbert–Samuel functions $h_{R/Ra}$ and h_R." We replace the last sentence of the proof by: "From this, the lemma follows by Proposition 12.5." Finally, we delete the last sentence from Theorem 12.8. The proof of the theorem remains unchanged. Observe that the Hilbert–Samuel polynomial is not used anywhere outside Chapter 12 in the book.

12.5. The elements $c_i := \frac{x_i + I}{1} \in A_{\mathfrak{m}} =: R$ generate the maximal ideal $\mathfrak{m}_{\mathfrak{m}}$ of R. By Exercise 12.4 we have $R/\mathfrak{m}_{\mathfrak{m}} \cong A/\mathfrak{m} \cong K$. By the discussion before Proposition 12.5, $\mathrm{gr}(R)$ is generated as a K-algebra by the elements $a_i := c_i t + (\mathfrak{m}_{\mathfrak{m}})_{R^*}$, and the a_i are homogeneous of degree 1. Let J be the kernel of the map $K[x_1, \dots, x_n] \to \mathrm{gr}(R)$, $x_i \mapsto a_i$. We are done if we can show that $J = I_{\mathrm{in}}$.

To prove that I_{in} is contained in J, take $f \in I \setminus \{0\}$ and write $\widehat{f} := f_{\mathrm{in}} - f$. So $f_{\mathrm{in}} \equiv \widehat{f} \mod I$, and every monomial in $\widehat{f}$ has degree larger than $\deg(f_{\mathrm{in}}) =: d$. Therefore

$$f_{\mathrm{in}}(c_1 t, \dots, c_n t) = f_{\mathrm{in}}(c_1, \dots, c_n) t^d = \widehat{f}(c_1, \dots, c_n) t^d \in \mathfrak{m}_{\mathfrak{m}}^{d+1} t^d \subseteq (\mathfrak{m}_{\mathfrak{m}})_{R^*},$$

where the last inclusion follows from the definition of R^*. We conclude $f_{\mathrm{in}} \in J$, so $I_{\mathrm{in}} \subseteq J$.

For proving the reverse inclusion, take $f \in J$. Since J is a homogeneous ideal, we may assume that f is homogeneous of some degree d, and $f \neq 0$. We have $0 = f(a_1, \dots, a_n) = f(c_1, \dots, c_n) t^d + (\mathfrak{m}_{\mathfrak{m}})_{R^*}$, so $f(c_1, \dots, c_n) \in \mathfrak{m}_{\mathfrak{m}}^{d+1}$ by the definition of R^*. This means that there exists $a \in A \setminus \mathfrak{m}$ such that $a \cdot (f + I) \in \mathfrak{m}^{d+1}$. We may write $a = h + I$ with $h \in K[x_1, \dots, x_n]$, so $hf + I \in \mathfrak{m}^{d+1}$. This means that there exists $g \in \mathfrak{n}^{d+1}$ with $hf - g \in I$. From $a \notin \mathfrak{m}$ we conclude $h \notin \mathfrak{n}$, so $h(0) \neq 0$ and $(hf)_{\mathrm{in}} = h(0) \cdot f$. The condition $g \in \mathfrak{n}^{d+1}$ means that every monomial of g has degree $> d$, so by the above

$$(hf - g)_{\mathrm{in}} = h(0) \cdot f.$$

We conclude that $h(0) \cdot f \in I_{\mathrm{in}}$, so also $f \in I_{\mathrm{in}}$. This completes the proof.

13.6.

(a) Since the polynomial $x_2^2 - x_1^2(x_1 + 1) \in K[x_1, x_2]$ is irreducible, $K[X]$ is an integral domain. Therefore the same holds for its localization R.

(b) By Exercise 13.5(d) there exists $f = \sum_{i=0}^{\infty} a_i x_1^i \in K[[x_1]]$ with $f^2 = x_1 + 1$. For $k \in \mathbb{N}_0$, form the polynomials

$$A_k := x_2 - x_1 \sum_{i=0}^{k} a_i x_1^i \quad \text{and} \quad B_k := x_2 + x_1 \sum_{i=0}^{k} a_i x_1^i \in K[x_1, x_2].$$

Clearly (A_k) and (B_k) are Cauchy sequences with respect to the Krull topology given bythe filtration $I_n := \mathfrak{n}^n$ with $\mathfrak{n} := (x_1, x_2)$, and the product sequence $(A_k \cdot B_k)$ converges to $x_2^2 - x_1^2(x_1+1)$. Also observe that none of the A_k or B_k lie in $\mathfrak{n}^2$. Applying the canonical map $K[x_1, x_2] \to R$ to the A_k and B_k yields Cauchy sequences in R whose product converges to 0, and no element of these sequences lies in $\mathfrak{m}^2$, the square of the maximal ideal of R. The sequences have limits, A and B, in the completion $\widehat{R}$. A and B must be nonzero, since the A_k and B_k lie outside $\mathfrak{m}^2$. Since the limit of the product sequence is 0, it follows with Exercise 13.4(c) that $A \cdot B = 0$. So $\widehat{R}$ has zero divisors.

14.11. Since $R := K[X]$ is a Dedekind domain and $\bar{l} \neq 0$, $(\bar{l})$ is a finite product of maximal ideals. A maximal ideal $\mathfrak{m} \in \operatorname{Spec}_{\max}(R)$ occurs in this product if and only if $\bar{l} \in \mathfrak{m}$, i.e., if and only if $\mathfrak{m}$ corresponds to a point in the intersection $L \cap X$. So the $\mathfrak{m}_i$ are precisely the maximal ideals occurring in the product. The difficulty lies in the fact that some $\mathfrak{m}_i$ may coincide, so we have to get the multiplicities right. If ξ_i has multiplicity n_i as a zero of f, we need to show that $(\bar{l}) \in \mathfrak{m}_i^{n_i}$, but $(\bar{l}) \notin \mathfrak{m}_i^{n_i+1}$. Fix an i. By a change of coordinates, we may assume that

$$P_i = (0,0), \;\; L = \{(\xi, 0) \mid \xi \in K\}, \;\; \xi_i = 0, \;\text{ and } l = x_2.$$

Then $g = x_2 \cdot h + f(x_1)$ with $h \in K[x_1, x_2]$ and $f \in K[t]$ as defined in the exercise. By definition, n_i is the maximal k such that x_1^k divides $f(x_1)$. With $\mathfrak{n} := (x_1, x_2) \in \operatorname{Spec}_{\max}(K[x_1, x_2])$ (so $\mathfrak{m}_i = \mathfrak{n}/(g)$), we need to show that n_i is the maximal k with

$$x_2 + (g) \in (\mathfrak{n}/(g))^k. \tag{S.14.1}$$

The condition (S.14.1) is equivalent to the existence of $u \in K[x_1, x_2]$ such that all monomials in $x_2 - ug$ have degree $\geq k$. First consider the case that $h(0,0) = 0$. Since X is nonsingular, it follows by the Jacobian criterion (Theorem 13.10) that $f'(0) \neq 0$, so $n_i = 1$. In this case, x_2 occurs as a monomial in $x_2 - ug = x_2(1 - uh) - uf(x_1)$ for every $u \in K[x_1, x_2]$, so the maximal k satisfying (S.14.1) is $1 = n_i$.

Now consider the case $h(0,0) \neq 0$. Then h is invertible as an element of the formal power series ring $K[[x_1, x_2]]$ (see Exercise 1.2(b)). In particular, there exists $u \in K[x_1, x_2]$ such that all monomials in $uh - 1$ have degree $\geq n_i$, so the same is true for $x_2 - ug = x_2(1 - uh) - uf(x_1)$. On the other hand, for every $u \in K[x_1, x_2]$, $x_2 - ug$ has monomials of degree $\leq n_i$, since x_2 occurs if $u(0,0) = 0$, and otherwise $x_1^{n_i}$ occurs. Therefore in this case the maximal k satisfying (S.14.1) is n_i again. This finishes the proof.

14.12.

(a) We have $L = K(\overline{x}_1, \overline{x}_2)$ with $\overline{x}_2^2 - \overline{x}_1^3 - a\overline{x}_1 - b = 0$, and $R = K[\overline{x}_1, \overline{x}_2]$. Let $\mathcal{O}$ be a place of L with maximal ideal $\mathfrak{p}$, and let ν: $L \to \mathbb{Z}$ be the corresponding discrete valuation. If ν were trivial on $K(\overline{x}_1)$, then $K(\overline{x}_1)$ would be contained in $\mathcal{O}$, so $\mathcal{O} = L$ since L is integral over $K(\overline{x}_1)$ and $\mathcal{O}$ is integrally closed in L. This contradiction shows that ν is nontrivial on $K(\overline{x}_1)$. Consider two cases.
(1) $\nu(\overline{x}_1) \geq 0$. Then by the results of Exercise 14.2, $K[\overline{x}_1] \subseteq \mathcal{O}$, and there exists $\xi_1 \in K$ such that $\overline{x}_1 - \xi_1 \in \mathfrak{p}$. We have $\overline{x}_2^2 \in \mathcal{O}$, so $\overline{x}_2 \in \mathcal{O}$ and hence $R \subseteq \mathcal{O}$. Choose $\xi_2 \in K$ with $\xi_2^2 = \xi_1^3 + a\xi_1 + b$. Then

$$(\overline{x}_2 - \xi_2)(\overline{x}_2 + \xi_2) = \overline{x}_1^3 + a\overline{x}_1 + b - \left(\xi_1^3 + a\xi_1 + b\right) \in \mathfrak{p},$$

so $\overline{x}_2 - \xi_2 \in \mathfrak{p}$ or $\overline{x}_2 + \xi_2 \in \mathfrak{p}$. By changing our choice of ξ_2, we may assume the first possibility. With $P := (\xi_1, \xi_2) \in E$, we get $\mathfrak{m}_P = (\overline{x}_1 - \xi_1, \overline{x}_2 - \xi_2)_R \subseteq \mathfrak{p}$, so $R \cap \mathfrak{p} = \mathfrak{m}_P$. This implies $R \backslash \mathfrak{m}_P \subseteq \mathcal{O} \backslash \mathfrak{p} = \mathcal{O}^\times$, so $R_P := R_{\mathfrak{m}_P} \subseteq \mathcal{O}$. But R_P is a place of L since E is nonsingular by Exercise 13.10. Therefore if R_P were strictly contained in $\mathcal{O}$, $\mathcal{O}$ would be equal to L. This contradiction shows that $\mathcal{O} = R_P$.
(2) $\nu(\overline{x}_1) < 0$. Then $\overline{y}_1 := 1/\overline{x}_1 \in \mathfrak{p}$. With $\overline{y}_2 := \overline{x}_1/\overline{x}_2$ we have the relation

$$\overline{y}_2^2 \cdot \left(1 + a\overline{y}_1^2 + b\overline{y}_1^3\right) = \overline{y}_1,$$

so $\overline{y}_2 \in \mathfrak{p}$. Therefore $S := K[\overline{y}_1, \overline{y}_2] \subseteq \mathcal{O}$, and $\mathfrak{m} := (\overline{y}_1, \overline{y}_2)_S \subseteq \mathfrak{p}$. Using the Jacobian criterion (Theorem 13.10), we conclude from the above relation that $S_\mathfrak{m}$ is regular. By the same argument as above, we obtain $\mathcal{O} = S_\mathfrak{m}$. So there exists exactly one place for which $\nu(\overline{x}_1) < 0$. We write this place as $\mathcal{O}_\infty$, and its maximal ideal as $\mathfrak{p}_\infty$.
We now show that $R \cap \mathfrak{p}_\infty = \{0\}$. It follows from the equation defining E that we have a K-automorphism φ of L mapping $\overline{x}_1$ to itself and $\overline{x}_2$ to $-\overline{x}_2$. If $f \in R$, then clearly $f \cdot \varphi(f) \in K[\overline{x}_1]$. Moreover, φ maps $\mathfrak{p}_\infty$ to itself, so if $f \in R \cap \mathfrak{p}_\infty$ we obtain

$$f \cdot \varphi(f) \in K[\overline{x}_1] \cap \mathfrak{p}_\infty = K[1/\overline{y}_1] \cap \mathfrak{p}_\infty = \{0\},$$

so $f = 0$.

(b) Let φ: $L \to L$ be as above. By the assumption on a and b, the polynomial $x_1^3 + ax_1 + b$ has three pairwise distinct zeros $\alpha_1, \alpha_2, \alpha_3 \in K$. With $P_i := (0, \alpha_i) \in E$, φ fixes the places $\mathcal{O}_{P_i}$. Looking at the results from (a), we see that φ also fixes $\mathcal{O}_\infty$.
Now we consider $K(x)$ and claim that for every K-automorphism ψ: $K(x) \to K(x)$ thereexist $\alpha, \beta, \gamma, \delta \in K$ with

$$\psi(x) = \frac{\alpha x + \beta}{\gamma x + \delta}.$$

(Notice that this gives an automorphism only if $\alpha\delta - \beta\gamma \neq 0$, but we do not need this here.) Indeed, if we write $\psi(x) = g/h$ with $g, h \in K[x]$ coprime, then $K(x) = K(g/h)$. We have $g(x) - \frac{g}{h} \cdot h(x) = 0$. With a new indeterminate t, the polynomial $g(x) - th(x) \in K[t,x]$ is irreducible, so it is also irreducible in $K(t)[x]$. Since g/h is transcendental over K, it follows that $g(x) - \frac{g}{h} \cdot h(x) = 0$ is a minimal equation for x over $K(g/h)$. So its degree must be one, and we get $g = \alpha x + \beta$ and $h = \gamma x + \delta$ as claimed. If $\psi = \text{id}$, then ψ fixes infinitely many places. Which places are fixed if $\psi \neq \text{id}$? The places of $K(x)$ are determined in Exercise 14.2. A place corresponding to a point $\xi \in K$ is fixed if and only if $\frac{\alpha\xi+\beta}{\gamma\xi+\delta} = \xi$, so at most two such places are fixed. In addition, the place corresponding to the point at infinity may be fixed, giving at most three fixed places. This concludes the proof of (b).

(c) If $(f)_R = \mathfrak{m}_P$, then $f \in R$, so $f \notin \mathfrak{p}_\infty$ by (a). On the other hand, if $(f)_R = \mathfrak{m}_P \cdot \mathfrak{m}_Q^{-1}$ and $f \in \mathfrak{p}_\infty$, then by interchanging P and Q and substituting f by f^{-1}, we also get $f \notin \mathfrak{p}_\infty$. (In fact, the latter case turns out to be impossible by the theory of divisors of projective curves.) So in both cases, $f \in \mathfrak{p}_P \setminus \mathfrak{p}_P^2$, and f does not lie in the maximal ideal of any place $\mathcal{O} \neq \mathcal{O}_P$ of L.

Since $f \notin K$, f is transcendental over K, so L is a finite field extension of $K(f)$. We are done if we can show that the degree $d := [L : K(f)]$ is one. Let A be the integral closure of $K[f]$ in L. By Lemma 8.27, A is finitely generated as a module over $K[f]$. Since A is torsion-free, the structure theorem for finitely generated modules over a principal ideal domain (see Lang [33, Chapter XV, Theorem 2.2]) tells us that A is free. Clearly A contains a basis of L over $K(f)$, and on the other hand no more than d elements of A can be linearly independent. So A is a free $K[f]$-module of rank d. This implies that $A/(f)_A$ has dimension d as a vector space over $K[f]/(f)_{K[f]} = K$.

From $f \in \mathfrak{p}_P$ it follows that $K[f] \subseteq \mathcal{O}_P$, so also $A \subseteq \mathcal{O}_P$ since $\mathcal{O}_P$ is integrally closed in L. Since $\mathcal{O}_P = R_P$, there is a map

$$\psi \colon A \to K, \ a \mapsto a(P),$$

which is clearly K-linear and surjective. We claim that $\ker(\psi) = (f)_A$. If we can prove this, then $A/(f)_A \cong K$, so $d = 1$, and we are done. Since $f \in \mathfrak{p}_P$, f lies in $\ker(\psi)$. Conversely, take $a \in \ker(\psi)$ and consider the quotient $b := a/f \in L$. We need to show that $b \in A$. This is true if $b \in A_\mathfrak{m}$ for every $\mathfrak{m} \in \operatorname{Spec}_{\max}(A)$, since then the ideal $\{c \in A \mid c \cdot b \in A\} \subseteq A$ is not contained in any maximal ideal. (One could also use Exercise 8.3 for this conclusion.)

So let $\mathfrak{m} \in \operatorname{Spec}_{\max}(A)$. Since A is a normal Noetherian domain of dimension 1, $A_\mathfrak{m}$ is a DVR. We also have $K \subseteq A_\mathfrak{m} \subseteq L$ and $L = \operatorname{Quot}(A_\mathfrak{m})$. Therefore $A_\mathfrak{m}$ is a place of L. If $A_\mathfrak{m} \neq \mathcal{O}_P$, then f does not lie in the maximal ideal of $A_\mathfrak{m}$, so $1/f \in A_\mathfrak{m}$. Since also $a \in A \subseteq A_\mathfrak{m}$, we get $b \in A_\mathfrak{m}$. On the other hand, if $A_\mathfrak{m} = \mathcal{O}_P$, then b lies in $A_\mathfrak{m}$ since $a \in \mathfrak{p}_P$ and $f \notin \mathfrak{p}_P^2$. So $b \in A_\mathfrak{m}$ for every $\mathfrak{m} \in \operatorname{Spec}_{\max}(A)$, and the proof is complete.

References

In the square brackets at the end of each reference we give the pages where the reference is cited.

1. William W. Adams, Phillippe Loustaunau, *An Introduction to Gröbner Bases*, Graduate Studies in Mathematics **3**, American Mathematical Society, Providence, 1994 [117].
2. Michael F. Atiyah, Ian Grant Macdonald, *Introduction to Commutative Algebra*, Addison-Wesley, Reading, 1969 [174].
3. Thomas Becker, Volker Weispfenning, *Gröbner Bases*, Springer, Berlin, 1993 [117, 118, 161].
4. David J. Benson, *Polynomial Invariants of Finite Groups*, Lond. Math. Soc. Lect. Note Ser. **190**, Cambridge University Press, Cambridge, 1993 [vii].
5. Wieb Bosma, John J. Cannon, Catherine Playoust, *The Magma algebra system I: The user language*, J. Symb. Comput. **24** (1997), 235–265 [127, 213].
6. Nicolas Bourbaki, *General Topology. Chapters 1–4*, Springer, Berlin, 1998 [2, 33].
7. Nicolas Bourbaki, *Algebra II, Chapters 4–7*, Elements of Mathematics, Springer, Berlin, 2003 [148].
8. Winfried Bruns, Jürgen Herzog, *Cohen-Macaulay Rings*, Cambridge University Press, Cambridge, 1993 [194].
9. Antonio Capani, Gianfranco Niesi, Lorenzo Robbiano, *CoCoA: A system for doing computations in commutative algebra*, available via anonymous ftp from `cocoa.dima.unige.it`, 2000 [126].
10. Luther Claborn, *Every abelian group is a class group*, Pacific J. Math. **18** (1966), 219–222 [210].
11. Thierry Coquand, Henri Lombardi, *A short proof for the Krull dimension of a polynomial ring*, Am. Math. Mon. **112** (2005), 826–829 [72].
12. David Cox, John Little, Donal O'Shea, *Ideals, Varieties, and Algorithms*, Springer, New York, 1992 [4, 117].
13. David Cox, John Little, Donal O'Shea, *Using Algebraic Geometry*, Springer, New York, 1998 [117].
14. Steven D. Cutkosky, *Resolution of Singularities*, vol. 63 of *Graduate Studies in Mathematics*, American Mathematical Society, Providence, 2004 [201].
15. Wolfram Decker, Christoph Lossen, *Computing in Algebraic Geometry. A quick start using SINGULAR*, vol. 16 of *Algorithms and Computation in Mathematics*, Springer, Berlin, 2006 [117].
16. Harm Derksen, Gregor Kemper, *Computing invariants of algebraic group actions in arbitrary characteristic*, Adv. Math. **217** (2008), 2089–2129 [118].

17. David Eisenbud, *Commutative Algebra with a View Toward Algebraic Geometry*, Springer, New York, 1995 [4, 15, 89, 117, 118, 160, 162, 171, 173–175, 184, 185, 194, 195, 199].
18. David Gale, *Subalgebras of an algebra with a single generator are finitely generated*, Proc. Am. Math. Soc. **8** (1957), 929–930.
19. Joachim von zur Gathen, Jürgen Gerhard, *Modern Computer Algebra*, Cambridge University Press, Cambridge, 1999 [127].
20. Robert Gilmer, *Multiplicative Ideal Theory*, Marcel Dekker, New York, 1972 [88, 213].
21. Daniel R. Grayson, Michael E. Stillman, *Macaulay 2, a software system for research in algebraic geometry*, available at `http://www.math.uiuc.edu/Macaulay2`, 1996 [126].
22. Gert-Martin Greuel, Gerhard Pfister, *A* **Singular** *Introduction to Commutative Algebra*, Springer, Berlin, 2002 [117, 160, 178].
23. Gert-Martin Greuel, Gerhard Pfister, Hannes Schönemann, *Singular version 1.2 user manual*, Reports On Computer Algebra **21**, Centre for Computer Algebra, University of Kaiserslautern, 1998, available at `http://www.mathematik.uni-kl.de/~zca/Singular` [127].
24. Paul R. Halmos, *Naive Set Theory*, Springer, New York, 1974 [11].
25. Joe Harris, *Algebraic Geometry. A First Course*, Springer, New York, 1992 [4].
26. Robin Hartshorne, *Algebraic Geometry*, Springer, New York, 1977 [4, 38, 205].
27. David Hilbert, *Über die vollen Invariantensysteme*, Math. Ann. **42** (1893), 313–370 [23].
28. Harry C. Hutchins, *Examples of commutative rings*, Polygonal Publishing House, Passaic, N.J., 1981 [107, 213].
29. Theo de Jong, *An algorithm for computing the integral closure*, J. Symb. Comput. **26** (1998), 273–277 [118].
30. Gregor Kemper, *The calculation of radical ideals in positive characteristic*, J. Symb. Comput. **34** (2002), 229–238 [118].
31. Martin Kreuzer, Lorenzo Robbiano, *Computational Commutative Algebra 1*, Springer, Berlin, 2000 [117].
32. Martin Kreuzer, Lorenzo Robbiano, *Computational Commutative Algebra 2*, Springer, Berlin, 2005 [117].
33. Serge Lang, *Algebra*, second edn., Addison-Wesley, Redwood City, 1984 [2, 9, 90, 101–103, 110, 171, 186–188, 199, 209, 216, 233].
34. Max D. Larsen, Paul J. McCarthy, *Multiplicative Theory of Ideals*, Academic, New York, 1971 [207].
35. Saunders Mac Lane, *Modular fields. I. Separating transcendence bases*, Duke Math. J. **5** (1939), 372–393 [186].
36. Ryutaroh Matsumoto, *Computing the radical of an ideal in positive characteristic*, J. Symb. Comput. **32** (2001), 263–271 [118].
37. Hideyuki Matsumura, *Commutative Algebra*, Mathematics Lecture Note Series **56**, Benjamin, Reading, 1980 [4, 107, 182, 184, 193].
38. Hideyuki Matsumura, *Commutative Ring Theory*, Cambridge Studies in Advanced Mathematics **8**, Cambridge University Press, Cambridge, 1986 [171, 174].
39. Ferdinando Mora, *An algorithm to compute the equations of tangent cones*, in: *Computer algebra (Marseille, 1982)*, Lecture Notes in Comput. Sci. **144**, pp. 158–165, Springer, Berlin 1982 [178].
40. Masayoshi Nagata, *On the closedness of singular loci*, Inst. Hautes Études Sci. Publ. Math. **1959** (1959), 29–36 [193].
41. Masayoshi Nagata, *Local Rings*, Wiley, New York, 1962 [89, 107, 110].
42. Jürgen Neukirch, *Algebraic Number Theory*, vol. 322 of *Grundlehren der Mathematischen Wissenschaften*, Springer, Berlin, 1999 [210].
43. Emmy Noether, *Der Endlichkeitssatz der Invarianten endlicher linearer Gruppen der Charakteristik* p, Nachr. Ges. Wiss. Göttingen (1926), 28–35 [111].

44. Vladimir L. Popov, Ernest B. Vinberg, *Invariant theory*, in: N.N. Parshin, I.R. Shafarevich, eds., *Algebraic Geometry IV*, Encyclopaedia of Mathematical Sciences **55**, Springer, Berlin, 1994 [145, 147].
45. J.L. Rabinowitsch, *Zum Hilbertschen Nullstellensatz*, Math. Ann. **102** (1930), 520–520 [12].
46. Igor R. Shafarevich, *Basic Algebraic Geometry*, Springer, Berlin, New York, 1974 [213].
47. Karen E. Smith, Lauri Kahanpää, Pekka Kekäläinen, William Traves, *An Invitation to Algebraic Geometry*, Springer, New York, 2000 [4].
48. Tonny A. Springer, *Invariant Theory*, Lecture Notes in Math. **585**, Springer, Berlin, 1977 [145].
49. Tonny A. Springer, *Aktionen reduktiver Gruppen auf Varietäten*, in: Hanspeter Kraft, Peter Slodowy, Tonny A. Springer, eds., *Algebraische Transformationsgruppen und Invariantentheorie, DMV Seminar* **13**, Birkhäuser, Basel 1987 [147].
50. Bernd Sturmfels, *Algorithms in Invariant Theory*, Springer, Wien, New York, 1993 [vii, 145].
51. Wolmer V. Vasconcelos, *Computational Methods in Commutative Algebra and Algebraic Geometry*, Algorithms and Computation in Mathematics **2**, Springer, Berlin, 1998 [117, 118].
52. Lawrence C. Washington, *Elliptic Curves: Number Theory and Cryptography*, Discrete Mathematics and Its Applications, Chapman & Hall, Boca Raton, 2003 [212].

Notation

Index

Please note that a boldface page number indicates the page on which the word or phrase is defined.